THERMODYNAMIC PROPERTIES OF ISOMERIZATION REACTIONS

THERMODYNAMIC PROPERTIES OF ISOMERIZATION REACTIONS

M. L. Frenkel
Thermodynamics Research Center
Texas A & M University

G. Ya. Kabo
Chemistry Department
Byelorussian State University

G. N. Roganov
Chemistry Department
Mogilev Institute of Technology

Translated by
P. Rappoport

⦿HEMISPHERE PUBLISHING CORPORATION
A member of the Taylor & Francis Group
Washington   Philadelphia   London

USA	Publishing Office:	Taylor & Francis 1101 Vermont Avenue, N.W., Suite 200 Washington, DC 20005-2531 Tel: (202) 289-2174 Fax: (202) 289-3665
	Distribution Center:	Taylor & Francis 1900 Frost Road, Suite 101 Bristol, PA 19007-1598 Tel: (215) 785-5800 Fax: (215) 785-5515
UK		Taylor & Francis Ltd. 4 John Street London WC1N 2ET, UK Tel: 071 405 2237 Fax: 071 831 2035

First edition published in 1988 by Minsk "Universitetkoc", under the title "Termodinamicheskie Kharakteristiki reacktsii izomerizatsii."

THERMODYNAMIC PROPERTIES OF ISOMERIZATION REACTIONS

Copyright © 1993 by Hemisphere Publishing Corporation. All rights reserved. Printed in the United States of America. Except as permitted under the United States Copyright Act of 1976, no part of this publication may be reproduced or distributed in any form or by any means, or stored in a database or retrieval system, without the prior written permission of the publisher.

1 2 3 4 5 6 7 8 9 0   BRBR   9 8 7 6 5 4 3 2

Cover design by Michelle Fleitz.
A CIP catalog record for this book is available from the British Library.
⊚ The paper in this publication meets the requirements of the ANSI Standard Z39.48-1984(Permanence of Paper).

Library of Congress Cataloging-in-Publication Data

Frenkel, M. L.
   [Termodinamicheskie kharakteristiki reaktsii izomerizatsii. English]
   Thermodynamic properties of isomerization reactions / M.L. Frenkel, G.Ya. Kabo, G.N. Roganov; translated by P. Rappoport.
    p. cm.
   Includes bibliographical references and index.
    1. Isozmerization.   2. Thermodynamics—Tables.   3. Organic compounds—Tables.   I. Kabo, G. Ya.   II. Roganov, G. N.
III. Title.
QD281.R35F7413   1992
547′.2—dc20   92-18703
                                                                                        CIP

ISBN 1-56032-111-3

CONTENTS

FOREWORD   vii

INTRODUCTION   1

   1. Specific Aspects of Determining the Thermodynamic
    Properties of Isomerization Reactions   1
   2. Qualitative Regularities of Changes in the Enthalpy and
    Entropy Functions of Isomerizations   4
   3. Description of Tables   12

**THERMODYNAMIC PROPERTIES OF ISOMERIZATION
REACTIONS**   16

REFERENCES   211

FORMULA INDEX   227

FOREWORD

Thermodynamic data that characterize the properties of individual substances and process parameters are essential for research as well as for technological and engineering calculations. The thermodynamic properties of individual substances are presented in a systematic form in the reference books *Thermodynamic Properties of Individual Substances*, V. P. Glushko, Ed. (Nauka, 1978); *Thermodynamic Properties of Organic Compounds*, D. Stull, E. Westrum, and G. Sinke (Wiley, 1969); *Computer Analyzed Thermochemical Data: Organic and Organometallic Compounds*, J. Pedley and J. Rylans (Nature, 1977); *Thermochemistry of Organometallic Compounds*, J. Cox and G. Pilcher (Academic Press, 1974); *Thermodynamic Properties of Oxygen-Containing Organic Compounds*, I. A. Vasilyev and V. M. Petrov (Khimiya, 1984) and also in the monograph *Thermochemistry of Vaporization of Organic Compounds* by Yu. A. Lebedev and E. A. Miroshnichenko (Nauka, 1981). These handbooks or monographs, however, contain but a limited amount of data on the thermodynamics of organic compound reactions and chemical equilibria, although such data offer considerable scope for carrying out thermodynamic calculations pertaining to chemical processes and individual substances, as well as providing additional information in terms of checking the reliability of thermodynamic data.

Progress in experimental techniques and the elaboration of rapid and precise methods for qualitative analysis of multicomponent mixtures of organic compounds with diverse structures provided an impetus to more vigorous employment of equilibrium studies as an independent technique for determining the thermodynamic functions of reactions. The results obtained in studying chemical equilibria and reported in numerous research papers have been meagerly used in compiling handbooks on thermodynamics, although the evaluation of thermodynamic properties from equilibria data is, in some instances, more accurate than from the results of calorimetric measurements.

This handbook has been compiled with a view to systematizing the available data on the thermodynamic parameters of a major type of organic compound conversions, namely izomerization reactions. Since isomerizations most clearly display the relationship between the structure and the properties of compounds and constitute the steps of industrial processes, systematic presentation of thermodynamic data for isomerization reactions merits particular interest.

Because there are general regularities in the variation of thermodynamic properties with the structure of molecules, it is deemed inexpedient, in our opinion, to systematize the thermodynamic properties with reference to isomerization reactions of definite classes of compounds. For this reason the present handbook includes the thermodynamic parameters that have been determined and published during the past 25 years for about 1000 isomerization reactions of organic compounds of different classes and containing the C, H, N, O, Cl, F, B, I, S, P, and other atoms.

This handbook contains primarily the results of investigations of the equilibria of isomers and, in part, calorimetric data. This approach is motivated by the fact that in the 1960s there commenced an extensive application of highly efficient physicochemical methods (gas-liquid chromatography, nuclear magnetic resonance, etc.) that expanded significantly the possibility of carrying out isomer studies and enhanced markedly the accuracy of analyses.

The results of experimental studies cited in the tables are consistent with many propositions advanced by the present authors in the monograph *Thermodynamics and Equilibrium of Isomers*. In our opinion, this handbook will be conducive to expanding the range of thermodynamic data pertaining to organic compounds. The results of equilibrium studies may also be useful for calculating the properties of individual substances and for inclusion in information banks.

The authors express their deep gratitude to Professors V. P. Belousov and A. M. Rozhnov for valuable suggestions made in the course of reviewing the manuscript of this handbook.

INTRODUCTION

1. SPECIFIC ASPECTS OF DETERMINING THE THERMODYNAMIC PROPERTIES OF ISOMERIZATION REACTIONS

The thermodynamic functions presented systematically in this handbook were obtained primarily from the results of equilibrium studies of isomerization reactions and by measurements of the heats of combustion of isomer groups by the calorimetric method. In thermodynamic investigations of isomerization processes, each of these methods has advantages and disadvantages. When dealing with chemical processes of a closely allied nature, the error in small values of the thermodynamic functions of isomerization reactions determined by equilibrium studies is as a rule less than that in the respective parameters determined from calorimetric experiments. The values of $\Delta_r H°$ calculated from calorimetric data on the heats of formation are the differences (mostly small ones) between high values of $\Delta_f H°$, whereas in chemical equilibrium studies the values of $\Delta_r H°$ and $\Delta_r S°$ are found directly from the relationship $K_p = f(T)$. Moreover, in calorimetric measurements, the presence of minor amounts of impurities is likely to result in significant errors in the values of enthalpies of formation used for determining $\Delta_r H°$, whereas the impurities present even in relatively high concentrations have practically no effect on the results of equilibrium studies.

In practice, however, the selection of an investigation technique is not infrequently governed by its feasibility. Thus, when in calorimetric measurements complete combustion cannot be attained, the technique of combustion calorimetry is inapplicable. On the other hand, the equilibrium method fails when individual reactions are nonselective, "false" equilibria take place, or the concentration of a reaction mixture component is low.

Procedure employed to study isomerization equilibria. Use was made of either a flow or a static system to investigate the equilibria in isomerization processes. It should be borne in mind that calculating the thermodynamic functions of gas-phase reactions from data derived from studies liquid-phase equilibria in a static system is, more often than not, based on the assumption that the activity coefficients of isomers are equal, although this is not invariably the case. Moreover, the static system is not necessarily ideal in terms of thermodynamics, conceivably because of the use of diverse solvents capable of forming associations. Accordingly, when solvents are used in static systems, caution should be exercised when the system is designated by a symbol denoting the standard state. The advantages of the static method are the possibility of performing the experiments expeditiously and the simplicity of the apparatus. This method is also useful where it is desired to extend the temperature range (ΔT) of studies conducted in flow systems.

Catalysts. Isomer equilibria are established in both the absence and presence of catalysts. Recourse is made to catalysts in order to attain isomerization equilibria by lowering high potential barriers involved in the conversions. The catalysts used may be in various states of aggregation: gaseous (e.g., NO in olefin isomerizations), liquid (e.g., a mixture of t-C_4H_9OH and t-C_4H_9ONa in isomerizations of unsaturated sulfur-containing compounds), and solid (e.g., mixed metal oxides supported on γ-Al_2O_3). In flow systems, solid catalysts are almost always employed, whereas in liquid systems both liquid and solid catalysts are useful. Gaseous catalysts find very limited application. In isomerizations, the requirements to be met by catalysts are the same as in other processes—selectivity, a broad operating temperature range, preparation simplicity, mechanical strength (for solid catalysts), etc. In some instances, the catalysts used should satisfy other requirements, dictated, for example, by the need to eliminate the formation of stereospecific complexes on the surface of solid catalysts that might be responsible for the dissimilarity in the equilibrium and final isomer ratios. In the absence of catalysts, facile equilibrium establishment takes place when potential barriers involved in conversions are low. This situation is observed in tautomerizations (keto-enol, amine-imine, lactim-lactam, and other tautomerisms). It should be noted that isomer equilibria are studied not only in isomerizations but also in other reversible reactions—hydrogenation-dehydrogenation, hydration-dehydration, hydrohalogenation-dehydrohalogenation, etc.

Analysis of equilibrium mixtures. The quantitative analysis of isomer mixtures was performed by spectral (NMR, IR, RS, and UV spectroscopy) and chromatographic (gas-liquid and gas-solid chromatography) methods, among which the NMR, GLC, and UV methods have of late been used most extensively. Spectral analysis techniques are rapid, do not affect the chemical composition of the mixtures being analyzed, and are, therefore, adapted for analyzing processes noted for their low potential barriers. Because of their sensitivity to minor structural variations, the spectral techniques are indispensable for equilibrium studies of stereoisomeric molecules. Thus, most of the tautomeric conversions were studied by the UV spectroscopic technique. NMR spectroscopy was used in a significant part of studies concerned with the equilibria of cyclic diastereomers and conformers. The analysis of multicomponent isomer mixtures and side-reaction products by spectral techniques is often ruled out or hindered because of spectral line overlapping, and in such cases gas-liquid chromatography is useful. The features of GLC are that a small amount of substance is required for analysis and in-line arrangement of the reactor and the analytical instrument is possible. The chromatographic method is generally more precise in determining quantitatively the concentrations of components than the spectral method, but it requires a preliminary "fixation" of the reaction and chemical stability of the mixture in the course of analysis. The GLC method is employed to analyze equilibrium reactions in position and stereoisomerizations.

Errors in thermodynamic values. Errors in thermodynamic properties found from the equilibrium data are governed primarily by factors that depend on the experimental conditions: the accuracy of maintaining the requisite temperature, the temperature interval (ΔT) used, equilibrium fixation rigidity, the reliability of proof of equilibrium attainment, the effect of side reactions, and the accuracy of equilibrium mixture analysis. As a rule, in the case of isomerization equilibrium studies, maintaining the temperature to $\pm 0.1 - \pm 1$ K is adequate in view of the fact that K_p is quite temperature insensitive.

Expanding the temperature interval of equilibrium investigations generally increases the reliability of determining the temperature dependence of K_p, decreases the effect of random errors, and, as a result, decreases the errors in the values of $\Delta_r H°$ and $\Delta_r S°$ in the processes being investigated. On the other hand, excessive broadening of ΔT might cause the temperature-dependent variations of $\Delta_r H°$ and $\Delta_r S°$ associated with the dissimilarity in heat capacities of the isomers to be greater than the inherent experimental errors. For the most part, in the case of reactions in which $\Delta_r H° \leq 10$ kJ mol^{-1}, ΔT equals 100–150 K. The cited values of $\Delta_r H°$ and $\Delta_r S°$ generally pertain to the average temperature of the investigated interval or are recalculated in terms of 298 K.

To ensure the reliability of chemical equilibrium fixation, use is made of such techniques as rapid and significant temperature lowering, catalyst decomposition or removal, and dilution to diminish the concentration of reactants. No need for fixation arises if continuous analysis in the reaction zone is used.

The validity of the attained reaction equilibrium is ascertained by establishing that the equilibrium has not resulted from an incomplete reaction course, is not the result of side factors, and is not a random phenomenon. Proof is provided by the fact that the equilibrium is attained in terms of both the starting substances and the reaction products. In this case, mixtures having various compositions of starting reactants and reaction products are used. In these experiments, the condition of equilibrium state onset is that the value of K_p, the equilibrium constant, should vary only within the range of a normal experimental error. The constancy of K_p is to remain when the time of reaction mixture contact with a catalyst and the duration of thermal action on reactants in non-catalytic processes are varied. For catalytic processes in static systems, the experiments are conducted by varying the catalyst volume, whereas in flow systems both the catalyst volume and the space velocity of feedstock are varied. In equilibrium investigations it is also essential to ascertain that the development of side reactions neither affects the value of K_p nor distorts it. Unfortunately, in many investigations no due attention is given to the proof of equilibrium validity.

Errors in $\Delta_r H°$ and $\Delta_r S°$ are found by the methods of linear regressive analysis from the linear relationship

$$\ln K_p = f\left(\frac{1}{T}\right), \quad \ln K_p = \frac{\Delta_r S°}{R} - \frac{1}{T}\frac{\Delta_r H°}{R}$$

The error in $\Delta_r G°$ of isomerization reactions is found as the root-mean-square deviation from the arithmetic mean value.

2. QUALITATIVE REGULARITIES OF CHANGES IN THE ENTHALPY AND ENTROPY FUNCTIONS OF ISOMERIZATIONS

According to the postulates of the classical theory of the structure of molecules, chemical isomers may be divided into two groups:

1. Structural isomers differing from one another by a dissimilar sequence of atomic bonds in molecules whose structure can be presented using two-dimensional molecular models, that is, chemical formulas.
2. Stereoisomers, which are not identical because of a dissimilar arrangement of atoms but have identical structural parameters.

Structural isomers, in turn, can be subdivided into three types:

1. Functional isomers that differ from one another in the valence state of atoms or in the type of chemical bonds in molecules.
2. Cyclic isomers, in which the molecules contain cycles of different size or differ from one another by the mode of joining the cycles.

3. Position isomers, which differ from one another in structure but have the same valence state of atoms, as well as the same bond types and cyclic characteristics of the molecules.

Among stereoisomers, two types are conventionally distinguished:

1. Configuration isomers, which differ only in the spatial arrangement of atoms and fail to become identical as a result of intramolecular rotations and rearrangements without bond rupture.
2. Conformation isomers, comprising molecules that correspond to certain potential minima in intramolecular motion.

The classification of isomers may serve as a basis for the classification of isomerization reactions, but it should be remembered that the conversions of isomers may be of a complex nature.

Generalization of experimental data listed in the tables and elucidation of regularities in the thermodynamic properties as a function of isomer structure are done most successfully in keeping with the principles of the classical theory of molecular structure. However, requisite experimental data for such an analysis are often absent, so only qualitative relationships in the variations of the thermodynamic properties of interest in isomerizations will be discussed below.

Functional isomerizations constitute the smallest group of isomer conversions investigated experimentally. The tables list, for the most part, tautomeric rearrangements of the functional isomerism type, such as lactim-lactam, imine-enamine, cyanide-isocyanide, and some other rearrangements. The accuracy of thermodynamic values for tautomeric rearrangements is generally not high because of difficulties inherent in determining the composition of interconvertible compounds and high sensitivity of equilibrium ratios of the principal components to the solvent and catalyst types used and the reaction conditions.

The thermodynamic quantities $\Delta_r H°$ for functional reactions of isomerization may likewise be calculated from the calorimetric data, but in the tables such values are listed only when the appropriate thermodynamic values of $\Delta_f H°$ for the isomers were cited in the same publication. To generalize regarding the reactions of functional isomerization presents difficulties because the available experimental data for this class of compounds are scarce and nonsystematic.

Cycle isomerization reactions involving cycle size changes are presented in the tables primarily by the isomer conversions of hydrocarbons. In many cases the value of K_p is very low due to marked differences in the thermodynamic stability of these compounds, which makes equilibrium studies of this type difficult. Accordingly, the majority of $\Delta_r H°$ values for these reactions were found by calorimetric measurements. The relative energy stability of cycloalkane isomers is in good agreement with the energies of ring straining in molecules, which decrease in the following order (kJ mol^{-1}):

△ — □ — ⬭ — ⬠ — ⬡ — ⬢

double bond positions above arrows

115.5   111.3   93.5   26.8   25.1   0.8

This series makes it possible to assess qualitatively the changes in isomerization enthalpy with cycle size variation: maximum stability is displayed by six-membered cycloalkanes, the three- and four-membered compounds being less stable:

methylcyclopentene ⇌ cyclohexene $\Delta_r H^\circ_{318} = -18.07$ kJ mol^{-1}

allylcyclobutane ⇌ vinylcyclohexane $\Delta_r H^\circ_{298} = -23.42$ kJ mol^{-1}

ethylcyclopentene ⇌ methylcyclohexene $\Delta_r H^\circ_{298} = -22.88$ kJ mol^{-1}

bicyclopentyl ⇌ decalin (trans) $\Delta_r H^\circ_{298} = -55.2$ kJ mol^{-1}

bicyclobutane ⇌ cyclobutene $\Delta_r H^\circ_{298} = -60.2$ kJ mol^{-1}

bicyclobutane ⇌ CH$_2$=CH—CH=CH$_2$ $\Delta_r H^\circ_{298} = -48.1$ kJ mol^{-1}

cubane ⇌ methylenecyclohexene $\Delta_r H^\circ_{298} = -125.45$ kJ mol^{-1}

Assessing the value of $\Delta_r H^\circ$ from the strain energies of rings presents greater difficulties when isomerizations involve simultaneous changes in the mode of cycle linkage with one another and in the size of cycles or, in the case of polycyclic compounds, involve a plurality of cycles. The dependence of isomerization enthalpy on the size of cycles is not manifest in the isomerizations of perfluorocyclic compounds:

[Structure: perfluorocyclobutene] ⇌ CF_2=CF—CF=CF_2 $\Delta_r H^\circ_{298}$ = 50.2 kJ mol^{-1}

[Structure: 1,2-bis(trifluoromethyl)perfluorocyclobutene] ⇌ [(CF$_3$)(CF$_2$)C—C(CF$_3$)(CF$_2$)] $\Delta_r H^\circ_{595}$ = 1.67 kJ mol^{-1}

The *cycle isomerization reactions involving changes in the mode of cycle linkage* comprise essentially the conversions of polycyclic compounds into spiro compounds and of compounds with two-membered cycles linked via a chain of atoms into conjugated cycles linked with one another via a single atom.

Many reactions in which polycyclic compounds are converted into spiro compounds are characterized by lowering of the energy stability of spirocyclic compounds compared to that of bicyclic compounds, in particular, of cycloalkanes:

[1-methylcyclobutene fused] ⇌ [methylenecyclobutane fused] $\Delta_r H^\circ_{298}$ = 3.76 kJ mol^{-1}

[1-ethylcyclopentene] ⇌ [ethylidenecyclopentane] $\Delta_r H^\circ_{298}$ = 1.63 kJ mol^{-1}

[1-methylcyclohexene] ⇌ [methylenecyclohexane] $\Delta_r H^\circ_{298}$ = 7.15 kJ mol^{-1}

[3-methyl-1-methylcyclohexene with CH$_3$] ⇌ [2-methyl-1-methylenecyclohexane with CH$_3$] $\Delta_r H^\circ_{298}$ = 7.95 kJ mol^{-1}

However, in three-membered cyclic compounds and in bicyclic compounds the formation of compounds with a two-membered cycle in the spiro position may result in stability enhancement:

[Reaction 1: methylcyclopropane ⇌ methylenecyclopropane]

$$\Delta_r H^\circ_{298} = -42.67 \text{ kJ mol}^{-1}$$

[Reaction 2: methylnorbornene ⇌ methylenenorbornane]

$$\Delta_r H^\circ_{298} = -8.53 \text{ kJ mol}^{-1}$$

The energy stability of molecules increases by 3–17 kJ mol^{-1} in the reactions in which compounds with two-membered cycles linked through a chain of atoms undergo conversion to conjugated cycles:

$$CH_2=C-(CH_2)_2-C=CH_2 \rightleftharpoons \begin{array}{c} CH_3 \\ | \\ CH_3O \end{array} C=C \begin{array}{c} H \\ \\ H \end{array} C=C \begin{array}{c} OCH_3 \\ \\ CH_3 \end{array}$$
$$\begin{array}{cc} | & | \\ CH_3O & OCH_3 \end{array}$$

$$\Delta_r H^\circ_{298} = -17.5 \text{ kJ mol}^{-1}$$

[Reaction with methoxy-substituted structures]

$$\Delta_r H^\circ_{298} = -4.0 \text{ kJ mol}^{-1}$$

For the majority of *position isomerization reactions*, $|\Delta_r H^\circ| < 10$ kJ mol^{-1} and $|\Delta_r S^\circ| < 10$ J · mol^{-1}·K^{-1}, provided the differences between the symmetry numbers of the isomers are not great. Therefore, in the moderate temperature range (300–500 K) such isomers form equilibrium mixtures in which their concentrations are of the same order of magnitude. Accordingly, there arise favorable conditions for the experimental determination of equilibrium constants within adequately broad temperature intervals. But a limited number of reactions of this type are characterized by $|\Delta_r H^\circ| > 20$ kJ mol^{-1}:

$$CH_3-(CH_2)_{10}-CH_3 \rightleftharpoons CH_3-\underset{\underset{CH_3}{|}}{\overset{\overset{CH_3}{|}}{C}}-CH_2-\underset{}{\overset{\overset{CH_3}{|}}{CH}}-CH_2-\underset{\underset{CH_3}{|}}{\overset{\overset{CH_3}{|}}{C}}-CH_3$$

$$\Delta_r H^\circ_{298} = -24.52 \text{ kJ mol}^{-1}$$

[structure] ⇌ [structure] $\Delta_r H^\circ_{298} = -23.88$ kJ mol^{-1}

The tables list the thermodynamic functions for the equilibria of position isomerization reactions in which changes occur in the degree of carbon chain branching or in the alkyl group substituents at the cycles or at double bonds (two-membered cycles), or else substituents migrate along the chain or in the cycles and cycles (mostly two-membered ones) undergo transposition in the alkyl chain.

In specific cases, such rearrangements cause no changes in the "species" of atoms in terms of their surroundings or bonds in the molecules. The reactions in question pertain to the type of position isomerizations in which the isomers belong to the same "family" in terms of atoms or bonds.

For such isomers the values of $\Delta_f H^\circ$ are close to each other, so the enthalpies of isomerizations are low.

$$C_2H_5-\underset{\underset{CH_3}{|}}{CH}-C_4H_9 \rightleftharpoons C_3H_7-\underset{\underset{CH_3}{|}}{CH}-C_3H_7 \quad \Delta_r H^\circ_{443} = -0.52 \text{ kJ mol}^{-1}$$

$$C_2H_5-\underset{\underset{Cl}{|}}{CH}-C_4H_9 \rightleftharpoons C_3H_7-\underset{\underset{Cl}{|}}{CH}-C_3H_7 \quad \Delta_r H^\circ_{400} = 0.23 \text{ kJ mol}^{-1}$$

$$CH_3-CH_2-CH=CH-C_3H_7 \rightleftharpoons CH_3-CH=CH-C_4H_9$$
(trans-) (trans-)
$$\Delta_r H^\circ_{298} = -0.21 \text{ kJ mol}^{-1}$$

The enthalpies of reactions involving conversions of isomers belonging to different families vary from ~20 kJ mol^{-1} to values close to zero, whereas entropy changes are markedly affected by the differences in the symmetry numbers of isomer molecules:

$$CH_3-(CH_2)_6-CH_3 \rightleftharpoons CH_3-\underset{\underset{CH_3}{|}}{\overset{\overset{CH_3}{|}}{C}}-\underset{\underset{CH_3}{|}}{\overset{\overset{CH_3}{|}}{C}}-CH_3 \quad \Delta_r H^\circ_{298} = -16.94 \text{ kJ mol}^{-1}$$

$$Cl-CH_2-CH_2-CH_3 \rightleftharpoons CH_3-\underset{\underset{Cl}{|}}{CH}-CH_3 \quad \Delta_r H^\circ_{471} = -13.39 \text{ kJ mol}^{-1}$$

$$CH_2=\overset{\underset{|}{CH_3}}{C}-CH_2-CH_2-CH_3 \rightleftharpoons CH_2=CH-CH_2-\overset{\underset{|}{CH_3}}{CH}-CH_3$$

$$\Delta_r H^\circ_{394} = 6.9 \text{ kJ mol}^{-1}$$

[cyclopentene isomerization figure] $\Delta_r H^\circ_{465} = 7.49 \text{ kJ mol}^{-1}$

[cyclopropane substituted isomerization figure] $\Delta_r H^\circ_{298} = -5.02 \text{ kJ mol}^{-1}$

[ethyl-benzene to methyl-propylbenzene figure] $\Delta_r H^\circ_{298} = 0.38 \text{ kJ mol}^{-1}$

Configurational isomerization is observed in both acyclic and cyclic compounds. The enthalpies of conversion of diastereomeric acyclic compounds are generally $|\Delta_r H^\circ| < 10 \text{ kJ mol}^{-1}$ and regarding their order of magnitude are close to the differences in the energy of rotational isomers in respective classes of compounds:

$$CH_3-\overset{\underset{|}{Cl}}{CH}-\overset{\underset{|}{CH_3}}{CH}-C_2H_5 \rightleftharpoons CH_3-\overset{\underset{|}{Cl}}{CH}-\overset{\underset{|}{CH_3}}{CH}-C_2H_5$$

(*threo-*)   (*erythro-*)

$$\Delta_r H^\circ_{348} = 2.38 \text{ kJ mol}^{-1}$$

$$CH_3-\overset{\underset{|}{Cl}}{CH}-\overset{\underset{|}{Cl}}{CH}-C_2H_5 \rightleftharpoons CH_3-\overset{\underset{|}{Cl}}{CH}-\overset{\underset{|}{Cl}}{CH}-C_2H_5$$

(*threo-*)   (*erythro-*)

$$\Delta_r H^\circ_{323} = 1.17 \text{ kJ mol}^{-1}$$

$$CH_3-\overset{\underset{|}{OH}}{CH}-\overset{\underset{|}{CH_3}}{CH}-C_2H_5 \rightleftharpoons CH_3-\overset{\underset{|}{OH}}{CH}-\overset{\underset{|}{CH_3}}{CH}-C_2H_5$$

(*threo-*)   (*erythro-*)

$$\Delta_r H^\circ_{516} = 0.3 \text{ kJ mol}^{-1}$$

Equilibrium conversions of such diastereomers occur in the presence of catalysts, e.g., $AlCl_3$, $AlBr_3$, or BF_3, or in the course of reversible hydrogenation-dehydrogenation, dehydrohalogenation-hydrohalogenation, etc.

The reactions of configurational isomerization of cyclic compounds listed in the tables include the reactions of *cis* ⇌ *trans* isomerization of carbocyclic compounds with three to six carbon atoms in the cycle. The values of $|\Delta_r H°|$ in such conversions may vary in a rather broad interval and should be interpreted in terms of conformational analysis.

$\Delta_r H°_{298} = 11.46$ kJ mol^{-1}

$\Delta_r G°_{298} = -1.17$ kJ mol^{-1}

$\Delta_r G°_{500} = 7.53$ kJ mol^{-1}

$\Delta_r H°_{563} = 3.9$ kJ mol^{-1}

$\Delta_r H°_{422} = -19.28$ kJ mol^{-1}

The thermodynamic functions of configurational isomerization reactions are generally determined from equilibrium data. Calorimetric measurements are made difficult by the complexity of separating pure specimens of configuration isomers. Moreover, the values of $\Delta_r H°$ for configurational conversions based on calorimetric data have, as a rule, low accuracy because of the closeness of the $\Delta_f H°$ values for this group of isomers. However, if pure specimens of configuration isomers are available, calorimetric determination of the enthalpies of isomerization may be quite effective, e.g., in the case of cis ⇌ trans isomerization of decalin:

(*trans*-)   (*cis*-)

$$\Delta_r H°_{298} = -12.93 \pm 3.22 \text{ kJ mol}^{-1}$$

The equilibrium of *cis* ⇌ *trans* (Z ⇌ E) isomerizations in relation to the double bond is readily attained in reversible catalytic reactions of addition to the double bond, followed by the elimination of hydrogen, halogens, hydrogen halides, water, etc. To obtain exact values of $\Delta_r H°$ and $\Delta_r S°$ for *cis* ⇌ *trans* isomerizations, carrying out equilibrium investigations in the 150–250 K range should suffice. Chromatographic analysis provides a high accuracy of determining the ratio of these configurational isomers.

An analysis of the data listed in the tables justifies the following conclusions:

1. Irrespective of the size of substituents at the double bond, *trans* (E) isomers in the alkene and alkadiene series are generally more stable energetically than *cis* (Z) isomers by 3–6 kJ mol^{-1}.
2. Where in the substituents at the alkene double bond the radicals display maximum branching, the energy content difference between E and Z isomers grows to as much as 16.5–44.0 kJ mol^{-1}.
3. In the E ⇌ Z isomerization of vinyl ethers, the position of —OR— substituents affects the difference in the energy content of isomers less than does the type of hydrocarbon radicals—substituents at the double bond.

3. DESCRIPTION OF TABLES

The tables list the calorimetric values $\Delta_r H°$ for isomerization reactions only when the cited paper contained the results of $\Delta_f H°$ measurements for more than one isomer. If the reference, in addition to equilibrium constants or equilibrium concentrations determined experimentally at one (preferably room) temperature, also cited the values of the standard Gibbs energy of formation, the experimental data were included in the tables. Otherwise, these data were disregarded because the reliability of the available thermodynamic information (e.g., equilibrium attainment from both sides, reproducibility) could not be assessed.

When the same reference cited data for both the liquid and gaseous states, the data for the gaseous state only were included in the tables. In this case, if a specific solvent is listed in column 7, it means that the experimental results were obtained for a solution and thereafter recalculated for the gas phase. When the same reference contains data on isomer conversions in solutions and cites the use of multiple solvents, the listed data were obtained in the presence of "nonselective" solvents, such as benzene, cyclohexane, or carbon tetrachloride. Reference to the solvent used is particularly essential for the presentation of thermodynamic functions of stereoisomeric conversions of alicyclic five- and six-membered systems, as well as in the case of structural tautomeric conversions.

The tables include no results for the thermodynamics of conformational and partially stereoisomeric conversions in carbocyclic five- and six-membered systems that were obtained before 1965, since the results were presented in a systematic manner in the monograph *Conformational Analysis* by E. Eliel et al. (Mir, 1969).

All experimentally investigated isomerization reactions are nonuniformly distributed among reaction types. The share of functional and functional-cyclic isomerizations is only about 9%, that of cyclic isomerization is under 18%, and the share of position isomerization reactions is about 40%. This lack of uniformity in equilibrium investigations is partly due to objective factors (high values of $\Delta_r G°$ and $\Delta_r H°$, considerable difficulties encountered in choosing active and selective catalysts that make for the attainment of equilibrium ratios of isomers, absence of reliable proof of equilibrium attainment, etc.). In tautomeric rearrangements, particularly facile isomerization renders experimental techniques excessively sophisticated.

In the case of position isomerization, where the nature of molecules undergoes no such pronounced changes, equilibrium ratios are attained more readily in adequately broad temperature ranges, so the thermodynamic parameters of numerous reactions of this type can be determined very accurately and the investigations are attractive to experimentalists. It should be noted, however, that studies concerned with a large number of similar reactions (reactions of the same family) are often unwarranted, because of their immaterial informative value regarding the effect of molecular structure on the thermodynamic properties of substances. This consideration likewise applies to the group of configurational (geometric) *cis* ⇌ *trans* isomerization reactions that hold a conspicuous place in the present handbook.

It must be admitted that the selection of objects for studying isomer equilibria was largely sporadic and not infrequently was dictated by the chemical aspects of the reactions. In studies of correlations between the thermodynamic properties of isomers and the structure of molecules, it appears expedient to resort to a combination of isomer equilibria investigations and calorimetric measurements. This approach will make it possible to carry out an independent check of thermodynamic quantities, to analyze subtle aspects of the dependence of thermodynamic properties on isomer structure, and to obtain the sum total of thermodynamic properties of compounds that incorporate all structural fragments required for bringing out, on the basis of the classical theory of molecular structure, the regularity inherent in the thermodynamic properties of interest.

Column 1 lists the molecular formulas for isomers using the following sequence of atom positions: the carbon atom (C) comes first, followed by hydrogen (H). Other atoms are arranged in alphabetic order using the first letter of the Latin name of the elements. A conventional molecular formula, $C_xH_yA_aB_bD_d$...., serves to illustrate the system of arranging the compounds in tables. Initially, the number of atoms in the last of the elements in the formula (e.g., D) is increased to its maximum value $(d + n)$, followed by increasing the number of atoms in the element B, etc., the last being the number of carbon atoms in the formula. An exception to this rule is provided by compounds that contain no carbon atoms (for example, the isomer conversions of the compounds of the molecular formula F_2N_2 are the first compounds listed in the table) or hydrogen (the compounds containing no hydrogen atoms are placed at the beginning of a list of carbon-containing compounds whose molecular formula has the same number of carbon atoms).

In column 2, isomer conversions are designated as follows:

1.1—functional structural isomerizations
1.2—cyclic structural isomerizations
1.3—position structural isomerizations
2.1—configurational stereoisomerization

In column 3, isomer types are specified in brackets by the prefixes *endo-*, *exo-*, *theor-*, *erythro-*, *meso-*, *syn-*, *anti-*, *cis-*, and *trans-*. The formation of a mixture of *d* and *l* isomers is designated by the symbols *d* and *l*. When several dissimilar substituents are present in a compound, the resulting isomers cannot be unambiguously referred to as *cis-* or *trans-* forms, so the prefixes *cis-* and *trans-* merely point to the formation of a single isomer with a structure defined by the listed chemical formula. The following notation is used for branched radicals: isopropyl, i-C_3H_7; isobutyl, i-C_4H_9; and tertiary butyl radical, -t-C_4H_9. All other hydrocarbon radicals are assumed to be primary, straight-chain radicals. A low-boiling isomer is denoted by the letters "l.b." and a high-boiling one by the letters "h.b."

Column 4 lists the absolute temperatures, to which respective thermodynamic properties are referred, and also defines the state of aggregation of isomers that undergo a given conversion: g, gaseous; l, liquid; and s, solid state.

Column 5 presents enthalpy values and, in brackets, changes in the Gibbs free energy for isomer conversions at a temperature T (dimensions of the quantity are kJ mol^{-1}). If the error of the listed values was determined, it is cited after the sign ±.

Column 6 contains the values of entropy change in isomerizations and specifies relevant measurement errors (dimensions of the quantity, J · mol^{-1} · K^{-1}.

Column 7 presents information about the analytical techniques, methods of determining the thermodynamic properties, and presence of a catalyst and solvent in the reaction mixture.

The methods of determining the thermodynamic properties are designated as follows:

SE—equilibrium in a static system
FE—equilibrium in a flow system
CC—combustion calorimetry
RC—reaction calorimetry
ST—statistical computation method
LT—low-temperature polarimetric jump method
TP—low-temperature heat capacity measurement method

Where several methods have been required to obtain the set of thermodynamic properties, mention is made of all the techniques used. The presence of a catalyst in the reaction mixture is designated by the CT symbol.

The techniques of equilibrium system analysis are designated as indicated below:

GLC—gas-liquid chromatography
NMR—nuclear magnetic resonance spectroscopy
IR—infrared spectroscopy
UV—ultraviolet spectroscopy
MS—mass spectrometry
RSS—Raman scattering spectroscopy

A chemical compound formula, if listed in column 7, pertains to the solvent used in the reaction system.

Thermodynamic Properties of Isomerization Reactions

Formula	Type	Reaction	T, K (phase)	$\Delta_r H_T^\circ$ ($\Delta_r G_T^\circ$)	$\Delta_r S_T^\circ$	Technique	Reference
1	2	3	4	5	6	7	8
F_2N_2	2.1	(cis-) F–N=N–F ⇌ F–N=N–F (trans-)	298 (g)	12.5 ±1.25	—	CC	1
			361 (g)	0.0	−18.70	SE, GLC	2
			298 (g)	—	−10.25	ST	3
H_8B_5Cl	1.3	⇌	398 (l)	2.18	21.80	SE, NMR $(C_2H_5)_2O$	4
CH_3NO_2	1.1	CH_3–O–N=O ⇌ CH_3–NO_2	298 (g)	−0.54	−0.29	CC, ST	5
C_2HClF_2	2.1	(cis-) ⇌ (trans-)	591 (g)	(0.34 ±0.01)	—	SE, CT, GLC	6

Formula		Structures			Methods	Ref
$C_2H_2Br_2$	1.3	$CH_2=CBr_2 \rightleftharpoons$ $\underset{H}{\overset{Br}{\diagdown}}C=C\underset{H}{\overset{Br}{\diagup}}$	591 (g)	—	(-0.50 ± 1.09)	ST	6
		$CH_2=CBr_2 \rightleftharpoons$ $\underset{H}{\overset{Br}{\diagdown}}C=C\underset{Br}{\overset{H}{\diagup}}$	628 (g)	-5.44 ± 1.05	1.05 ± 0.17	FE, CT, GLC	7
	2.1	$\underset{H}{\overset{Br}{\diagdown}}C=C\underset{H}{\overset{Br}{\diagup}} \rightleftharpoons \underset{H}{\overset{Br}{\diagdown}}C=C\underset{Br}{\overset{H}{\diagup}}$	628 (g)	-4.31 ± 0.33	4.60 ± 0.33	FE, CT, GLC	7
C_2H_2ClF	2.1	$\underset{H}{\overset{Cl}{\diagdown}}C=C\underset{H}{\overset{F}{\diagup}}$ (cis-) $\rightleftharpoons \underset{H}{\overset{Cl}{\diagdown}}C=C\underset{F}{\overset{H}{\diagup}}$ (trans-)	577 (g)	1.13 ± 0.08	3.18 ± 0.33	FE, CT, GLC	7
$C_2H_2Cl_2$	2.1	$\underset{H}{\overset{Cl}{\diagdown}}C=C\underset{H}{\overset{Cl}{\diagup}}$ (cis-) $\rightleftharpoons \underset{H}{\overset{Cl}{\diagdown}}C=C\underset{Cl}{\overset{H}{\diagup}}$ (trans-)	615 (g)	3.26 ± 0.08	0.88 ± 0.17	SE, CT, GLC	250
$C_2H_2F_2$	2.1	$\underset{H}{\overset{F}{\diagdown}}C=C\underset{H}{\overset{F}{\diagup}}$ (cis-) $\rightleftharpoons \underset{H}{\overset{F}{\diagdown}}C=C\underset{F}{\overset{H}{\diagup}}$ (trans-)	298 (g)	2.59	0.0	SE, ST, CT	9
			619 (g)	3.88	5,60	SE, CT, GLC	10
$C_2H_2I_2$	2.1	$\underset{H}{\overset{I}{\diagdown}}C=C\underset{H}{\overset{I}{\diagup}}$ (cis-) $\rightleftharpoons \underset{H}{\overset{I}{\diagdown}}C=C\underset{I}{\overset{H}{\diagup}}$ (trans-)	595 (g)	0.0	4.60	SE, IR	11

Table (Continued)

Formula	Type	Reaction	T, K (phase)	$\Delta_r H_T^\circ$ ($\Delta_r G_T^\circ$)	$\Delta_r S_T^\circ$	Technique	Reference
$C_2H_3Cl_3$	1.3	$H-\underset{\underset{Cl}{\mid}}{\overset{\overset{Cl}{\mid}}{C}}-CH_2-Cl \rightleftarrows Cl-\underset{\underset{Cl}{\mid}}{\overset{\overset{Cl}{\mid}}{C}}-CH_3$	423 (l)	8.37	−9.20	SE, CT, GLC	12
C_2H_3N	1.1, 1.2	$CH_3-N=C \rightleftarrows CH_3-C\equiv N$	300 (g)	−99.14 ±0.59	—	RSS	13
$C_2H_4Br_2$	1.3	$CH_3-\underset{\underset{Br}{\mid}}{\overset{\overset{Br}{\mid}}{C}}-H \rightleftarrows Br-CH_2-CH_2-Br$	484 (g)	12.13 ±1.25	5.44 ±0.42	FE, CT, GLC	14
$C_2H_4Cl_2$	1.3	$CH_3-\underset{\underset{Cl}{\mid}}{\overset{\overset{Cl}{\mid}}{CH}} \rightleftarrows Cl-CH_2-CH_2-Cl$	385 (g)	−9.62 ±1.38	6.19 ±1.59	SE, CT, GLC	15
$C_3H_3Cl_3$	2.1	$\underset{(cis\text{-})}{\overset{Cl-CH_2}{\underset{Cl}{>}}C=C\overset{H}{\underset{Cl}{<}}} \rightleftarrows \underset{(trans\text{-})}{\overset{Cl-CH_2}{\underset{Cl}{>}}C=C\overset{Cl}{\underset{H}{<}}}$	358 (g)	−1.13	−0.84	SE, CT, GLC	16
$C_3H_4F_2$	1.3	$CH_2=CH-CHF_2 \rightleftarrows CH_3-CH=CF_2$	396 (g)	10.46	1.67	FE, CT, GLC	17
C_3H_5Br	1.3	$CH_2=CH-CH_2-Br \rightleftarrows CH_3-CH=CH-Br$ (cis- + trans-)	473 (g)	−3.01 ±0.42	—	SE, CT, GLC	18
	2.1	$\underset{(cis\text{-})}{\overset{CH_3}{\underset{H}{>}}C=C\overset{Br}{\underset{H}{<}}} \rightleftarrows \underset{(trans\text{-})}{\overset{CH_3}{\underset{H}{>}}C=C\overset{H}{\underset{Br}{<}}}$	473 (g)	3.05 ±0.42	—	SE, CT, GLC	18

Formula		Reaction	T, K (state)	ΔH	ΔG	Method	Ref.
C_3H_5Cl	1.3	$CH_2=CH-CH_2-Cl \rightleftarrows$ $CH_3\diagdown C=C \diagup Cl$ $H \diagup \diagdown H$ (cis-)	631 (l)	1.84 ±0.10	−2.84 ±0.17	SE, GLC	19
		$CH_2=CH-CH_2-Cl \rightleftarrows$ $CH_3\diagdown C=C \diagup H$ $H \diagup \diagdown Cl$ (trans-)	552 (g)	−14.64 ±0.84	−12.55 ±1.25	SE, CT, GLC	20
		$CH_2=CH-CH_2-Cl \rightleftarrows CH_3-CH=CH-Cl$ (cis- + trans-)	552 (g)	−10.88 ±0.84	−11.71 ±1.25	SE, CT, GLC	20
	2.1	$CH_2=CH-CH_2-Cl \rightleftarrows$ $CH_3\diagdown C=C \diagup Cl$ $H \diagup \diagdown H$ (cis-)	473 (g)	−9.96 ±0.42	—	SE, CT, GLC	18
		$CH_3\diagdown C=C \diagup H$ $H \diagup \diagdown Cl$ (trans-)	473 (g)	3.18 ±0.13	—	SE, CT, GLC	18
C_3H_5F	1.3	$CH_2=CH-CH_2-F \rightleftarrows$ $CH_3\diagdown C=C \diagup F$ $H \diagup \diagdown H$ (cis-)	483 (g, l)	2.93 ±0.27	−0.13 ±0.71	SE, FE, CT, GLC	21
		$CH_2=CH-CH_2-F \rightleftarrows$ $CH_3\diagdown C=C \diagup H$ $H \diagup \diagdown F$ (trans-)	387 (g)	−13.97	−5.86	FE, CT, GLC	17
		$CH_2=CH-CH_2-F \rightleftarrows$ $CH_3\diagdown C=C \diagup H$ $H \diagup \diagdown F$ (trans-)	387 (g)	−11.21	−7.11	FE, CT, GLC	17
		$CH_2=CH-CH_2-F \rightleftarrows CH_3-CH=CH-F$ (cis- + trans-)	473 (g)	−17.35 ±0.42	—	SE, CT, GLC	18

Table (Continued)

Formula	Type	Reaction	T, K (phase)	$\Delta_r H_T^\circ$ ($\Delta_r G_T^\circ$)	$\Delta_r S_T^\circ$	Technique	Reference
C_3H_5F	2.1	CH₃\C=C/F (cis-) ⇌ CH₃\C=C/H (trans-) H/ \H H/ \F	473 (g)	3.14 ±0.13	—	SE, CT, GLC	18
C_3H_5I	1.3	$CH_2=CH-CH_2-I$ ⇌ CH₃\C=C/I (cis-) H/ \H	387 (g)	2.68	−1.25	FE, CT, GLC	17
		$CH_2=CH-CH_2-I$ ⇌ CH₃\C=C/H (trans-) H/ \I	546 (g)	(−4.35)	—	SE, CT, GLC	20
			546 (g)	(0.84)	—	SE, CT, GLC	20
C_2H_5N	1.1, 1.2	$C_2H_5-N=C$ ⇌ $C_2H_5-C≡N$	300 (g)	−89.93 ±4.18	—	RSS	13
$C_3H_6Br_2$	1.3	Br-CH₂-CH₂-CH₂-Br ⇌ Br-CH₂-CH(Br)-CH₃	338 (l)	−4.73 ±0.13	3.14 ±0.42	SE, CT, GLC	251
$C_3H_6Cl_2$	1.3	Cl-CH₂-CH₂-CH₂-Cl ⇌ Cl-CH₂-CH(Cl)-CH₃	303 (l)	−6.07 ±0.84	1.67 ±2.51	SE, CT, GLC	22
		Cl-CH₂-CH₂-CH₂-Cl ⇌ Cl₂CH-CH₂-CH₃	464 (g)	16.42 ±3.05	−3.10 ±8.20	FE, CT, GLC	23

Formula		Equilibrium	T (K) (state)			Method	Ref.
C_3H_6O	1.1	$CH_3-\underset{\underset{O}{\parallel}}{C}-CH_3 \rightleftarrows CH_2=\underset{\underset{OH}{\mid}}{C}-CH_3$	298 (l)	35.72	—	RC	24
C_3H_7Br	1.3	$\underset{\underset{CH_2-CH_2-CH_3}{\mid}}{Br} \rightleftarrows \underset{\underset{CH_3-CH-CH_3}{\mid}}{Br}$	481 (g, l)	−12.13 ±1.59	−13.97 ±1.97	SE, FE, CT, IR	25
C_3H_7Cl	1.3	$\underset{\underset{CH_2-CH_2-CH_3}{\mid}}{Cl} \rightleftarrows \underset{\underset{CH_3-CH-CH_3}{\mid}}{Cl}$	471 (g, l)	−13.39 ±0.84	−13.26 ±1.46	SE, FE, CT, GLC	26
C_3H_7I	1.3	$\underset{\underset{CH_2-CH_2-CH_3}{\mid}}{I} \rightleftarrows \underset{\underset{CH_3-CH-CH_3}{\mid}}{I}$	344 (l)	−12.13 ±4.18	−27.19 ±7.95	SE, CT, IR	27
$C_3H_7N_5$	1.1 1.2	(imidazoline-azide tautomerism)	293 (l)	−5.0 ±1.3	−31.0 ±4.2	SE, NMR $CH_3-\underset{\underset{O}{\parallel}}{C}-CH_3$	70
C_4F_6	1.2	$CF_2=CF-CF=CF_2 \rightleftarrows$ (perfluorocyclobutene)	298 (g)	50.2	—	IR	28
C_4F_8	2.1	$\underset{CF_3}{\overset{F}{>}}C=C\underset{CF_3}{\overset{F}{<}}$ (cis-) $\rightleftarrows \underset{F}{\overset{CF_3}{>}}C=C\underset{F}{\overset{CF_3}{<}}$ (trans-)	468 (g)	−3.42 ±0.18	2.03 ±0.11	SE, GLC	29

Table (Continued)

Formula	Type	Reaction	T, K (phase)	$\Delta_r H^\circ_T$ ($\Delta_r S^\circ_T$)	$\Delta_r S^\circ_T$	Technique	Reference
$C_4H_4F_2$	1.3	(difluoromethylene cyclopropane ⇌ difluoromethyl cyclopropene)	488 (l)	−7.95 ±0.42	—	SE, NMR CH_3—CH_2—CH_3	30
$C_4H_5Cl_2NO$	1.1	(Cl$_2$C=CH—NH—C(O)CH$_3$ ⇌ Cl$_2$CH—CH=N—C(O)CH$_3$)	333 (l)	(18.8)	—	SE, NMR C_2H_5OH	31
C_4H_6	1.2	(methylenecyclopropane ⇌ methylcyclopropene)	298 (g)	42.67 ±2.09	—	CC	32
		(cyclobutene ⇌ bicyclobutane)	298 (g)	−60.2 ±2.1	—	CC	32
		(cyclobutene ⇌ CH$_2$=CH—CH=CH$_2$)	298 (g)	−48.1 ±2.1	—	CC	32

Formula		Equilibrium	T (K), phase			Method	Ref.
$C_4H_6O_2$	1.2	 (methyl-1,3-dioxole) ⇌ (methylene-1,3-dioxolane)	298 (g)	−0.5 ±0.6	7.5 ±1.4	SE, CT, GLC	33
			298 (l)	1.1 ±0.5	9.6 ±1.3	SE, CT, GLC, 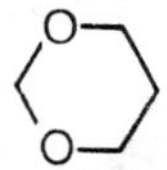 1,4-dioxane	33
C_4H_7Br	2.1	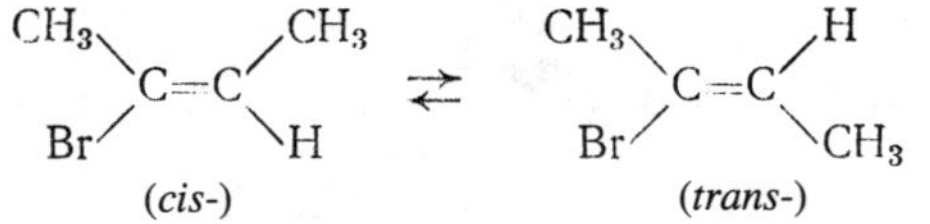 trans-CH$_3$CH=CHBr(CH$_3$) ⇌ cis-	366 (l)	4.81 ±0.21	0.84 ±0.84	SE, CT, GLC, C_6H_{12}	34
		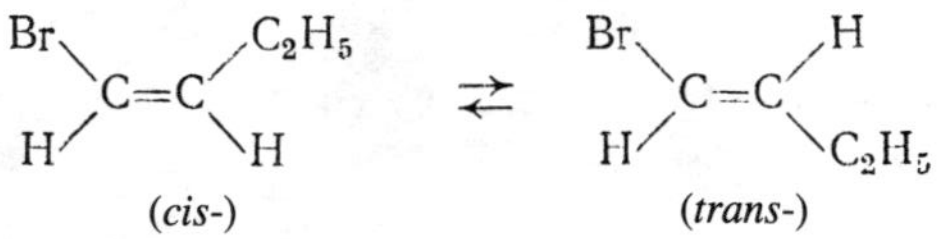 trans-BrCH=CH(C$_2$H$_5$) ⇌ cis-	366 (l)	0.00 ±0.13	−4.89 ±0.42	SE, CT, GLC, C_6H_{12}	34
C_4H_7BrO	2.1	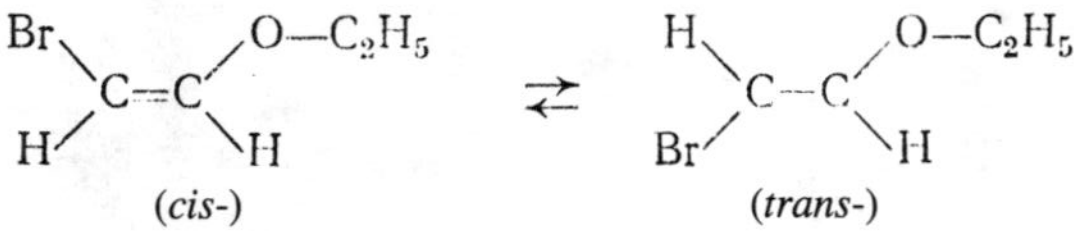 trans-(H,O-C$_2$H$_5$/Br,H) ⇌ cis-	298 (l)	6.1 ±1.0	7.0 ±3.0	SE, CT, GLC, 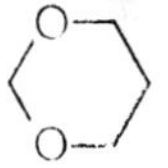 1,4-dioxane	35
C_4H_7Cl	1.3	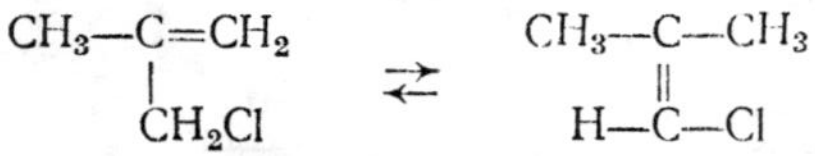 CH$_3$-C(CH$_2$Cl)=CH$_2$ ⇌ CH$_3$-C(CH$_3$)=CHCl	408 (g)	−7.95	−4.18	FE, CT, GLC	36
		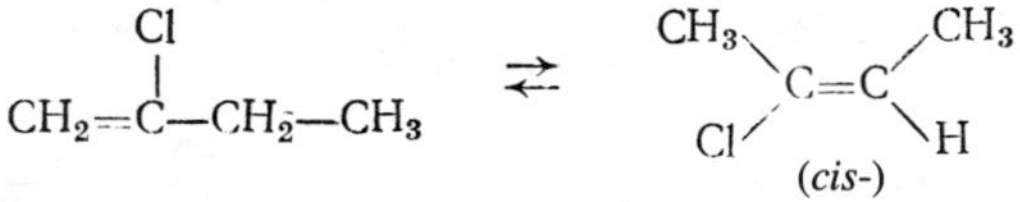 CH$_2$=C(Cl)-CH$_2$-CH$_3$ ⇌ cis-CH$_3$-C(Cl)=CH-CH$_3$	568 (g)	−8.78 ±0.10	3.35 ±0.13	FE, CT, GLC	37

23

Table (Continued)

Formula	Type	Reaction	T, K (phase)	$\Delta_r H_T^\circ$ ($\Delta_r G_T^\circ$)	$\Delta_r S_T^\circ$	Technique	Reference
C_4H_7Cl	1.3	$CH_2=C(Cl)-CH_2-CH_3 \rightleftharpoons CH_3-C(Cl)=C(H)-CH_3$ (trans-)	498 (g, l)	−8.78	−3.35	SE, FE, CT, GLC	38
		$CH_2=CH-CH_2-CH_2-Cl \rightleftharpoons CH_2=CH-CH(Cl)-CH_3$	568 (g)	−13.80 ±0.08	−4.81 ±0.13	FE, CT, GLC	37
		$CH_2=CH-CH(Cl)-CH_3 \rightleftharpoons CH_3-CH=CH-CH_2Cl$	498 (g, l)	−13.80	−4.68	SE, FE, CT, GLC	38
		$CH_3-CH(Cl)-CH=CH_2$ ... $\rightleftharpoons CH_3-C(Cl)=C(H)-CH_3$ (cis-/trans-)	548 (l)	−16.73	−7.53	SE, CT, GLC	38
	2.1	$CH_3-C(Cl)=C(H)-CH_3$ (cis-) $\rightleftharpoons$ (trans-)	523 (g)	−3.76	−4.81	SE, CT, GLC	38
		$Cl-C(C_2H_5)=C(H)-H$ (cis-) $\rightleftharpoons$ (trans-)	568 (g)	−5.02 ±0.04	−1.46 ±0.08	FE, CT, GLC	37
C_4H_7ClO	1.3	$Cl-CH_2-C(OCH_3)=CH_2 \rightleftharpoons CH_3-C(OCH_3)=C(H)-Cl$ (cis-)	498 (g, l)	2.84 ±0.30	0.42 ±0.50	SE, FE, CT, GLC	39
			298 (l)	5.2 ±0.6	−2.1 ±1.5	SE, CT, GLC	35

		Cl–CH$_2$\\CH$_3$–O/C=CH$_2$ ⇌ CH$_3$–O\\CH$_3$/C=C/Cl\\H (trans-)	298 (l)	−1.7 ±0.2	−4.5 ±0.5	SE, CT, GLC ⬡	35
	2.1	CH$_3$\\CH$_3$–O/C=C/H\\Cl (cis-) ⇌ CH$_3$\\CH$_3$–O/C=C/Cl\\H (trans-)	298 (l)	−7.0 ±0.3	−2.8 ±0.8	SE, CT, GLC ⬡	35
			298 (l)	5.3 ±2.2	4.3 ±0.5	SE, CT, GLC	35
		Cl\\H/C=C/O–C$_2$H$_5$\\H (cis-) ⇌ Cl\\H/C=C/H\\O–C$_2$H$_5$ (trans-)	311 (l)	2.77 ±0.57	−3.26 ±1.84	SE, CT, GLC	40
C$_4$H$_8$	1.3	CH$_2$=CH–CH$_2$–CH$_3$ ⇌ CH$_3$\\H/C=C/CH$_3$\\H (cis-)	420 (g)	−8.16	−8.74	FE, CT, GLC	41
		CH$_2$=CH–CH$_2$–CH$_3$ ⇌ CH$_3$\\H/C=C/H\\CH$_3$ (trans-)	540 (g)	−12.97 ±0.84	−15.06 ±1.67	SE, CT, GLC	42
			420 (g)	−11.50	−11.29	FE, CT, GLC	41
			500 (g)	−11.71 ±0.84	−12.97 ±1.67	FE, CT, GLC	43
			544 (g)	−12.07	−14.81	SE, CT, GLC	44

Table (Continued)

Formula	Type	Reaction	T, K (phase)	$\Delta_r H_T^\circ$ ($\Delta_r G_T^\circ$)	$\Delta_r S_T^\circ$	Technique	Reference
C_4H_8	2.1	$\begin{array}{c} CH_3\diagdown\diagup CH_3 \\ C=C \\ H\diagup\diagdown H \\ (cis\text{-}) \end{array} \rightleftarrows \begin{array}{c} CH_3\diagdown\diagup H \\ C=C \\ H\diagup\diagdown CH_3 \\ (trans\text{-}) \end{array}$	300 (g)	−12.51	−15.06	SE, CT, GLC	45
			540 (g)	−2.09 ±0.84	0.0 ±1.67	SE, CT, GLC	42
			420 (g)	−3.35	−2.53	FE, CT, GLC	41
			500 (g)	−4.81 ±0.42	−4.60 ±0.84	SE, FE, CT, GLC	43
			544 (g)	−1.84	−1.04	SE, CT, GLC	44
			300 (g)	−4.14	−4.27	SE, CT, GLC	45
			298 (g)	−4.34 ±0.04	−5.10 ±0.12	FE, CT, GLC	46
			298 (g)	−4.18 ±0.08	−4.85 ±0.29	SE, CT, GLC	47
			400 (g)	−4.11	−4.81	FE, CT, GLC	48

$C_4H_8Br_2$	1.3	$CH_2-CH_2-CH-CH_3$ $\quad$ Br $\quad$ Br $\quad\rightleftarrows\quad$ $CH_3-CH-CH-CH_3$ $\quad$ Br $\quad$ Br	414 (g, l)	−2.13 ±0.08	−6.11 ±0.21	SE, CT, GLC	49
		$CH_2-CH-CH_2-CH_3$ $\quad$ Br $\quad$ Br $\quad\rightleftarrows\quad$ $CH_3-CH-CH-CH_3$ $\quad$ Br $\quad$ Br	414 (g, l)	−8.32 ±0.29	−9.20 ±0.08	SE, CT, GLC	49
		$CH_2-CH_2-CH_2-CH_2$ $\quad$ Br $\qquad\qquad$ Br $\quad\rightleftarrows\quad$ $CH_2-CH-CH_2-CH_3$ $\quad$ Br $\quad$ Br	414 (g, l)	−1.76 ±0.42	2.26 ±1.25	SE, CT, GLC	50
		$CH_2-CH_2-CH_2-CH_2$ $\quad$ Br $\qquad\qquad$ Br $\quad\rightleftarrows\quad$ $CH_2-CH_2-CH-CH_3$ $\quad$ Br $\qquad$ Br	414 (g, l)	−7.74 ±0.42	−0.84 ±1.25	SE, CT, GLC	50
		$CH_2-CH-CH_2-CH_3$ $\quad$ Br $\quad$ Br $\quad\rightleftarrows\quad$ $CH_2-CH_2-CH-CH_3$ $\quad$ Br $\qquad$ Br	414 (g, l)	−5.98 ±0.25	−3.10 ±0.63	SE, CT, GLC	50
		$CH_3-C-CH_2-CH_3$ $\quad$ Br $\quad$ Br $\quad\rightleftarrows\quad$ $CH_3-CH-CH-CH_3$ $\quad$ Br $\quad$ Br	401 (g, l)	−13.39 ±0.84	4.81 ±2.30	SE, FE, CT, GLC	51
		$CH_2-CH-CH_2$ $\quad$ Br $\quad$ Br $\quad$ CH_3 $\quad\rightleftarrows\quad$ $CH_2-CH_2-CH-CH_3$ $\quad$ Br $\qquad\qquad$ CH_3	413 (g, l)	−8.78 ±0.25	−7.19 ±0.59	SE, CT, GLC	52
	2.1	(meso-) $\quad\rightleftarrows\quad$ (d,l)	308 (l)	−1.80 ±0.17	−2.34 ±0.63	SE, CT, GLC	53

Table (Continued)

Formula	Type	Reaction	T, K (phase)	$\Delta_r H_T^\circ$ ($\Delta_r G_T^\circ$)	$\Delta_r S_T^\circ$	Technique	Reference
$C_4H_8Cl_2$	1.3	$CH_3-CHCl-CHCl-CH_3 \rightleftharpoons CH_3-CCl_2-CH_2-CH_3$	348 (l)	14.10 ±0.33	4.35 ±2.09	SE, CT, GLC	54
C_4H_8O	2.1	(meso-) $\rightleftharpoons$ (d, l)	311 (l)	−1.13 ±0.21	−2.89 ±0.71	SE, CT, GLC CCl_4	55
	2.1	(cis-) $\rightleftharpoons$ (trans-) methyl propenyl ether	311 (l)	−3.82 ±0.24	−6.19 ±0.79	SE, CT, GLC	40
			298 (l)	(−0.50 ±0.13)	—	SE, CT, GLC	56
C_4H_8OS	1.3	$CH_2=CH-CH_2-S(=O)-CH_3 \rightleftharpoons$ (cis-) sulfoxide	293 (l)	(3.6)	—	SE, CT, GLC $t-C_4H_9OH$	57

Formula		Equilibrium	T, K (state)	ΔH		Methods	Ref.
$C_4H_8O_2$	2.1	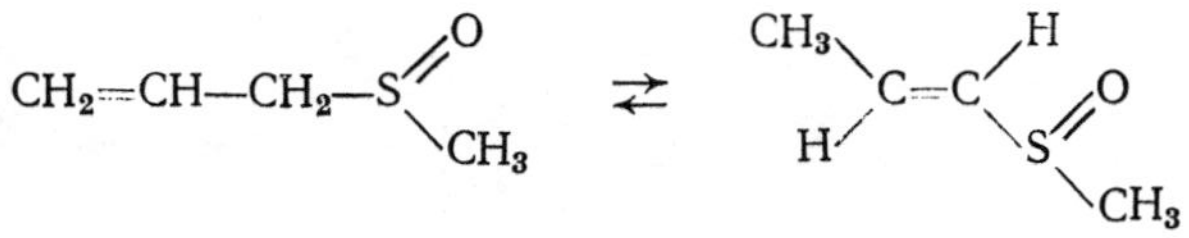 $CH_2=CH-CH_2-S(=O)-CH_3 \rightleftharpoons CH_3-C(H)=C(H)-S(=O)-CH_3$ (trans-)	293 (l)	(−3.0)	—	SE, CT, GLC $t-C_4H_9OH$	57
		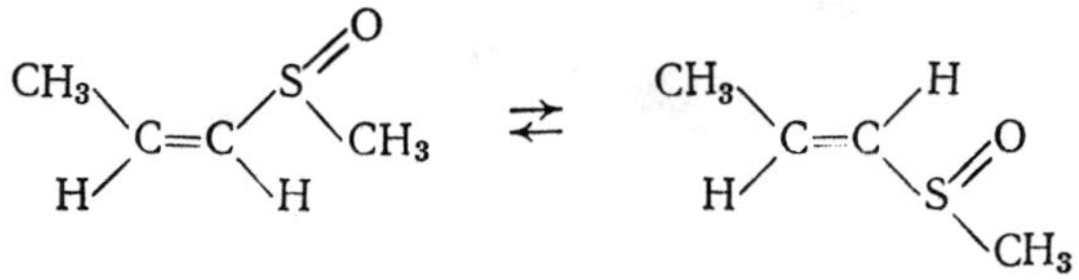 $\rightleftharpoons$ (cis- isomer of methyl propenyl sulfoxide) (trans-)	293 (l)	(−6.5)	—	SE, CT, GLC $t-C_4H_9OH$	57
	1.3	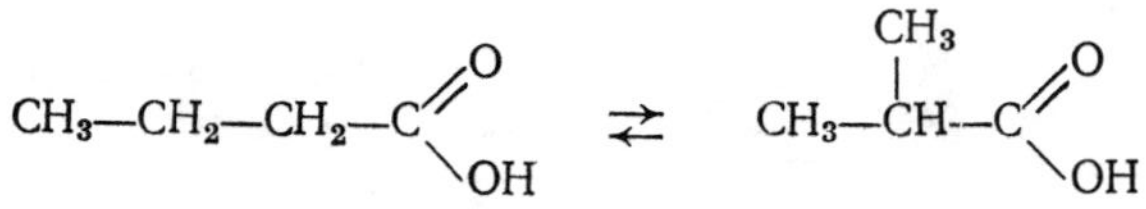 $CH_3-CH_2-CH_2-C(=O)OH \rightleftharpoons (CH_3)_2CH-C(=O)OH$	298 (l)	−7.78	—	CT	58
		 1,3-dioxane $\rightleftharpoons$ 1,4-dioxane	298 (g)	21.9 ±1.33	—	CT	59
	2.1	 $CH_3O-C(OCH_3)=CH_2$ (cis-) $\rightleftharpoons$ (trans-) dimethoxyethene	391 (g)	6.04 ±0.23	6.94 ±0.63	SE, CT, GLC	60
		same (l)	375 (l)	6.48 ±0.08	5.61 ±0.21	SE, CT, GLC	60

Table (*Continued*)

Formula	Type	Reaction	T, K (phase)	$\Delta_r H_T^\circ$ ($\Delta_r G_T^\circ$)	$\Delta_r S_T^\circ$	Technique	Reference
$C_4H_8O_2S$	1.3	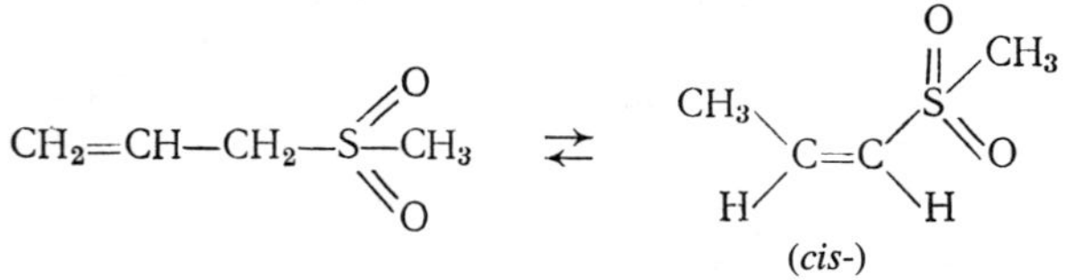 $CH_2{=}CH{-}CH_2{-}S(O)_2{-}CH_3 \rightleftarrows$ cis-[CH₃,H/C=C/S(O)₂CH₃,H]	293 (l)	(11.2)	—	SE, CT, GLC $t{-}C_4H_9OH$	57
		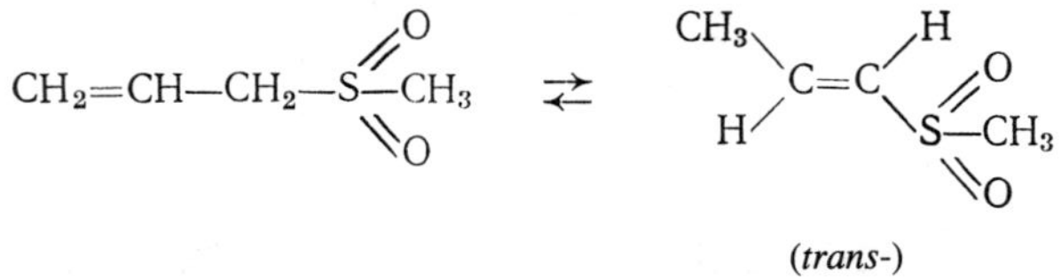 $CH_2{=}CH{-}CH_2{-}S(O)_2{-}CH_3 \rightleftarrows$ trans-[CH₃,H/C=C/H,S(O)₂CH₃]	293 (l)	(−2.1)	—	SE, CT, GLC $t{-}C_4H_9OH$	57
	2.1	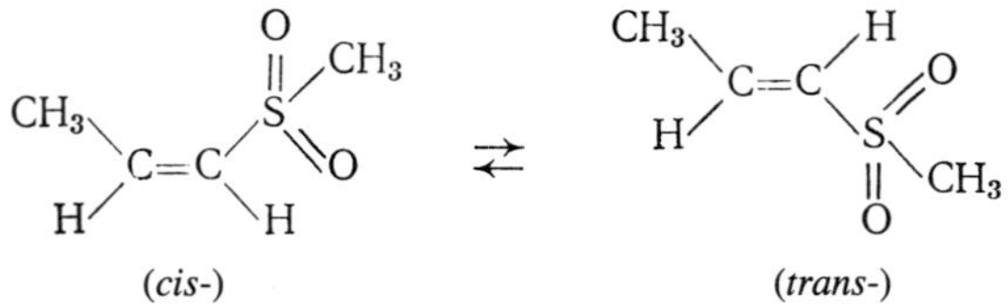 cis- $\rightleftarrows$ trans-	293 (l)	(−13.3)	—	SE, CT, GLC $t{-}C_4H_9OH$	57
C_4H_8S	1.3	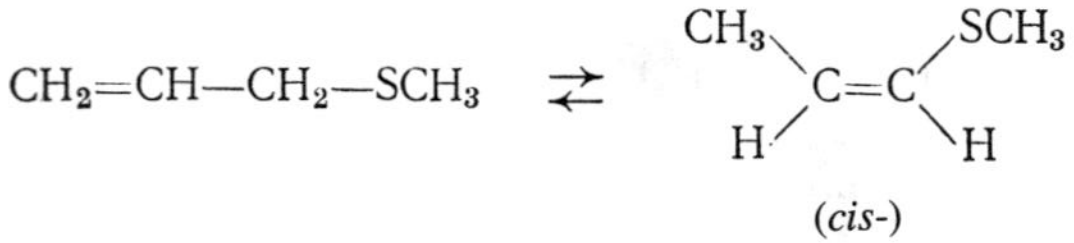 $CH_2{=}CH{-}CH_2{-}SCH_3 \rightleftarrows$ cis-[CH₃,H/C=C/SCH₃,H]	293 (l)	(−15.5)	—	SE, CT, GLC $t{-}C_4H_9OH$	57
	1.3	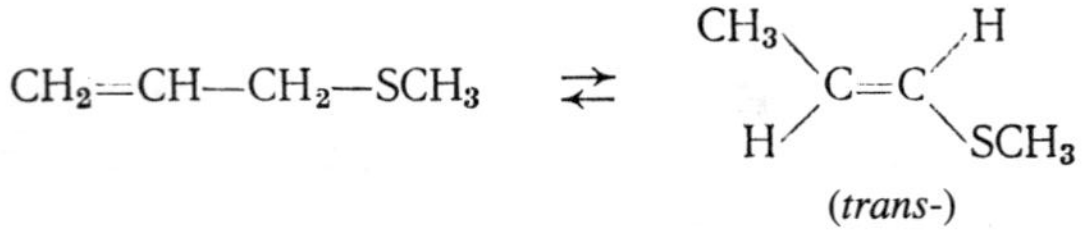 $CH_2{=}CH{-}CH_2{-}SCH_3 \rightleftarrows$ trans-[CH₃,H/C=C/H,SCH₃]	293 (l)	(−14.7)	—	SE, CT, GLC $t{-}C_4H_9OH$	57

Formula	Ratio	Equilibrium	T (K), state	ΔH	ΔS	Method	Ref
C_4H_9Br	2.1	CH_3-$C(SCH_3)$=CH-H (cis-) ⇌ CH_3-$C(H)$=CH-SCH_3 (trans-)	293 (l)	(0.75)	—	SE, CT, GLC t—C_4H_9OH	57
	1.3	CH_3-$CHBr$-CH_2-C_2H_5 ⇌ $BrCH_2$-CH_2-CH_2-C_2H_5	491 (g)	12.97 ±0.63	7.19 ±1.05	FE, CT, GLC	61
		CH_3-$CHBr$-$CH(CH_3)$-CH_3 ⇌ CH_3-$CBr(CH_3)$-CH_3	400 (l)	−15.90 ±0.67	−24.26 ±2.09	SE, CT, GLC	62
C_4H_9NO	1.1	CH_3-$C(OCH_3)$=NCH_3 ⇌ CH_3-$C(=O)$-$N(CH_3)_2$	298 (g)	−68.18 ±10.46	—	RC	63
C_4H_9NSi	1.1 1.2	CH_3-$Si(CH_3)_2$-$C≡N$ ⇌ CH_3-$Si(CH_3)_2$-$N=C$	298 (g)	16.75 ±0.17	2.34	SE, ST IR, NMR	64
C_5H_4BrNO	1.1 1.2	2-hydroxy-6-bromopyridine ⇌ 6-bromo-2(1H)-pyridone	323 (l)	−6.12	−11.25	SE, UV C_2H_5OH–H_2O	65
C_5H_4ClNO	1.1 1.2	2-hydroxy-6-chloropyridine ⇌ 6-chloro-2(1H)-pyridone	323 (l)	−4.39	−9.04	SE, UV C_2H_5OH–H_2O	65

Table (Continued)

Formula	Type	Reaction	T, K (phase)	$\Delta_r H_T^\circ$ ($\Delta_r G_T^\circ$)	$\Delta_r S_T^\circ$	Technique	Reference
C_5H_4ClNO	1.1 1.2	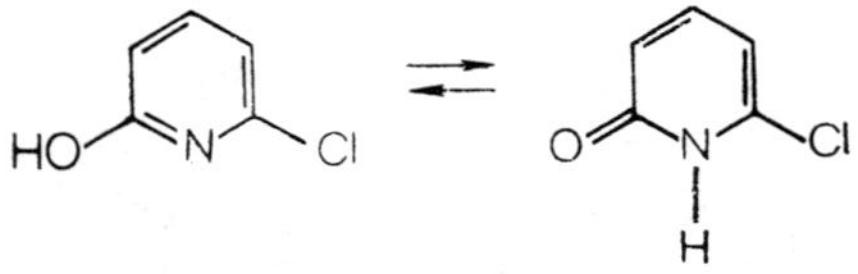 2-chloro-6-hydroxypyridine ⇌ 6-chloro-pyridin-2(1H)-one	298 (l)	(6.7)	—	SE, UV	66
		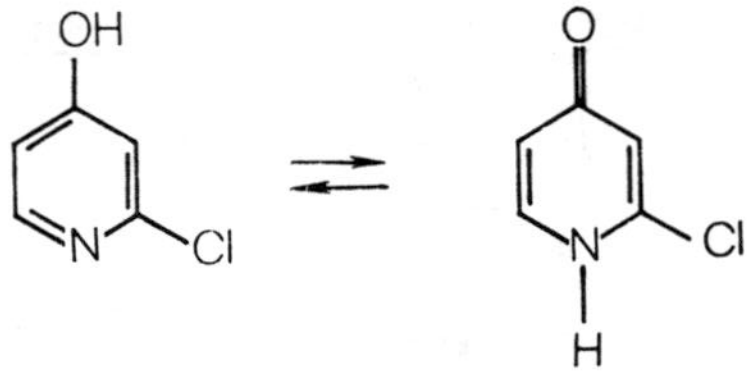 2-chloro-4-hydroxypyridine ⇌ 2-chloro-pyridin-4(1H)-one	298 (l)	(4.2)	—	SE, UV CH_3OH	66
C_5H_4ClNS	1.1 1.2	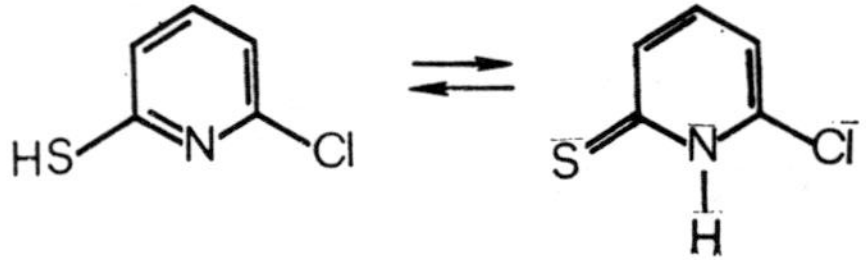 2-chloro-6-mercaptopyridine ⇌ 6-chloro-pyridine-2(1H)-thione	298 (l)	(7.9)	—	SE, UV $CHCl_3$	66
		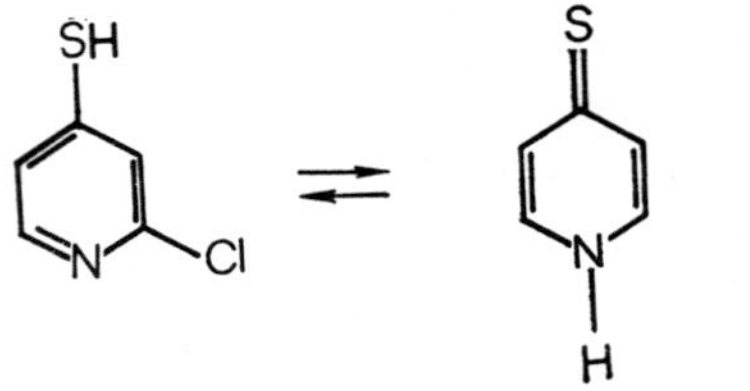 2-chloro-4-mercaptopyridine ⇌ 2-chloro-pyridine-4(1H)-thione	298 (l)	(8.8)	—	SE, UV	66

Formula	1.1/1.2	Tautomers	T (K), state	value	value	Method	Ref
C_5H_5NO	1.1 / 1.2	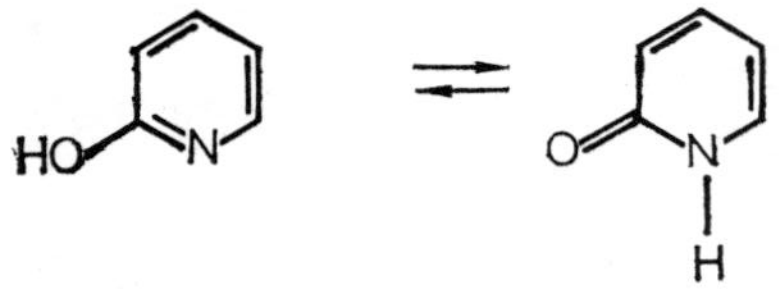 2-hydroxypyridine ⇌ 2-pyridone (N–H)	298 (l)	(17.2)	—	SE, UV	67
			508 (g)	-1.25 ± 1.25	12.5	SE, UV	68
C_5H_5NS	1.1 / 1.2	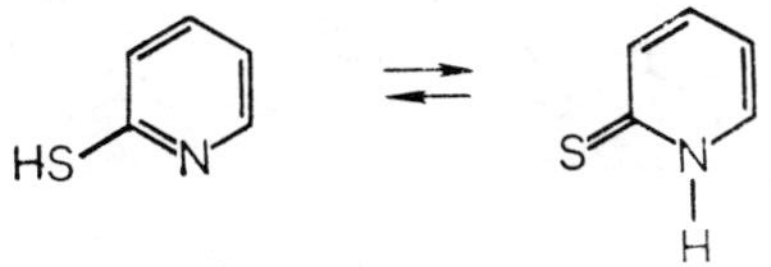 2-mercaptopyridine ⇌ 2-thiopyridone (N–H)	298 (l)	(28.0)	—	SE, UV	67
			298 (l)	(5.9)	—	SE, UV (benzene)	66
$C_5H_3N_3O_2$	1.1 / 1.2	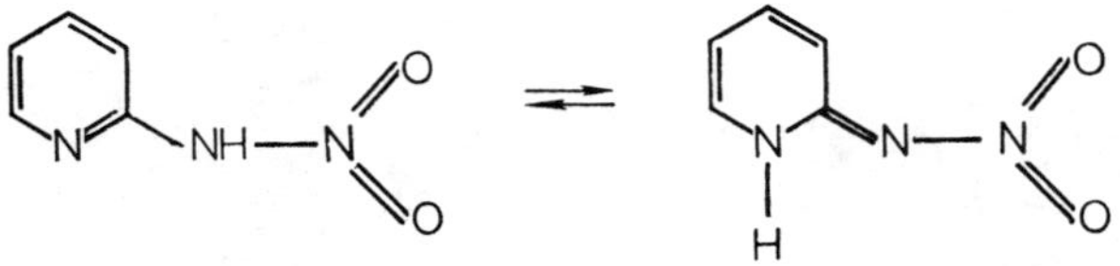 2-nitraminopyridine ⇌ 2-nitrimino-N–H pyridine	331 (l)	-4.52	-11.44	SE, UV, H_2O	69
$C_5H_5N_5O$	1.1 / 1.2	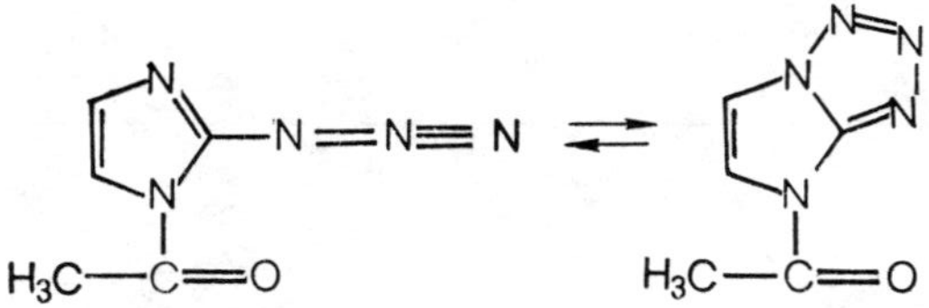 2-azido-1-acetylimidazole ⇌ acetyltetrazoloimidazole	293 (l)	-5.4 ± 1.3	-33.9 ± 4.6	SE, NMR, $CH_3\text{--}C(\!=\!O)\text{--}CH_3$	70

Table (*Continued*)

Formula	Type	Reaction	T, K (phase)	$\Delta_r H°_T$ ($\Delta_r G°_T$)	$\Delta_r S°_T$	Technique	Reference
$C_5H_6N_2$	1.3	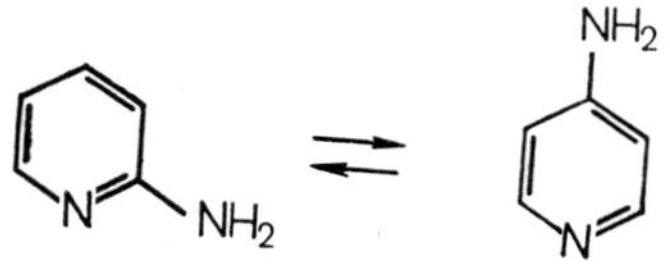 4-aminopyridine ⇌ 2-aminopyridine	298 (g)	11.8 ±1.8	—	CC	71
			298 (g)	—	−6.98	ST	252
		 3-aminopyridine ⇌ 2-aminopyridine	298 (g)	—	−1.33	ST	252
			298 (g)	26.1 ±1.8	—	CC	71

$C_5H_6N_2O$	1.1 1.2	2-aminopyridine ⇌ 2-iminopyridine	298 (l)	(−35.6)	—	SE, UV	67
	1.1 1.2	2-amino-6-hydroxypyridine ⇌ 6-amino-2(1H)-pyridinone	323 (l)	−4.66	−20.12	SE, UV	65
C_5H_8	2.1	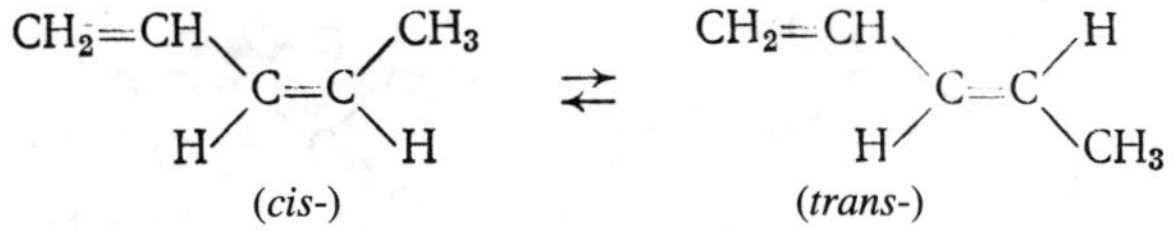 cis-/trans-1,3-pentadiene	533 (g)	−4.35 ±0.13	−0.59 ±0.21	SE, CT, GLC	72
	1.2	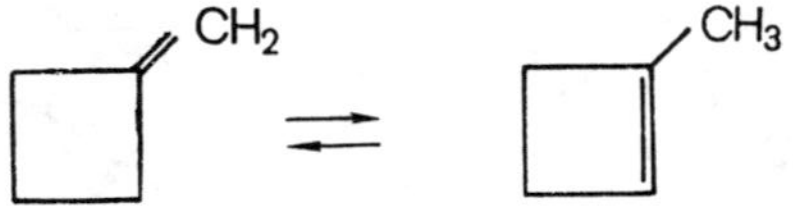 methylenecyclobutane ⇌ methylcyclobutene	298 (l)	(4.31 ±0.08)	—	SE, CT, GLC	73
			298 (g)	−3.76	—	CC	32
C_5H_8O	2.1	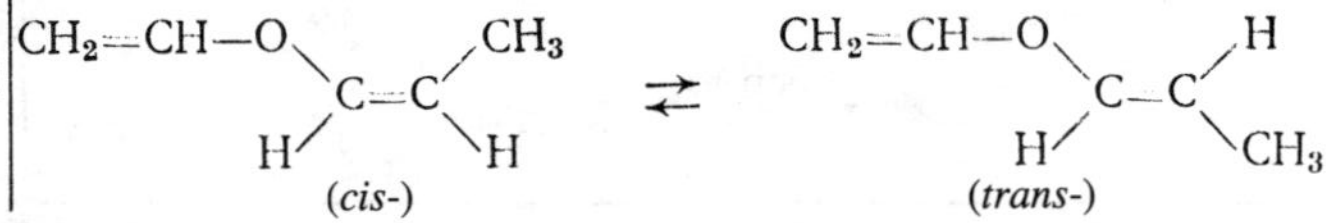 cis-/trans-ethyl vinyl ether	298 (g)	2.3 ±0.4	−1.5 ±0.6	SE, CT, GLC	74

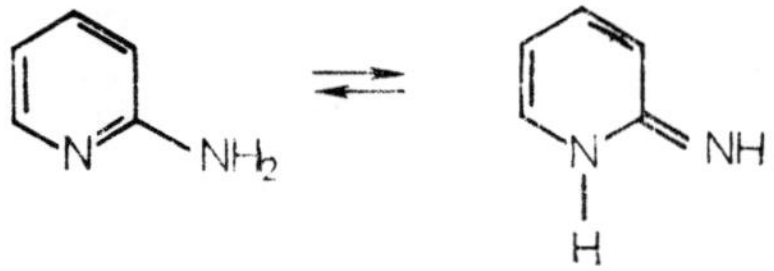

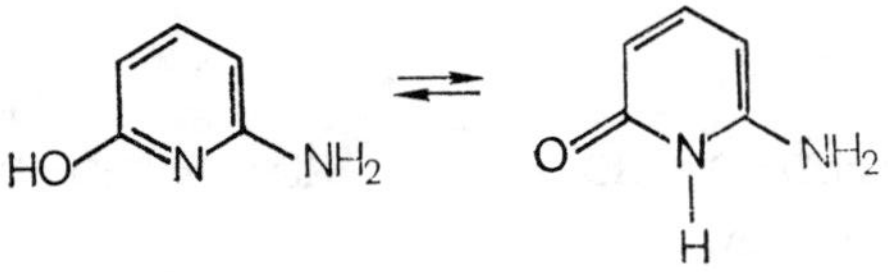

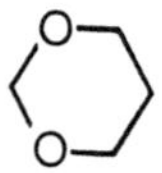

Table (Continued)

Formula	Type	Reaction	T, K (phase)	$\Delta_r H°_T$ ($\Delta_r G°_T$)	$\Delta_r S°_T$	Technique	Reference
C_5H_8O	1.2	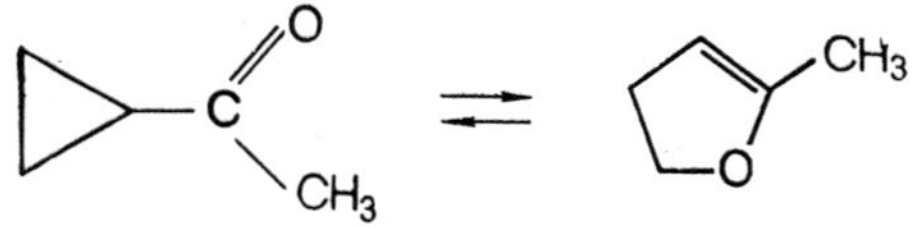 cyclopropyl methyl ketone ⇌ 2-methyl-2,3-dihydrofuran	702 (g)	−9.6 ±3.3	−18.4 ±4.8	SE, GLC	75
		 1-cyclopentenol ⇌ cyclopentanone	298 (l)	−32.92	—	RC	76
$C_5H_8O_2$	2.1	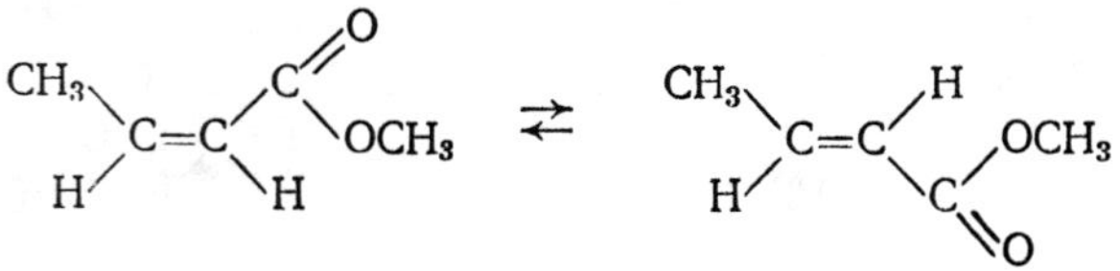 methyl (cis)-2-butenoate ⇌ methyl (trans)-2-butenoate	390 (l)	(−5.86 ±0.21)	—	SE, CT, GLC CH_3OH	56
	1.1	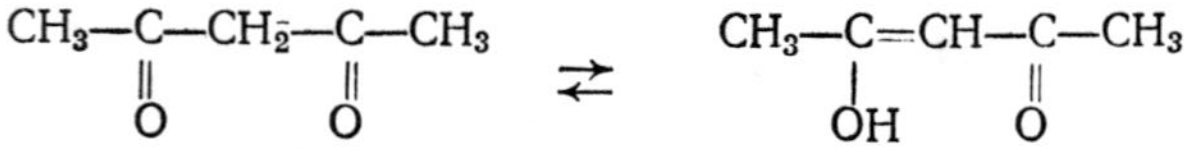 $CH_3-C-CH_2-C-CH_3$ (diketo) ⇌ $CH_3-C=CH-C-CH_3$ (enol)	436 (g)	−16.8	—	SE, IR	77
			409 (g)	−18.0	—	SE, UV	78
			376 (g)	−7.9 ±0.4	—	SE, UV	79
	1.2		282 (l)	−10.0 ±0.8	−21.8 ±2.5	SE, NMR	80

$C_5H_8O_3$	2.1	(structure: CH$_3$O-C(=C(H)-C(=O)OCH$_3$)-H (cis-) ⇌ trans-)	298 (g)	(−12.18 ±0.29)	—	SE, IR	83
		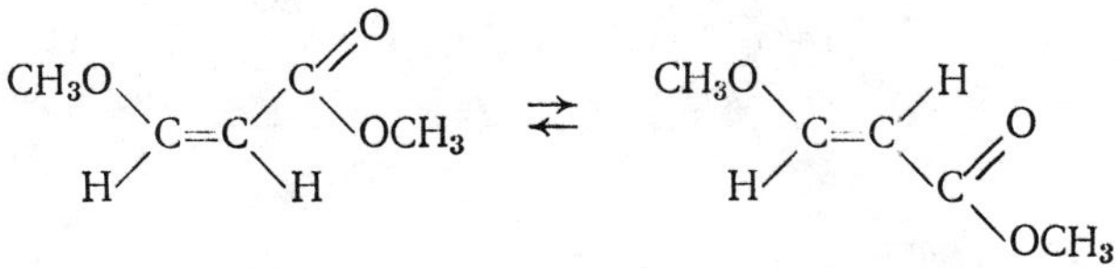	373 (l)	(−15.06 ±2.09)	—	SE, CT, GLC 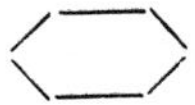 ⬡	56
C_5H_9BrO	1.3	 (structure: Br-CH$_2$-C(OC$_2$H$_5$)=CH$_2$ (cis-) ⇌ trans-)	298 (l)	1.4 ±1.0	−8.0 ±4.0	SE, CT, NMR  ⬡	35
$C_5H_9BrO_2$	2.1	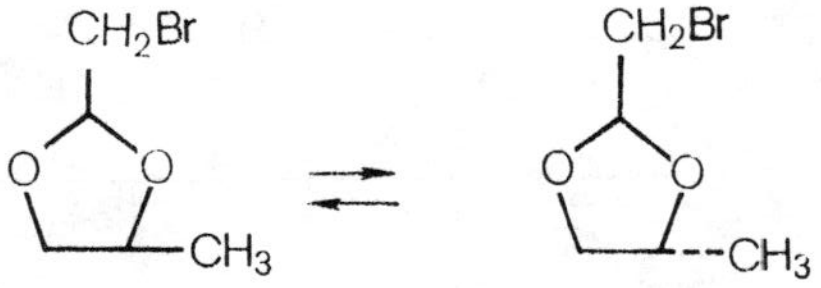 (structure: 2-bromomethyl-4-methyl-1,3-dioxolane cis- ⇌ trans-)	341 (l)	0.0	−0.84	SE, CT, GLC, NMR CCl$_4$	84

(Additional rows above:)

	1.3	(3-methoxy-2,5-dihydrofuran ⇌ isomer)	298 (l)	7.4 ±0.4	7.6 ±1.1	SE, CT, GLC ⬡	81
	1.2	(methylene-1,3-dioxane ⇌ 5-methyl-1,3-dioxine)	298 (g)	−18.3 ±0.9	10.2 ±2.4	SE, CT, GLC	82

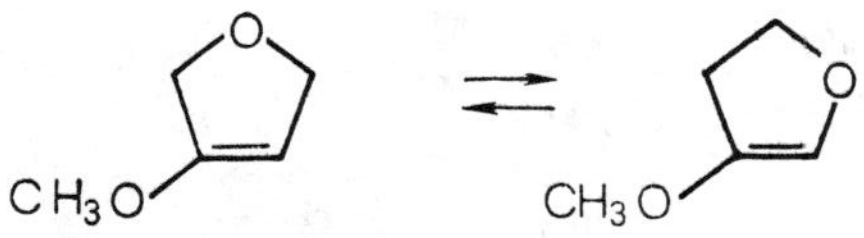

37

Table (Continued)

Formula	Type	Reaction	T, K (phase)	$\Delta_r H_T^\circ$ ($\Delta_r G_T^\circ$)	$\Delta_r S_T^\circ$	Technique	Reference
C_5H_9ClO	1.3	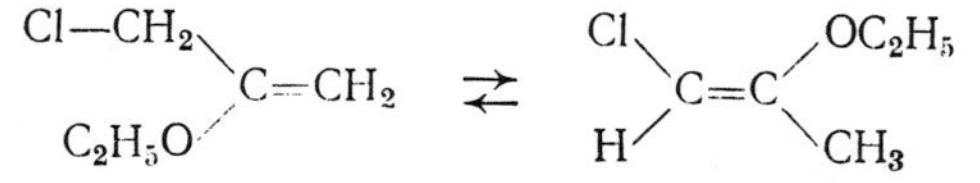 Cl–CH$_2$–C(OC$_2$H$_5$)=CH$_2$ ⇌ (Cl)(H)C=C(OC$_2$H$_5$)(CH$_3$) (cis-)	298 (l)	5.2 ±0.5	−0.3 ±1.4	SE, CT, GLC  ⬡	35
		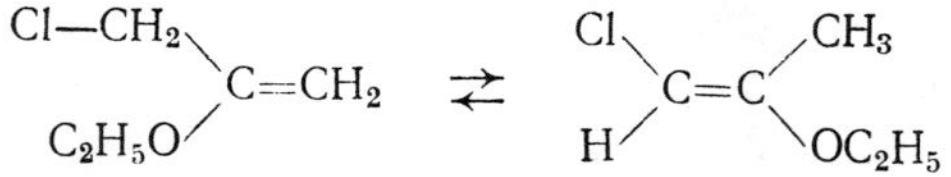 Cl–CH$_2$–C(OC$_2$H$_5$)=CH$_2$ ⇌ (Cl)(H)C=C(CH$_3$)(OC$_2$H$_5$) (trans-)	298 (l)	−1.9 ±0.1	−4.8 ±0,4	SE, CT, GLC  ⬡	35
		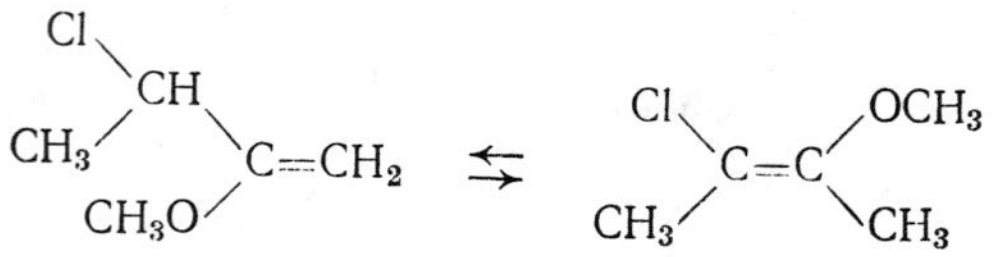 (Cl)(CH$_3$)CH–C(OCH$_3$)=CH$_2$ ⇌ (Cl)(CH$_3$)C=C(OCH$_3$)(CH$_3$) (cis-)	298 (l)	8.9 ±0.2	4.8 ±0.6	SE, CT, GLC  ⬡	35
		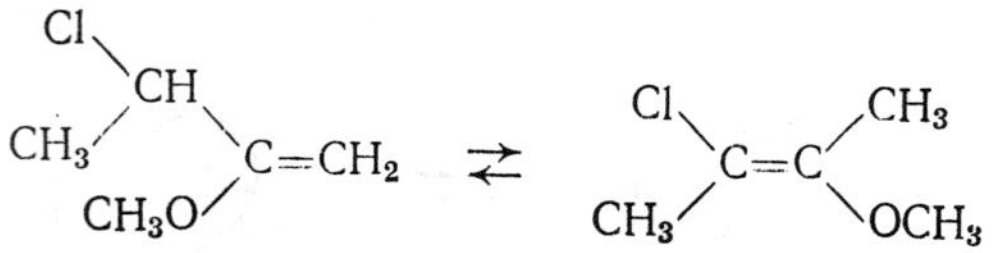 (Cl)(CH$_3$)CH–C(OCH$_3$)=CH$_2$ ⇌ (Cl)(CH$_3$)C=C(CH$_3$)(OCH$_3$) (trans-)	298 (l)	5.0 ±0.9	5.4 ±2.5	SE, CT, GLC ⬡	35
	2.1	 (Cl)(H)C=C(OC$_2$H$_5$)(CH$_3$) ⇌ (Cl)(H)C=C(CH$_3$)(OC$_2$H$_5$) (trans-)	298 (l)	−7.1 ±0.6	−4.7 ±1.8	SE, CT, GLC ⬡	35

		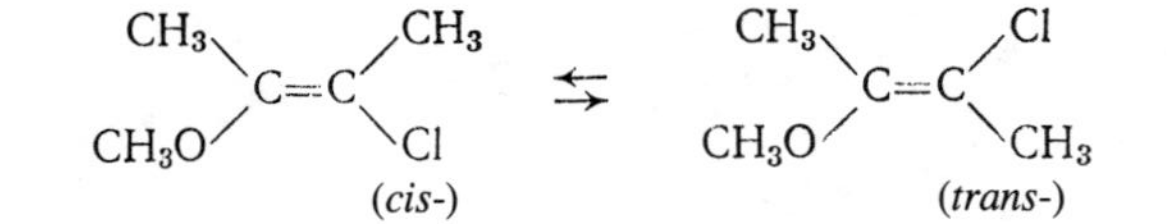 CH₃\C=C/Cl / \CH₃O CH₃ (cis-) ⇌ CH₃\C=C/Cl / \CH₃O CH₃ (trans-)	298 (l)	−4.1 ±0.8	0.0 ±2.2	SE, CT, GLC ⬡	35
C₅H₉FO	1.3	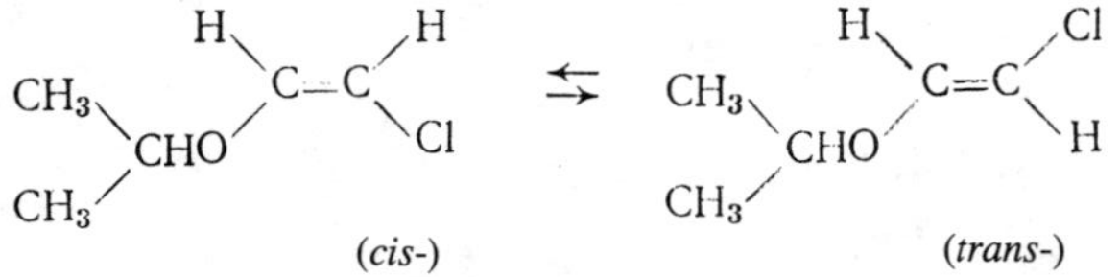 H\C=C/Cl / \CH₃ CHO CH₃ (cis-) ⇌ H\C=C/CHO / \CH₃ CH₃ (trans-)	298 (l)	6.5 ±0.2	4.6 ±0.5	SE, CT, GLC ⬡	35
		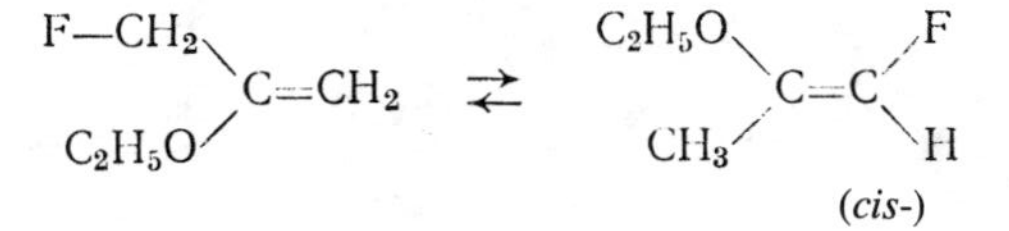 F−CH₂\C=CH₂ / C₂H₅O ⇌ C₂H₅O\C=C/F / \CH₃ H (cis-)	298 (l)	9.9 ±1.1	4.0 ±3.0	SE, CT, GLC ⬡	35
		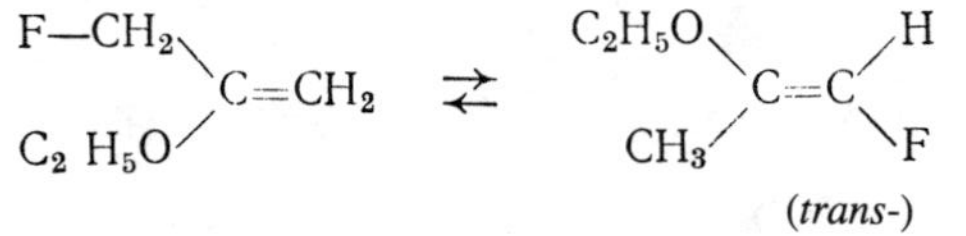 F−CH₂\C=CH₂ / C₂H₅O ⇌ C₂H₅O\C=C/H / \CH₃ F (trans-)	298 (l)	7.3 ±0.4	1.4 ±1.1	SE, CT, GLC ⬡	35
	2.1	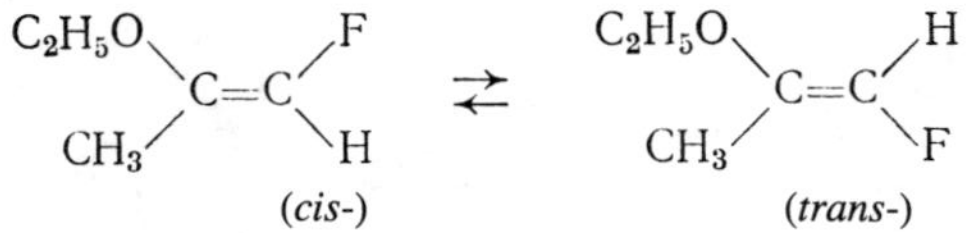 C₂H₅O\C=C/F / \CH₃ H (cis-) ⇌ C₂H₅O\C=C/H / \CH₃ F (trans-)	298 (l)	−2.7 ±1.8	3.0 ±5.0	SE, CT, GLC ⬡	35
C₅H₉NO	1.1 1.2	[piperidinone tautomers]	298 (l)	(44.3)	—	SE, UV	67

Table (Continued)

Formula	Type	Reaction	T, K (phase)	$\Delta_r H_T^\circ$ ($\Delta_r G_T^\circ$)	$\Delta_r S_T^\circ$	Technique	Reference
C_5H_9NS	1.1 1.2	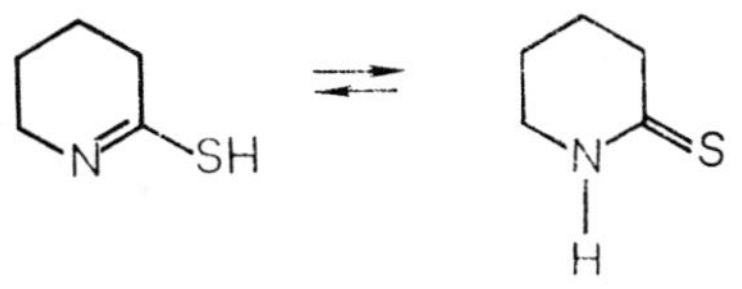 2-mercaptopyridine (SH form) ⇌ 2-thiopiperidone (NH, C=S form)	298 (l)	(50.2)	—	SE, UV	67
C_5H_{10}	1.3	$CH_2=CH-CH(CH_3)-CH_3$ ⇌ $CH_3-C(CH_3)=CH-CH_3$	622 (g)	-14.22 ± 1.46	-1.25 ± 1.75	FE, CT, GLC	85
			298 (l)	-16.47 ± 1.28	—	CC	86
		$CH_2=C(CH_3)-CH_2-CH_3$ ⇌ $CH_3-C(CH_3)=CH-CH_3$	562 (g)	-8.07 ± 0.48	-5.02 ± 0.84	FE, CT, GLC	85
			298 (l)	9.36 ± 0.97	—	CC	86
		$CH_2=CH-CH_2-C_2H_5$ ⇌ $CH_3-C(C_2H_5)=CH-H$ (cis-)	461 (g)	-9.37 ± 0.25	-9.08 ± 0.5	FE, CT, GLC	87
			298 (l)	-6.52 ± 0.84	—	CC	86

	CH$_2$=CH—CH$_2$—C$_2$H$_5$ ⇌ CH$_3$\C=C/H with H, C$_2$H$_5$ (trans-)	461 (g)	−11.29 ±0.29	−6.65 ±0.59	FE, CT, GLC	87
	CH$_3$\CH$_2$=C—CH$_2$—CH$_3$ ⇌ CH$_3$\C=C/H with H, C$_2$H$_5$ (trans-)	298 (l)	2.98 ±1.07	—	CC	86
2.1	CH$_3$\C=C/H with C$_2$H$_5$, H (cis-) ⇌ CH$_3$\C=C/H with H, C$_2$H$_5$ (trans-)	400 (g)	−4.09	−1.88	FE, CT, GLC	48
		461 (g)	−2.93 ±0.42	2.38 ±0.42	FE, CT, GLC	87
		298 (g)	−4.81 ±0.13	−3.38 ±0.42	SE, CT, GLC	47
		298 (l)	−4.49 ±0.92	—	CC	86
		298 (g)	−4.90 ±0.08	−3.56 ±0.28	FE, CT, GLC	46
C$_5$H$_{10}$Br$_2$ 1.3	Br—CH$_2$—CH$_2$—CH—CH$_2$—CH$_3$ with Br ⇌ Br—CH$_2$—CH$_2$—CH$_2$—CH—CH$_3$ with Br	413 (l)	−2.51 ±0.08	0.00 ±0.13	SE, CT, GLC	88
	Br, Br—CH$_2$—CH—CH$_2$—C$_2$H$_5$ ⇌ Br—CH$_2$—CH$_2$—CH—C$_2$H$_5$ with Br	413 (l)	−6.11 ±0.42	−1.59 ±1.25	SE, CT, GLC	88
	Br—CH$_2$—CH$_2$—CH$_2$—CH$_2$ with Br ⇌ Br—CH$_2$—CH$_2$—CH$_2$—CH—CH$_3$ with Br	413 (l)	−9.62 ±2.93	−0.54 ±5.85	SE, CT, GLC	88

Table (Continued)

Formula	Type	Reaction		T, K (phase)	$\Delta_r H^\circ_T$ ($\Delta_r G^\circ_T$)	$\Delta_r S^\circ_T$	Technique	Reference
$C_5H_{10}Br_2$	1.3	$\underset{Br}{\overset{Br}{\mid}}$ CH$_2$—CH—CH$_2$—CH$_2$—CH$_2$	⇌ CH$_3$—CH—CH$_2$—CH—CH$_3$ (meso-) with Br, Br	428 (l)	7.99	11.80	SE, CT, GLC	89
	2.1	CH$_3$—CH—CH—CH$_2$—CH$_3$ (meso-) Br, Br	⇌ CH$_3$—CH—CH—CH$_2$—CH$_3$ (erythro-) Br, Br	428 (l)	7.99	11.80	SE, CT, GLC	89
		CH$_3$—CH—CH—CH$_2$—CH$_3$ (threo-) Br, Br	⇌ CH$_3$—CH—CH—CH$_2$—CH$_3$ (erythro-) Br, Br	428 (l)	1.88	2.34	SE, CT, GLC	89
		CH$_3$—CH—CH$_2$—CH—CH$_3$ (meso-) Br, Br	⇌ CH$_3$—CH—CH$_2$—CH—CH$_3$ (d,l) Br, Br	428 (l)	−2.38	−1.34	SE, CT, GLC	89
$C_5H_{10}Cl_2$	2.1	CH$_3$—CH—CH—CH$_2$—CH$_3$ (threo-) Cl, Cl	⇌ CH$_3$—CH—CH—CH$_2$—CH$_3$ (erythro-) Cl, Cl	323 (l)	1.17 ±0.21	3.14 ±0.71	SE, CT, GLC	90
		CH$_3$—CH—CH$_2$—CH—CH$_3$ (meso-) Cl, Cl	⇌ CH$_3$—CH—CH$_2$—CH—CH$_3$ (d,l) Cl, Cl	323 (l)	−2.43 ±0.29	−1.55 ±0.92	SE, CT, GLC	90

$C_5H_{10}N_2$	1.2	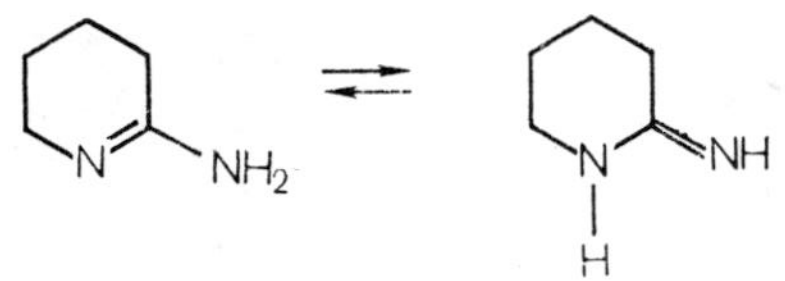 ![structure: 2-iminopiperidine ⇌ 2-aminopyridine tautomers]	298 (l)	(0.0)	—	SE, UV	67
$C_5H_{10}O$	1.3	$CH_2=C(OCH_3)-CH_2-CH_3$ ⇌ cis isomer with CH_3O, CH_3	298 (g)	0.45 ±0.38	−1.6 ±0.8	SE, CT, GLC	91
		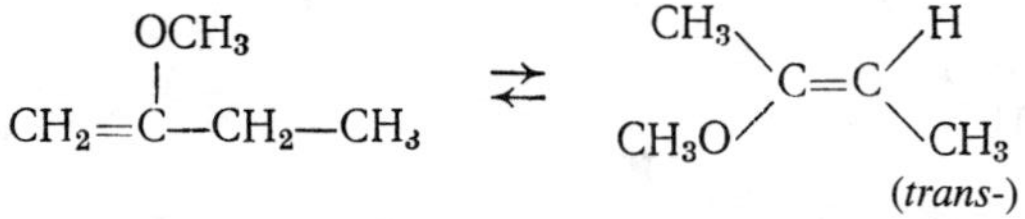 $CH_2=C(OCH_3)-CH_2-CH_3$ ⇌ trans isomer	298 (g)	10.30 ±0.60	11.8 ±1.5	SE, CT, GLC	91
	2.1	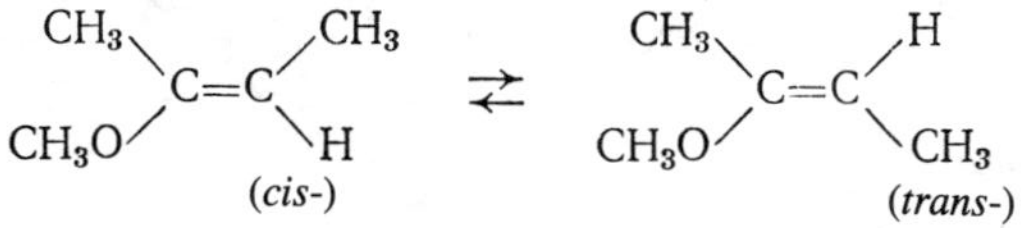 cis-2-methoxy-2-butene ⇌ trans-2-methoxy-2-butene	298 (g)	9.85 ±0.41	13.4 ±0.9	SE, CT, GLC	91
		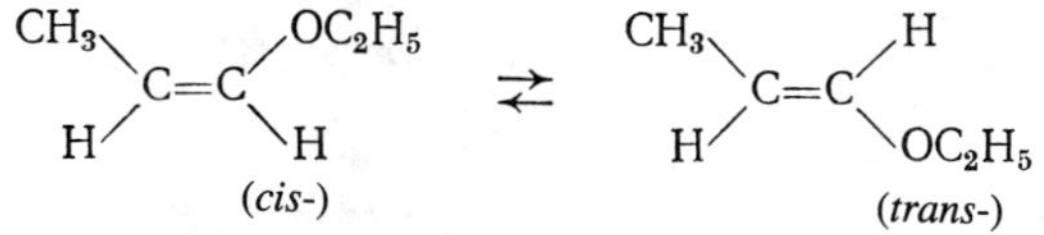 cis ethoxypropene ⇌ trans ethoxypropene	311 (l)	−1.55 ±0.08	−7.53 ±0.42	SE, CT, GLC	40
		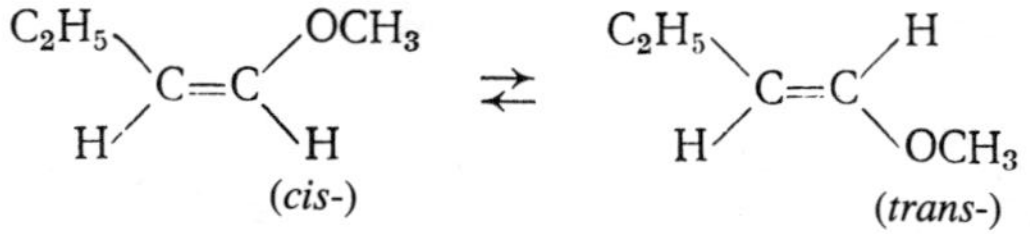 cis 1-ethyl-1-methoxyethene ⇌ trans isomer	298 (g)	−2.2 ±0.6	−4.9 ±1.5	SE, CT, GLC	74
		![cyclohexene structure]	298 (l)	(−1.51 ±0.13)	—	SE, CT, GLC 	56

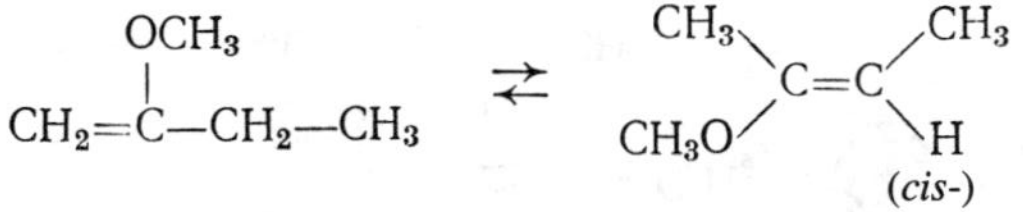

43

Table (*Continued*)

Formula	Type	Reaction	T, K (phase)	$\Delta_r H_T^\circ$ ($\Delta_r G_T^\circ$)	$\Delta_r S_T^\circ$	Technique	Reference
$C_5H_{10}OS$	1.3	$CH_2=CH-CH_2-S-C_2H_5$ with O double bond $\rightleftarrows$ cis isomer with CH_3, H, C=C, S(=O)–C_2H_5, H	293 (l)	(1.7)	—	SE, CT, GLC t–C_4H_9OH	57
		$CH_2=CH-CH_2-S-C_2H_5$ with O double bond $\rightleftarrows$ trans isomer with CH_3, H, C=C, H, S(=O)–C_2H_5	293 (l)	(−4.7)	—	SE, CT, GLC t–C_4H_9OH	57
		cis sulfoxide $\rightleftarrows$ trans sulfoxide	293 (l)	(−6.4)	—	SE, CT, GLC t–C_4H_9OH	57
$C_5H_{10}O_2$	1.3	$C_4H_9-C(=O)-OH \rightleftarrows i-C_4H_9-C(=O)-OH$	298 (l)	−5.73	—	CC	58
		$C_4H_9-C(=O)-OH \rightleftarrows CH_3-CH(C_2H_5)-C(=O)-OH$	298 (l)	−7.91	—	CC	58
		$C_4H_9-C(=O)-OH \rightleftarrows t-C_4H_9-C(=O)-OH$	298 (l)	−15.85	—	CC	58

Formula		Equilibrium	T (K) (phase)	ΔH	ΔS	Method, solvent	Ref.
	2.1	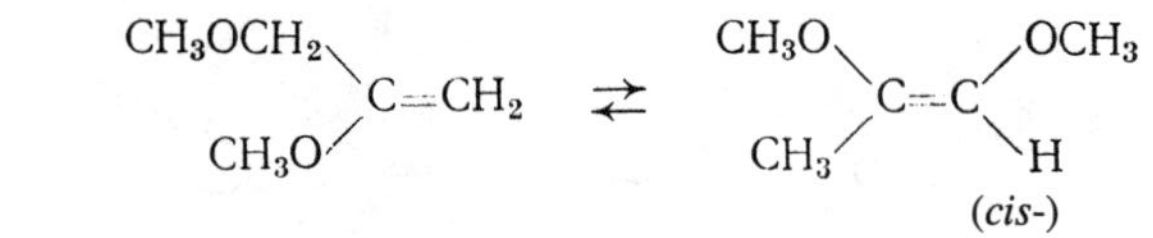 CH₃OCH₂\\C=CH₂ / CH₃O ⇌ CH₃O\\C=C /OCH₃\\H (cis-) / CH₃	298 (l)	6.2 ±1.1	8.4 ±3.5	SE, CT, GLC 	81
		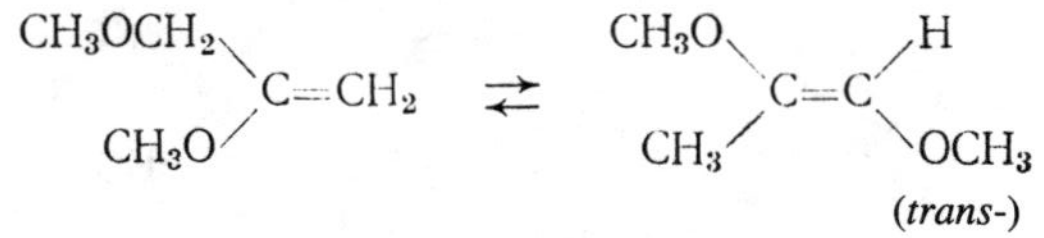 CH₃OCH₂\\C=CH₂ / CH₃O ⇌ CH₃O\\C=C/H\\OCH₃ (trans-) / CH₃	298 (l)	0.6 ±0.4	3.2 ±1.0	SE, CT, GLC 	81
		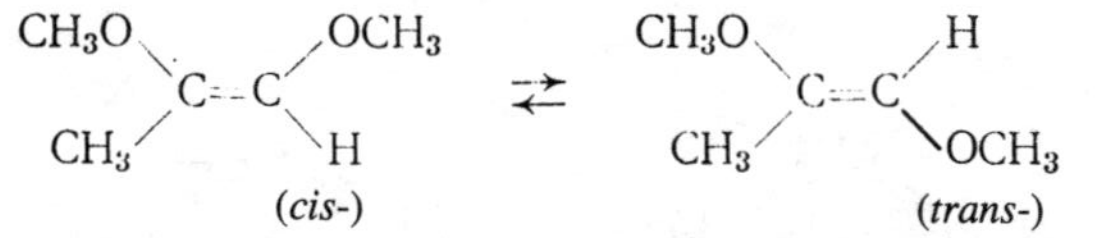 CH₃O\\C=C/OCH₃\\H (cis-) / CH₃ ⇌ CH₃O\\C=C/H\\OCH₃ (trans-) / CH₃	298 (l)	−5.6 ±0.8	−5.2 ±2.5	SE, CT, GLC	81
		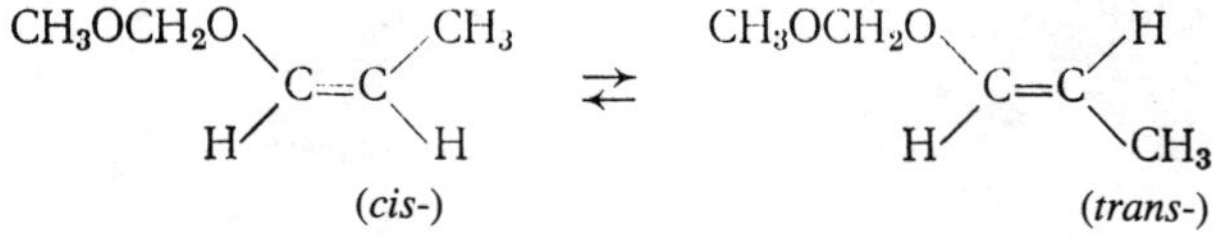 CH₃OCH₂O\\C=C/CH₃\\H (cis-) ⇌ CH₃OCH₂O\\C=C/H\\CH₃ (trans-)	298 (l)	2.3 ±0.2	0.7 ±0.5	SE, CT, GLC	92
		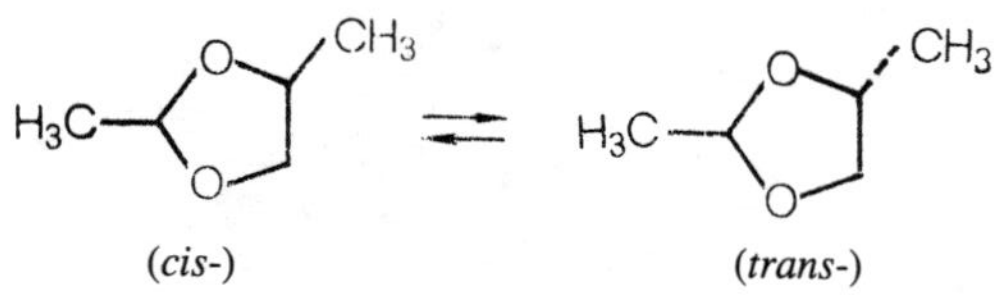 H₃C⟨O—O⟩CH₃ (cis-) ⇌ H₃C⟨O—O⟩CH₃ (trans-)	298 (l)	(1.15 ±0.03)	—	SE, CT, GLC (C₂H₅)₂O	93
C₅H₁₀O₂S	1.3	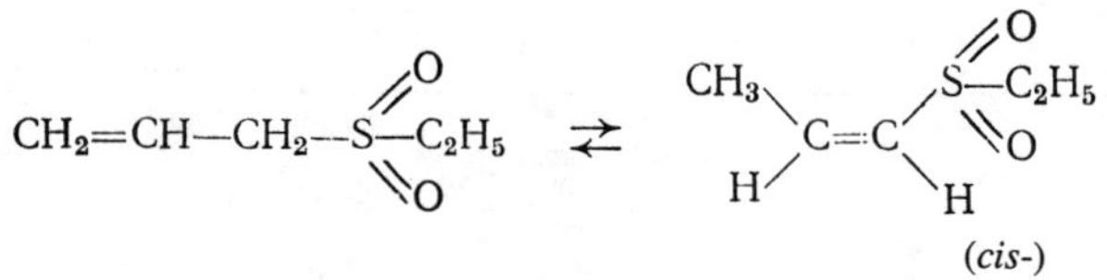 CH₂=CH—CH₂—S(=O)(=O)—C₂H₅ ⇌ CH₃\\C=C/S(=O)(=O)—C₂H₅\\H (cis-)	293 (l)	(10.1)	—	SE, CT, GLC t-C₄H₉OH	57

45

Table (*Continued*)

Formula	Type	Reaction	T, K (phase)	$\Delta_r H_T^\circ$ ($\Delta_r G_T^\circ$)	$\Delta_r S_T^\circ$	Technique	Reference
$C_5H_{10}O_2S$	1.3	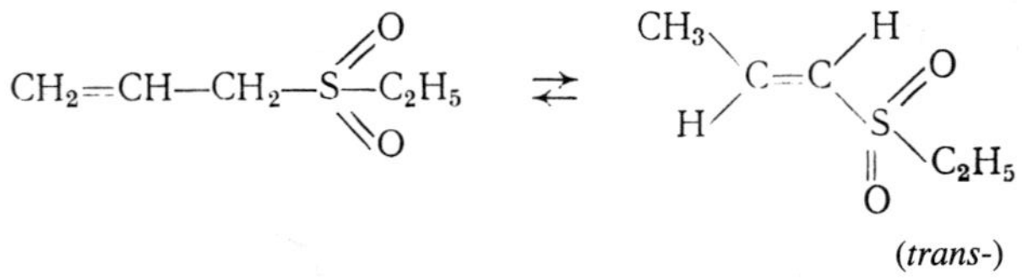 $CH_2=CH-CH_2-S(=O)(=O)-C_2H_5 \rightleftharpoons$ CH$_3$\C=C/H (with S(=O)(O)C$_2$H$_5$) (*trans-*)	293 (l)	(−3.6)	—	SE, CT, GLC t—C_4H_9OH	57
	2.1	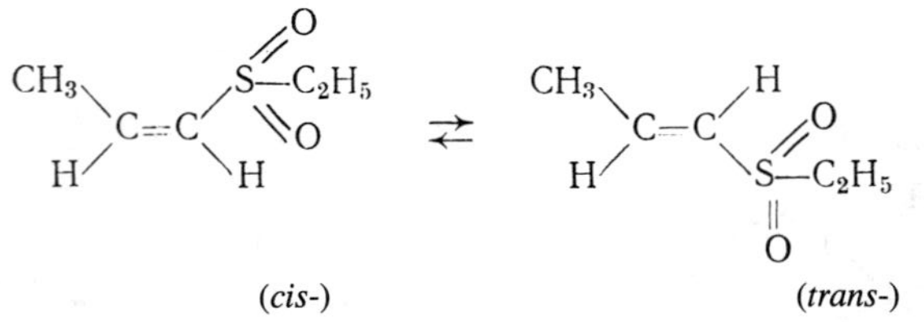 (*cis-*) $\rightleftharpoons$ (*trans-*)	293 (l)	(−13.6)	—	SE, CT, GLC t—C_4H_9OH	57
$C_5H_{10}S$	1.3	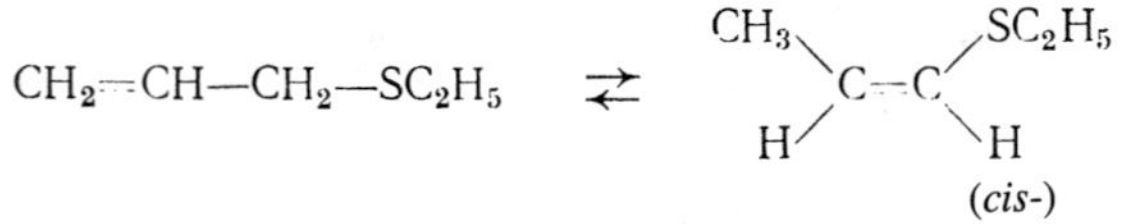 $CH_2=CH-CH_2-SC_2H_5 \rightleftharpoons$ (*cis-*)	293 (l)	(−15.0)	—	SE, CT, GLC t—C_4H_9OH	57
		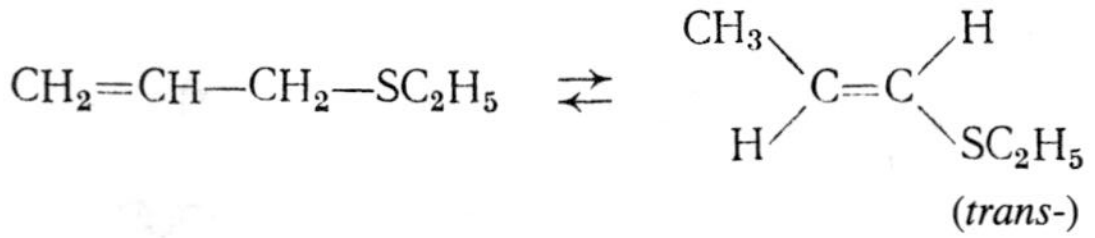 $CH_2=CH-CH_2-SC_2H_5 \rightleftharpoons$ (*trans-*)	293 (l)	(−15.3)	—	SE, CT, GLC t—C_4H_9OH	57
	2.1	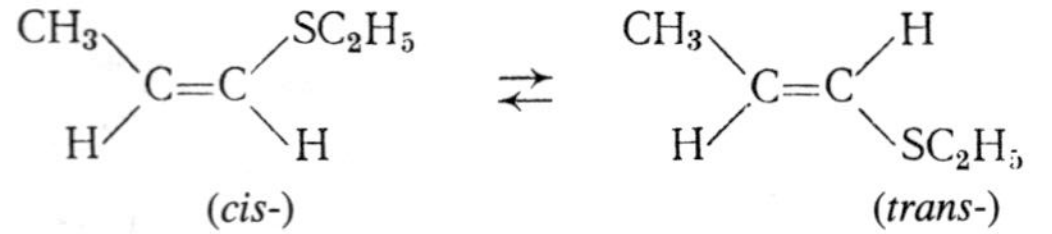 (*cis-*) $\rightleftharpoons$ (*trans-*)	293 (l)	(−0.25)	—	SE, CT, GLC t—C_4H_9OH	57

Formula		Reaction	T, K (state)	ΔH ±	ΔS ±	Method	Ref.
$C_5H_{11}Br$	1.3	(2-methylthiolane) ⇌ (3-methylthiolane)	298 (g)	3.72 ±1.00	—	CC	94
	1.3	$CH_3-CH(Br)-CH_2-C_2H_5$ ⇌ $CH_3-CH_2-CH(Br)-C_2H_5$	328 (l)	−0.05 ±0.42	−3.97 ±1.67	SE, CT, GLC	95
$C_5H_{11}Cl$	1.3	$CH_3-CH(Cl)-CH_2-C_2H_5$ ⇌ $CH_3-CH_2-CH(Cl)-C_2H_5$	461 (g)	0.0 ±0.42	−4.85 ±1.42	FE, CT, GLC	8
		$CH_3-C(Cl)(CH_3)-CH_2-CH_3$ ⇌ $CH_3-CH(Cl)-CH(CH_3)-CH_3$	407 (g, l)	−0.54 ±0.42	−6.15 ±1.25	SE, FE, CT, GLC	96
		$CH_3-CH(Cl)-CH_2-CH_2-CH_3$ ⇌ $CH_3-CH(CH_3)-CH(Cl)-CH_3$	323 (l)	7.91 ±0.21	7.32 ±0.84	SE, CT, GLC	97
C_5H_{12}	1.3	$CH_3-CH_2-CH_2-CH_2-CH_3$ ⇌ $C(CH_3)_4$	298 (g)	−21.38 ±1.42	—	⬡	98
		$CH_3-CH(CH_3)-CH_2-CH_3$ ⇌ $C(CH_3)_4$	298 (g)	−14.26 ±0.81	—	CC	99
		$CH_3-CH_2-CH_2-CH_2-CH_3$ ⇌ $CH_3-CH(CH_3)-CH_2-CH_3$	298 (g)	−7.03 ±0.46	—	CC	98

Table (Continued)

Formula	Type	Reaction	T, K (phase)	$\Delta_r H_T^\circ$ ($\Delta_r G_T^\circ$)	$\Delta_r S_T^\circ$	Technique	Reference
C_5H_{12}	1.3		298 (g)	−6.90 ±0.78	—	CC	99
$C_5H_{12}GeO$	1.1 1.2	$CH_3-Ge-CH_2-C\overset{O}{\underset{H}{\diagup}}$ with CH_3 groups $\rightleftarrows$ $CH_3-Ge-O-C\overset{CH_2}{\underset{H}{\diagup}}$ with CH_3 groups	363 (l)	11.04 ±0.21	8.78 ±0.42	SE, NMR	100
$C_5H_{12}S$	1.3	SH−CH$_2$−CH−CH$_2$−CH$_3$ (CH$_3$) $\rightleftarrows$ SH−CH$_2$−CH$_2$−CH−CH$_3$ (CH$_3$)	298 (g)	0.08 ±1.42	—	CC	94
		SH−CH$_3$−CH−CH−CH$_3$ (CH$_3$) $\rightleftarrows$ CH$_2$−C−CH$_3$ (SH, CH$_3$, CH$_3$)	298 (g)	−7.74 ±1.23	—	CC	94
		SH−CH$_2$−CH−CH$_2$−CH$_3$ (CH$_3$) $\rightleftarrows$ SH−CH$_3$−CH−CH−CH$_3$ (CH$_3$)	298 (g)	−6.32 ±1.42	—	CC	94
C_6F_{10}	1.2	$CF_3-C=C-CF_3$ / F_2C-CF_2 $\rightleftarrows$ CF_3-C-CF_2 / $F_2C=CF_2$ with CF$_3$	595 (g)	1.67 ±0.84	20.37	SE, GLC	101
$C_6H_2Cl_4$	1.3	(1,2,3,5-tetrachlorobenzene) $\rightleftarrows$ (1,2,4,5-tetrachlorobenzene)	298 (g)	—	−6.15	ST	102

48

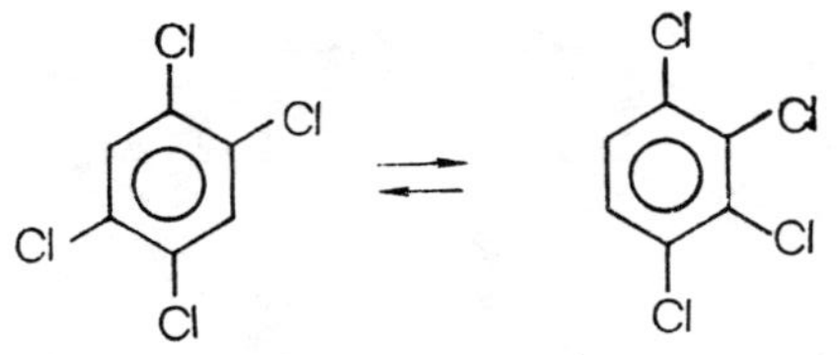

					Reaction	
102	ST	3.68	—	298 (g)	1,2,3-C₆H₃Cl₃ ⇌ 1,3,5-C₆H₃Cl₃	
253	CC	—	9.5	298 (g)	1,2,3-C₆H₃Cl₃ ⇌ 1,2,4-C₆H₃Cl₃	
102	ST	−2.84	—	298 (g)	1,2,3-C₆H₃F₃ ⇌ 1,2,4-C₆H₃F₃	C₆H₂F₄ 1.3
102	ST	−2.93	—	298 (g)	1,2,3,4-C₆H₂F₄ ⇌ 1,2,3,5-C₆H₂F₄	

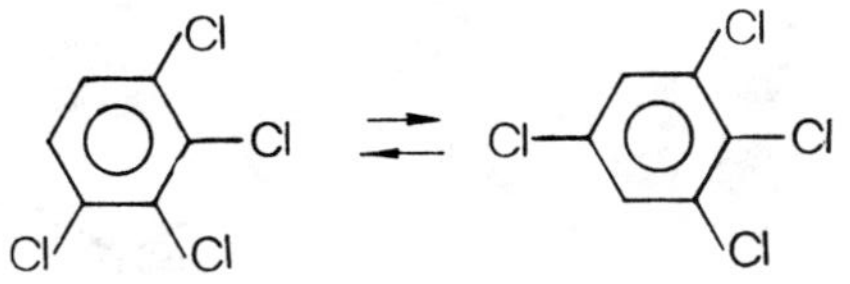

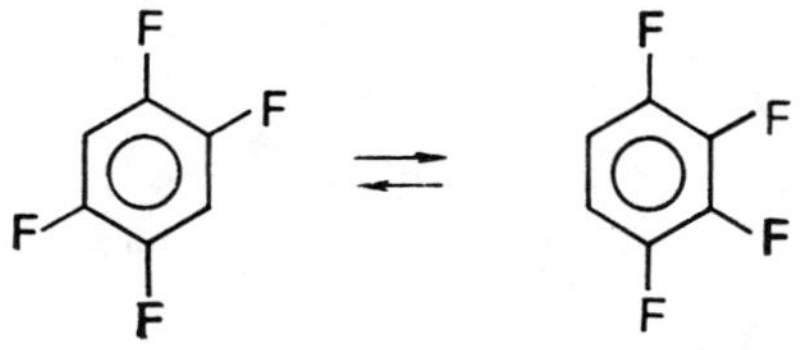

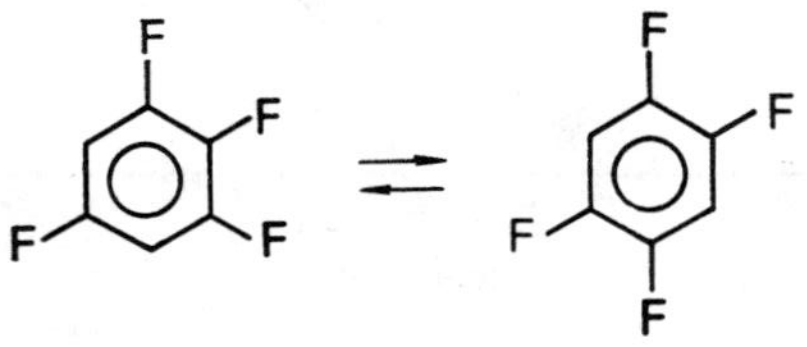

Table (*Continued*)

Formula	Type	Reaction	T, K (phase)	$\Delta_r H°_T$ ($\Delta_r G°_T$)	$\Delta_r S°_T$	Technique	Reference
$C_6H_3Cl_3$	1.3	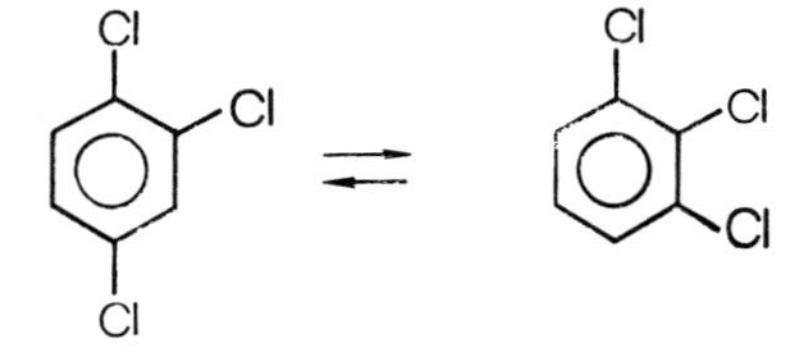 1,2,4-trichlorobenzene ⇌ 1,2,3-trichlorobenzene	298 (g)	—	−6.15	ST	102
		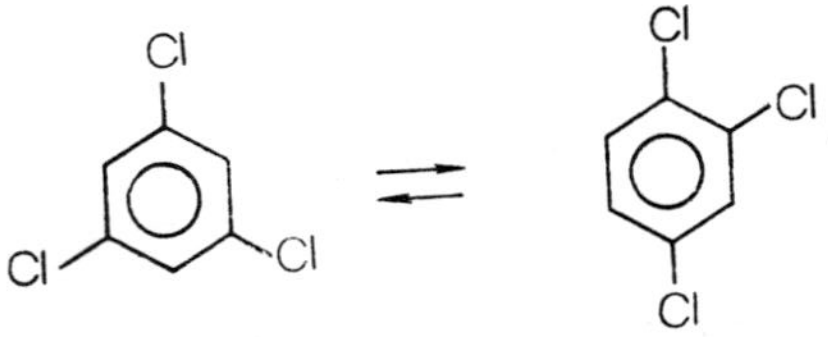 1,3,5-trichlorobenzene ⇌ 1,2,4-trichlorobenzene	298 (g)	—	13.80	ST	102
			298 (g)	−5.31	—	CC	253
$C_6H_3F_3$	1.3	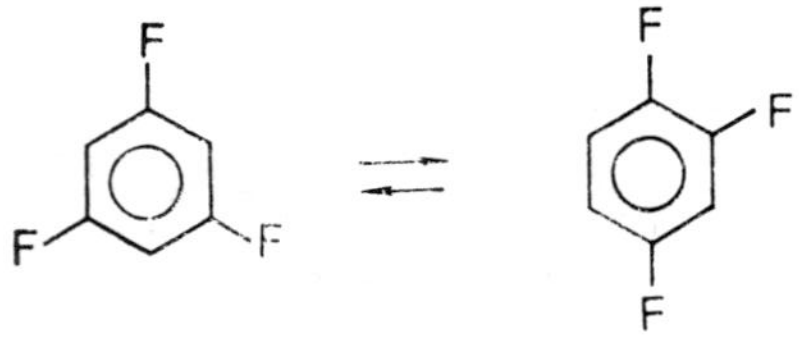 1,3,5-trifluorobenzene ⇌ 1,2,4-trifluorobenzene	298 (g)	—	16.15	ST	102

50

Formula		Equilibrium	T (K) (state)			Method	Ref.
$C_6H_4Br_2$	1.3	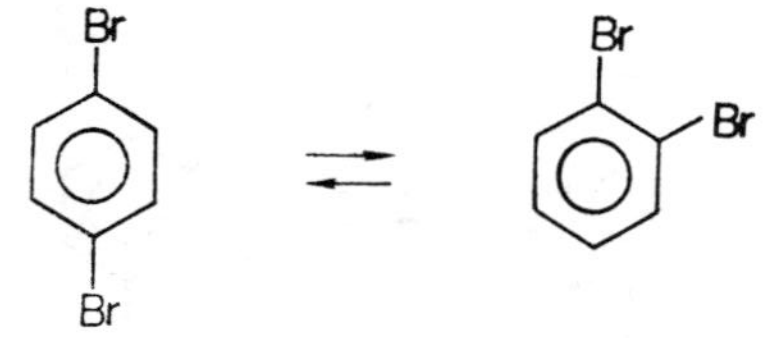 1,2-$C_6H_4Br_2$ ⇌ 1,4-$C_6H_4Br_2$	298 (g)	—	0.42	ST	102
		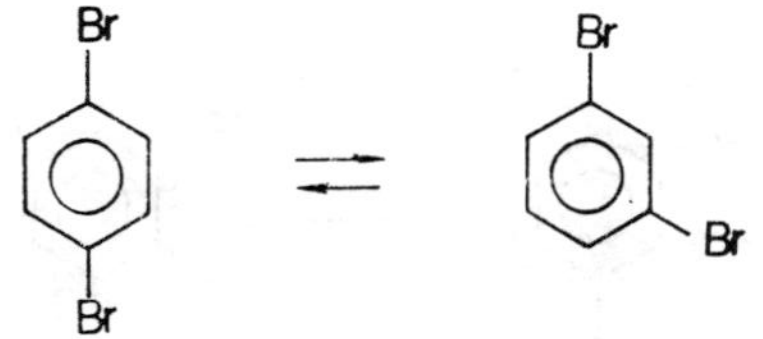 1,3-$C_6H_4Br_2$ ⇌ 1,4-$C_6H_4Br_2$	298 (g)	—	5.69	ST	102
$C_6H_4Cl_2$	1.3	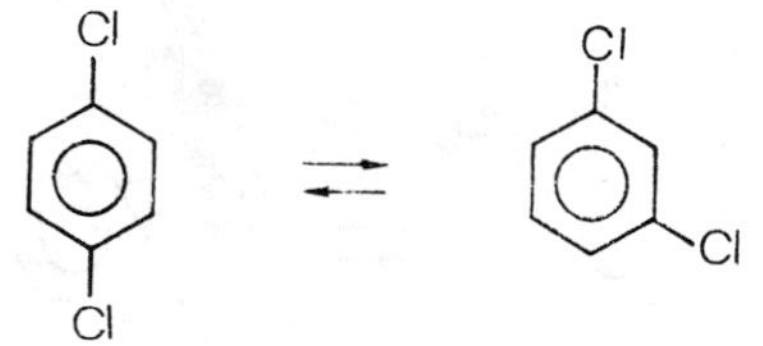 1,3-$C_6H_4Cl_2$ ⇌ 1,4-$C_6H_4Cl_2$	298 (g)	—	6.73	ST	102
			298 (g)	4.22	—	CC	103
		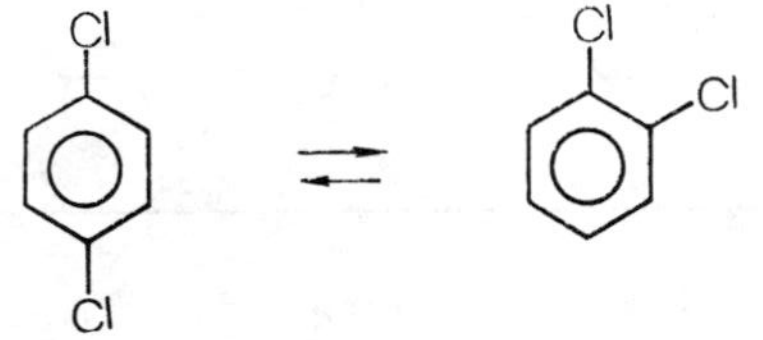 1,2-$C_6H_4Cl_2$ ⇌ 1,4-$C_6H_4Cl_2$	298 (g)	—	5.02	ST	102
			298 (g)	8.75	—	CC	103

Table (*Continued*)

Formula	Type	Reaction	T, K (phase)	$\Delta_r H_T^\circ$ ($\Delta_r G_T^\circ$)	$\Delta_r S_T^\circ$	Technique	Reference
$C_6H_4Cl_2$	1.3	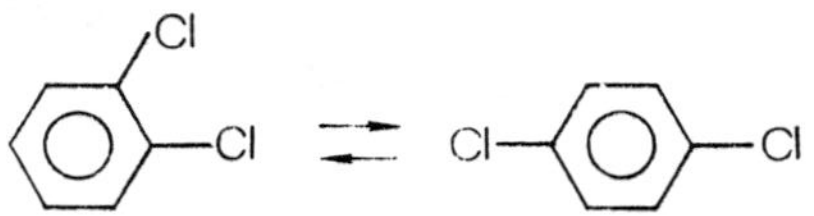 1,2-dichlorobenzene ⇌ 1,4-dichlorobenzene	298 (g)	−8.4	—	CC	254
		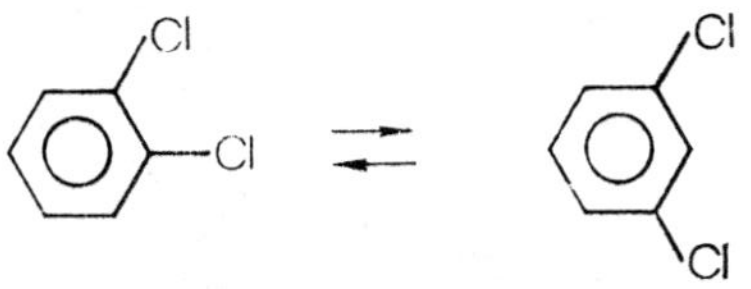 1,2-dichlorobenzene ⇌ 1,3-dichlorobenzene	298 (g)	−4.9	—	CC	254
$C_6H_4F_2$	1.3	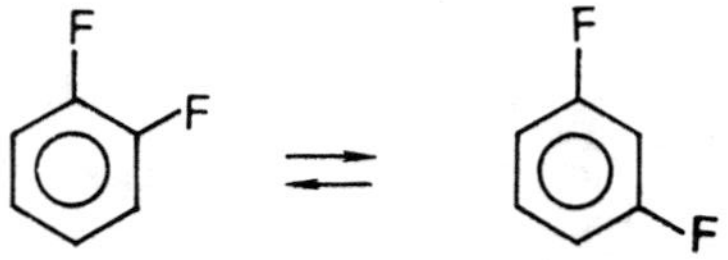 1,2-difluorobenzene ⇌ 1,3-difluorobenzene	298 (g)	—	−1.84	ST	102
		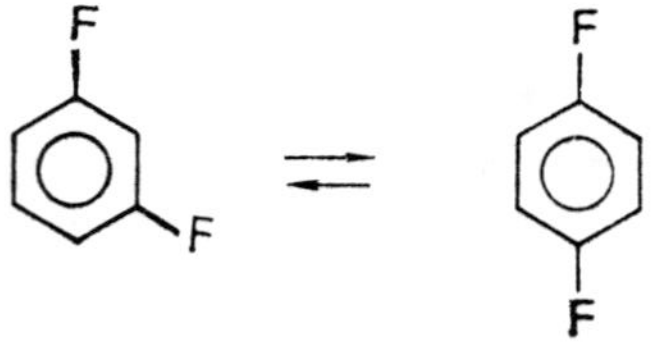 1,3-difluorobenzene ⇌ 1,4-difluorobenzene	298 (g)	—	−5.02	ST	102

Formula		Equilibrium	T, K (phase)	ΔG	ΔH	Method	Ref.
$C_6H_4N_2$	1.3	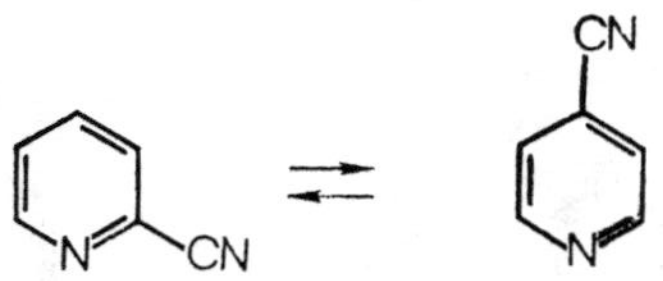 2-cyanopyridine ⇌ 4-cyanopyridine	298 (g)	2.8 ±2.5	—	CC	71
		 2-cyanopyridine ⇌ 3-cyanopyridine	298 (g)	−2.8 ±2.4	—	CC	71
$C_6H_6O_2$	1.3	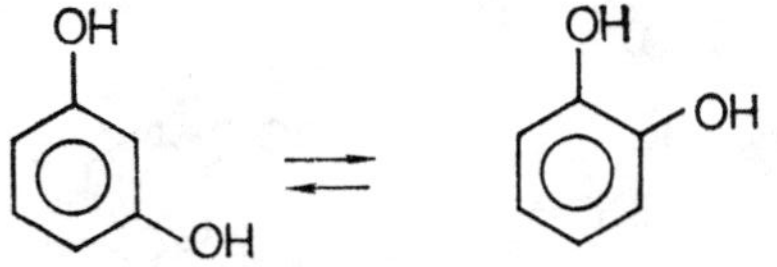 1,3-dihydroxybenzene ⇌ 1,2-dihydroxybenzene	298 (g)	2.93 ±0.84	—	CC	104
		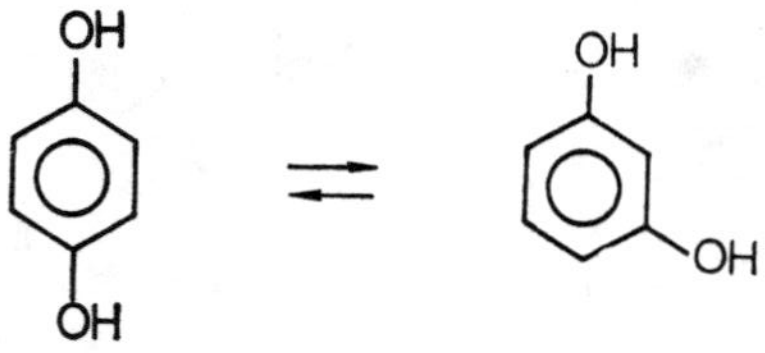 1,4-dihydroxybenzene ⇌ 1,3-dihydroxybenzene	298 (g)	0.84 ±0.70	—	CC	104
$C_6H_7F_3O_3$	1.1	$CF_3-\underset{O}{\overset{\parallel}{C}}-CH_2-\underset{O}{\overset{\parallel}{C}}-OC_2H_5$ ⇌ $CF_3-\underset{OH}{\overset{\mid}{C}}=CH-\underset{O}{\overset{\parallel}{C}}-OC_2H_5$	313 (l)	−14.89	35.14	SE, NMR	105

Table (Continued)

Formula	Type	Reaction	T, K (phase)	$\Delta_r H_T^\circ$ ($\Delta_r G_T^\circ$)	$\Delta_r S_T^\circ$	Technique	Reference
C_6H_7N	1.1 1.2	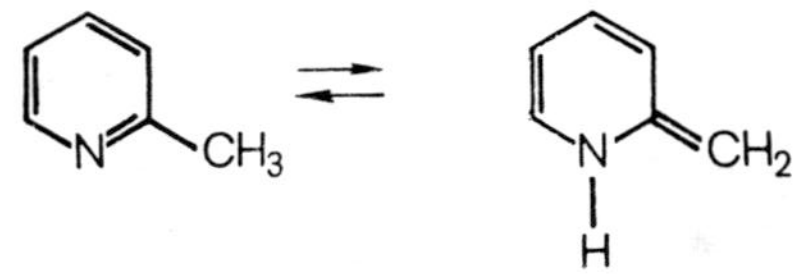 2-methylpyridine ⇌ 2-methylene-1H-pyridine	298 (l)	(−79.5)	—	SE, UV	67
C_6H_7NO	1.1 1.2	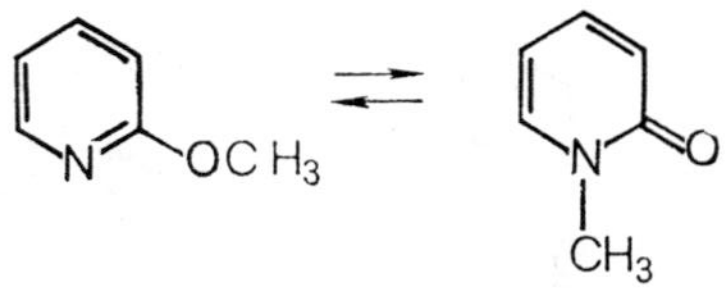 2-methoxypyridine ⇌ 1-methyl-2-pyridone	403 (l)	−51.87 ±3.3	—	CC	106
		4-methoxypyridine ⇌ 1-methyl-4-pyridone	508 (g)	−32.2 ±9.62	—	SE, UV	68
	1.1	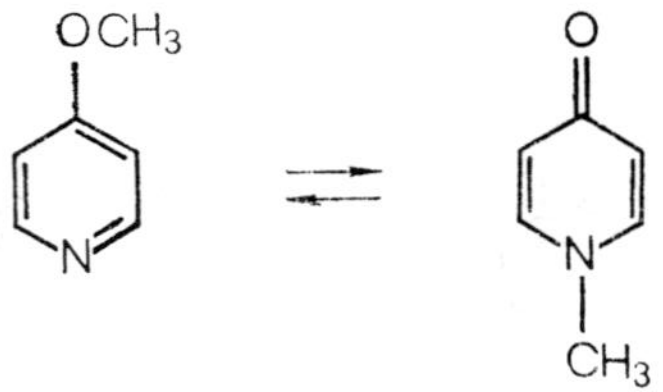	403 (l)	−36.81	—	CC	106

Formula		Equilibrium	T (state)			Method / Solvent	Ref.
C_6H_7NS	1.1 1.2	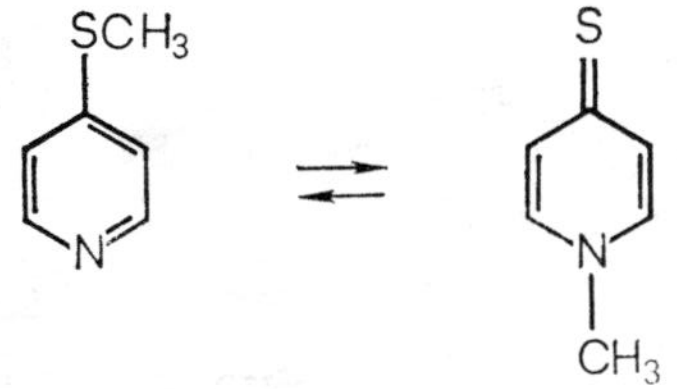 SCH₃–pyridine ⇌ N-CH₃ pyridinethione	298 (g)	116.7 ±16.3	—	CC	107
$C_6H_8N_2O_2S$	1.1 1.2	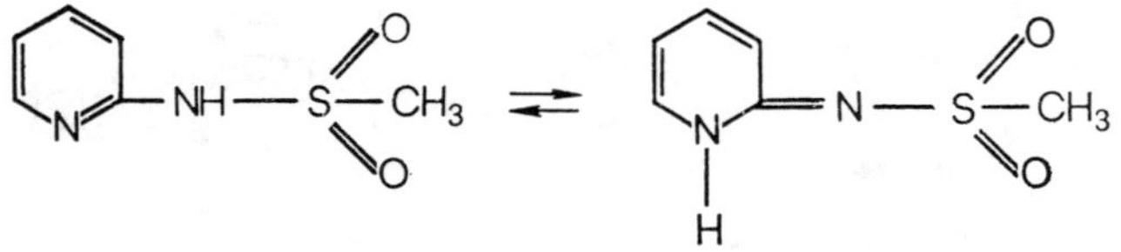 NH–SO₂–CH₃ pyridyl ⇌ N=S(O)(CH₃)=O pyridyl-H	331 (l)	−7.24	−9.20	SE, UV H_2O	69
$C_6H_9ClO_2$	1.3	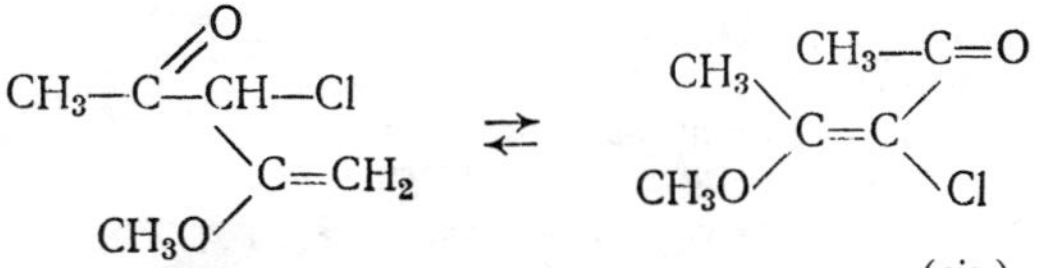 CH₃–CO–CH(Cl)–C(OCH₃)=CH₂ ⇌ CH₃–C(=O)–C(Cl)=C(CH₃)(OCH₃) (cis-)	298 (g)	−10.5 ±1.0	−10.0 ±2.7	SE, NMR, GLC  (benzene)	108
$C_6H_9ClO_3$	1.3	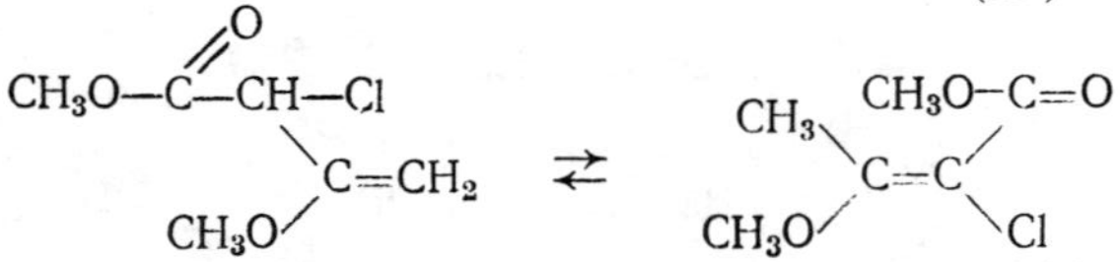 CH₃O–CO–CH(Cl)–C(OCH₃)=CH₂ ⇌ CH₃O–C(=O)–C(Cl)=C(CH₃)(OCH₃) (cis-)	298 (g)	−14.4 ±1.0	−21.5 ±2.7	SE, NMR, GLC (benzene)	108
		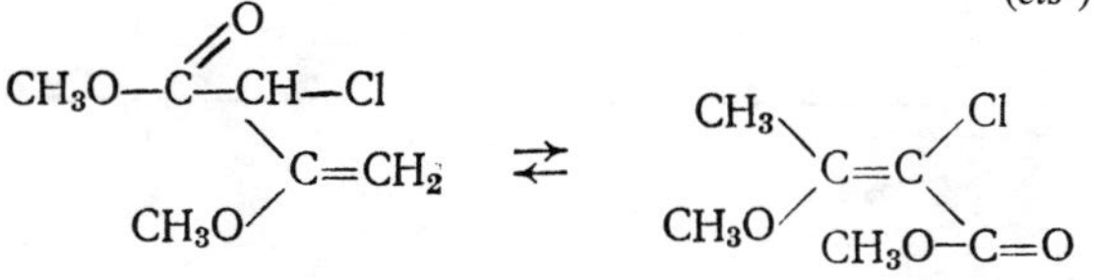 CH₃O–CO–CH(Cl)–C(OCH₃)=CH₂ ⇌ CH₃–C(Cl)=C(OCH₃)(CH₃O–C=O) (trans-)	298 (g)	−0.7 ±0.1	−4.4 ±0.2	SE, NMR, GLC CCl_4	108

Table (Continued)

Formula	Type	Reaction	T, K (phase)	$\Delta_r H°_T$ ($\Delta_r G°_T$)	$\Delta_r S°_T$	Technique	Reference
$C_6H_9ClO_3$	2.1	$\underset{(cis\text{-})}{\overset{CH_3}{\underset{CH_3O}{C}}=\overset{CH_3O-C=O}{\underset{Cl}{C}}} \rightleftarrows \underset{(trans\text{-})}{\overset{CH_3}{\underset{CH_3O}{C}}=\overset{Cl}{\underset{CH_3O-C=O}{C}}}$	298 (g)	13.8 ±1.3	17.2 ±3.4	SE, NMR, GLC CCl_4	108
C_6H_{10}	1.3	$CH_2=CH-CH=CH-CH-C_2H_5 \rightleftarrows CH_3-CH=CH-CH=CH-CH_3$ (cis- + trans-) (cis-,cis- + cis-,trans- + trans-,cis- + trans-,trans-)	548 (g)	−11.59 ±0.75	−6.32 ±0.17	FE, CT, GLC	109
		$\underset{H}{\overset{CH_2=CH}{C}}=\underset{H}{\overset{C_2H_5}{C}} \rightleftarrows \underset{H}{\overset{CH_3}{C}}=\underset{H}{\overset{C=C}{\underset{H}{\overset{CH_3}{C}}}}$ (trans-) (trans-,trans-)	548 (g)	−12.80 ±0.88	−12.55 ±0.17	FE, CT, GLC	109
	2.1	$\underset{H}{\overset{CH_2=CH}{C}}=\underset{H}{\overset{C_2H_5}{C}} \rightleftarrows \underset{H}{\overset{CH_2=CH}{C}}=\underset{H}{\overset{H}{\underset{C_2H_5}{C}}}$ (cis-) (trans-)	548 (g)	−4.52 ±0.63	1.72 ±0.33	FE, CT, GLC	109
		$\underset{(cis\text{-})}{CH_2=CH-CH_2\overset{CH_3}{\underset{H}{C}}=\underset{H}{C}} \rightleftarrows \underset{(trans\text{-})}{CH_2=CH-CH_2\overset{H}{\underset{H}{C}}=\underset{CH_3}{C}}$	573 (g)	−3.01 ±0.46	1.17 ±0.25	FE, CT, GLC	109

Reaction	T (K) (state)	ΔH	ΔG	Method	Ref.
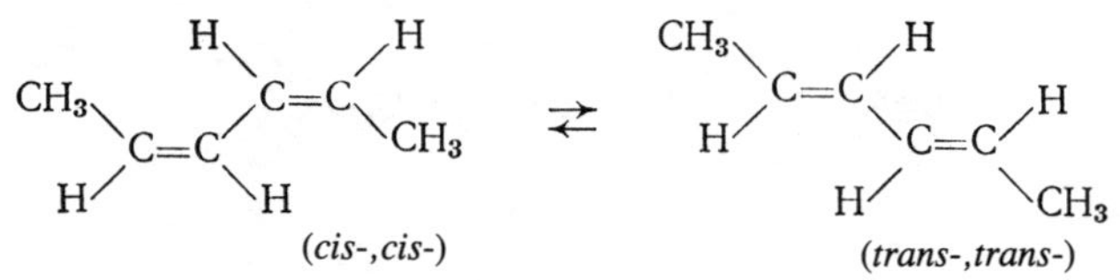 CH₃-CH=CH-CH=CH-CH₃ (cis-,cis-) ⇌ (trans-,trans-)	548 (g)	−3.93 ±0.42	4.89 ±0.21	FE, CT, GLC	109
(cis-,trans-) ⇌ (trans-,trans-)	318 (l)	−7.11	−7.11	SE, CT, GLC, C₇H₁₆	110
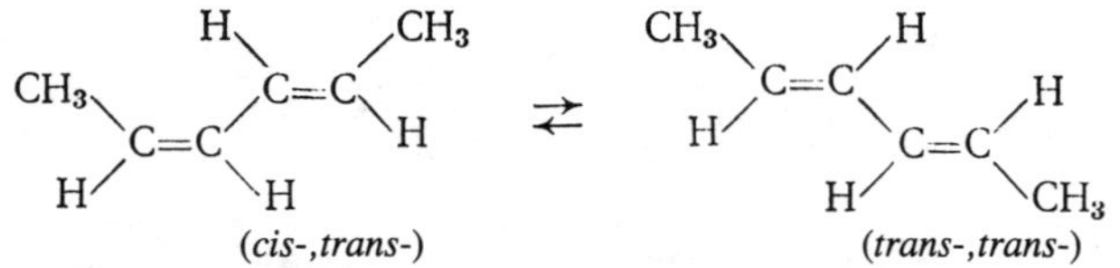 CH₂=C(CH₃)-CH(CH₃)-... (cis-) ⇌ (trans-)	548 (g)	−3.51 ±0.63	−5.73 ±0.17	FE, CT, GLC	109
	318 (l)	−3.60	−5.86	SE, CT, GLC, C₇H₁₆	110
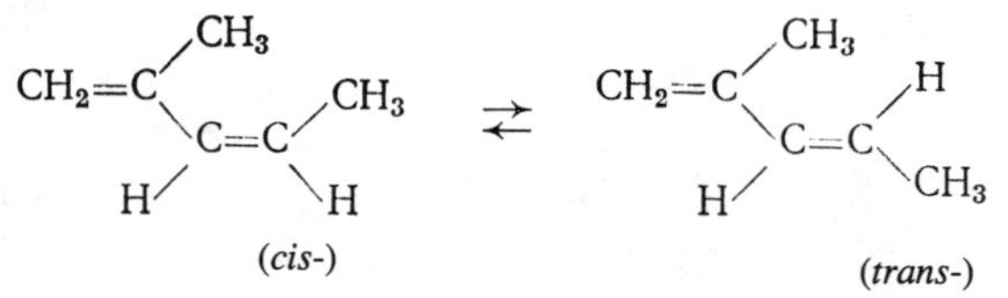 CH₃-C(CH₃)=CH-CH=CH₂ ⇌ (cis-)	548 (g)	−17.20 ±0.79	−11.92 ±0.29	FE, CT, GLC	255
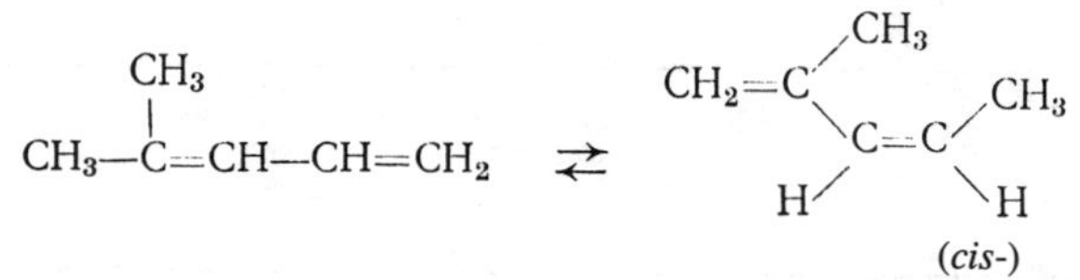	548 (g)	16.11 ±0.71	12.30 ±0.21	FE, CT, GLC	255

1.3

Table (Continued)

Formula	Type	Reaction	T, K (phase)	$\Delta_r H_T^\circ$ ($\Delta_r G_T^\circ$)	$\Delta_r S_T^\circ$	Technique	Reference
C_6H_{10}	1,3	$CH_3-\overset{\underset{\displaystyle CH_3}{\mid}}{C}=CH-CH=CH_2 \rightleftarrows \underset{H}{\overset{CH_3}{C}}=C\underset{CH_3}{\overset{H}{\diagdown}}$ (trans-)	548 (g)	−1.09 ±1.03	0.38 ±0.21	FE, CT, GLC	255
		$CH_3-C\equiv C-CH_2-C_2H_5 \rightleftarrows CH\equiv C-CH_2-CH_2-C_2H_5$	298 (l)	(6.14)	—	SE, CT, GLC CH_3-S-CH_3 $\parallel$ O	111
		$CH_3-CH_2-C\equiv C-C_2H_5 \rightleftarrows CH_3-C\equiv C-CH_2-C_2H_5$	298 (l)	14.64 ±2.43	—	RC, CT C_6H_{14}	112
		$CH_3-CH_2-C\equiv C-C_2H_5 \rightleftarrows CH_3-C\equiv C-CH_2-C_2H_5$	298 (l)	(−5.00)	—	SE, CT, GLC CH_3-S-CH_3 $\parallel$ O	111
		$CH_3-CH_2-C\equiv C-C_2H_5 \rightleftarrows CH\equiv C-CH_2-CH_2-C_2H_5$	298 (l)	−2.26 ±2.90	—	RC, CT C_6H_{14}	112
		$CH_3-CH_2-C\equiv C-C_2H_5 \rightleftarrows CH\equiv C-CH_2-CH_2-C_2H_5$	298 (l)	(1.13)	—	SE, CT, GLC	111
		3-methylcyclopentene ⇌ 1-methylcyclopentene	466 (g, l)	7.49 ±0.21	−8.12 ±0.42	SE, FE, CT, GLC	113

523 (g)	(11.04) ±0.17	—	FE, CT, GLC	73
298 (l)	14.10 ±0.86	—	CC	114
298 (g)	10.96 ±0.98	—	CC	115
523 (g)	(10.92) ±0.17	—	FE, CT, GLC	73
466 (g, l)	8.07 ±0.29	−1.63 ±0.42	SE, FE, CT, GLC	113
298 (l)	6.11 ±1.12	—	CC	114
411 (g, l)	1.17	—	SE, FE, CT, GLC	116
 466 (g, l)	15.8 ±0.8	−6.7 ±1.7	SE, FE, CT, GLC	113

[Structures: 1-methylcyclopentene ⇌ 3-methylcyclopent-2-ene (3 pairs); last pair shows methylenecyclopentane on right]

1.2
1.3

Table (Continued)

Formula	Type	Reaction	T, K (phase)	$\Delta_r H_T°$ ($\Delta_r G_T°$)	$\Delta_r S_T°$	Technique	Reference
C_6H_{10}	1.2, 1.3	 1-methylcyclopentene ⇌ methylenecyclopentane	298 (g)	16.31	—	CC	32
	1.2	3-methylcyclopentene ⇌ cyclohexene	298 (l)	17.69 ±1.03	—	CC	114
			523 (g)	(19.99 ±0.17)	—	FE, CT, GLC	73
	1.3		298 (l)	−21.17 ±1.09	—	CC	114
$C_6H_{10}O$	1.3	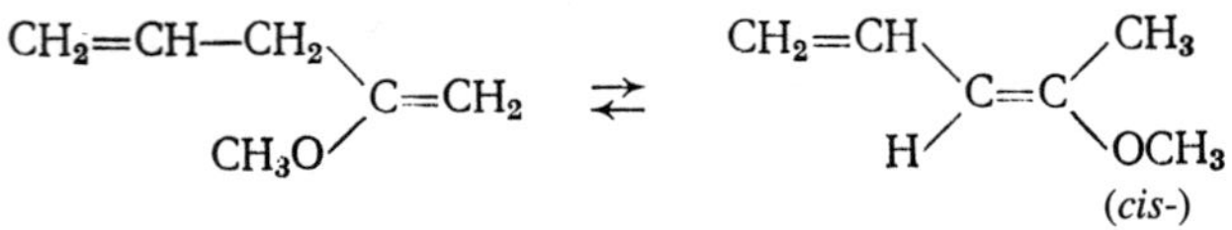 $CH_2=CH-CH_2-C(OCH_3)=CH_2$ ⇌ cis-$CH_2=CH-CH=C(CH_3)(OCH_3)$	298 (l)	−29.2 ±1.9	−25.0 ±5.0	SE, CT, GLC	117
		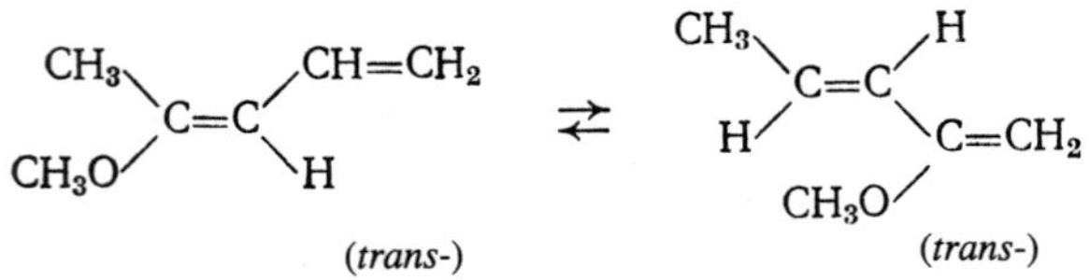 trans-$CH_3-C(OCH_3)=CH-CH=CH_2$ ⇌ cis-$CH_3-C(OCH_3)=CH-CH=CH_2$	298 (l)	−2.6 ±0.7	−7.0 ±2.0	SE, CT, GLC 	117
	2.1	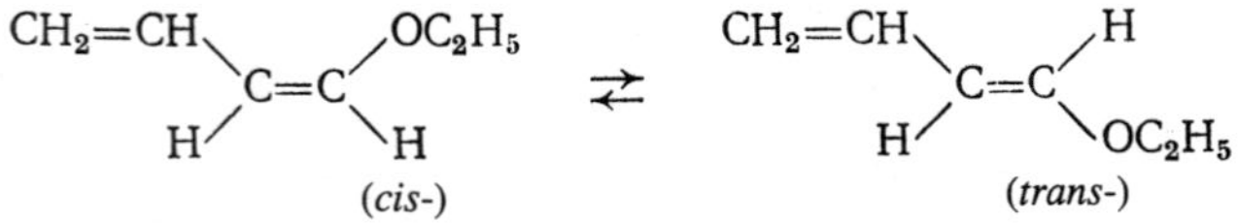 trans-$CH_2=CH-C(OC_2H_5)=CH$ ⇌ cis-$CH_2=CH-C(OC_2H_5)=CH$	311 (l)	−3.87 ±0.25	−6.23 ±0.79	SE, CT, GLC	40

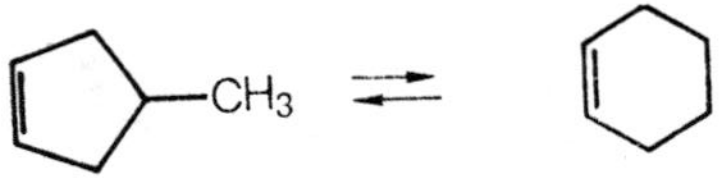

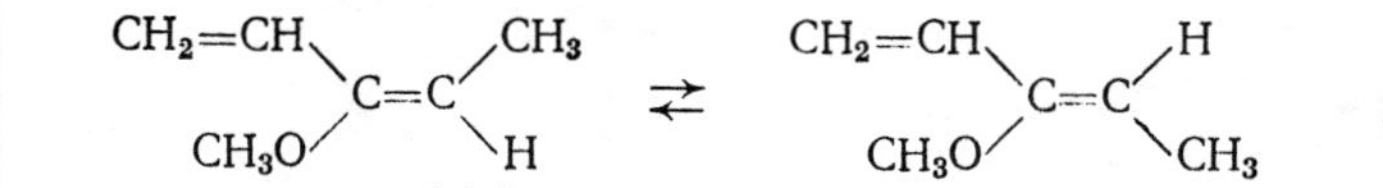

Equilibrium	T (K), phase	ΔH	ΔG	Method	Ref.
CH₂=CH–C(CH₃O)=C(CH₃)(H) (cis-) ⇌ CH₂=CH–C(CH₃O)=C(H)(CH₃) (trans-)	298 (l)	5.2 ±0.6	9.9 ±1.7	SE, CT, GLC	118
CH₃–C(CH₃O)=C(H)(CH=CH₂) (cis-) ⇌ CH₃–C(CH₃O)=C(CH=CH₂)(H) (trans-)	298 (l)	−11.2 ±1.1	−10.0 ±3.0	SE, CT, GLC	117
H–C(CH₃)=C(CH₃O)(C=CH₂) (cis-) ⇌ CH₃O–C(H)=C(CH₃)(C=CH₂) (trans-)	298 (l)	−16.0 ±1.1	−15.0 ±3.0	SE, CT, GLC	117
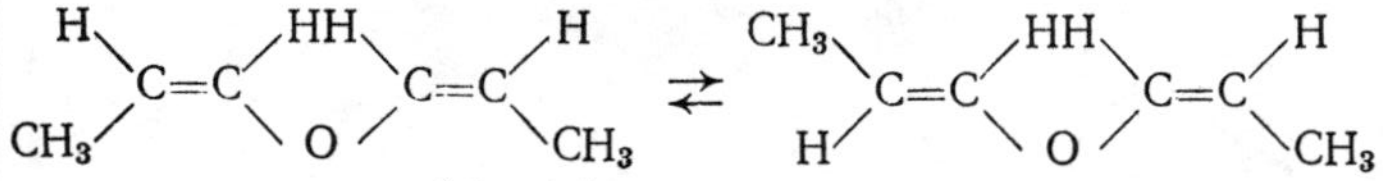 (cis-,cis-) ⇌ (trans-,cis-) dimethyl dioxine	298 (g)	2.9 ±0.3	7.2 ±0.6	SE, CT, GLC	74
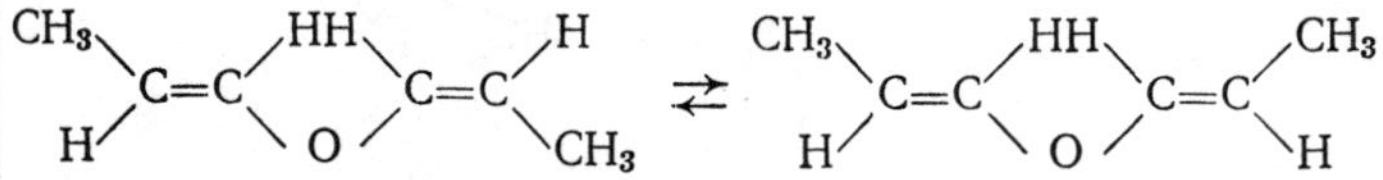 (trans-,cis-) ⇌ (trans-,trans-) dimethyl dioxine	298 (g)	2.3 ±0.4	−5.8 ±0.9	SE, CT, GLC	74
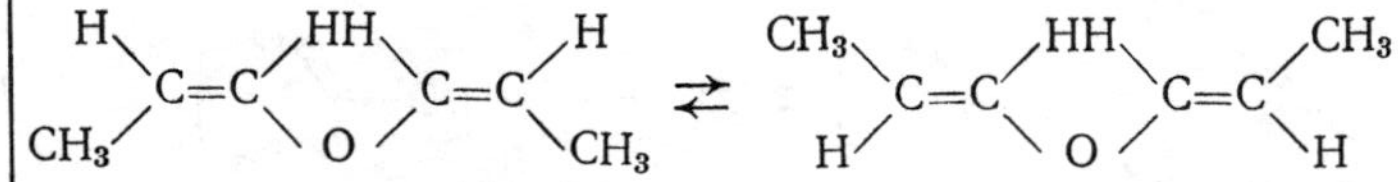 (cis-,cis-) ⇌ (trans-,trans-) dimethyl dioxine	298 (g)	5.1 ±0.4	1.4 ±0.9	SE, CT, GLC	74
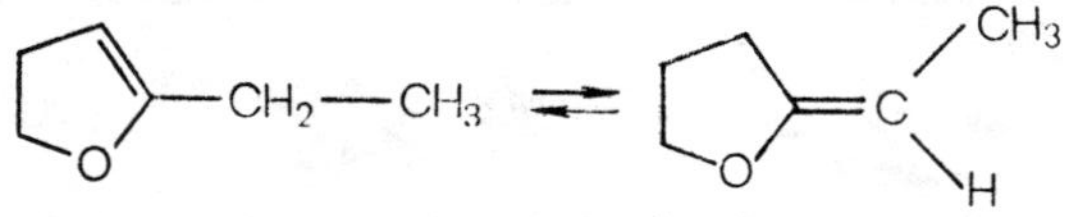 2-ethyl-4,5-dihydrofuran ⇌ (exo-) 2-ethylidene tetrahydrofuran	298 (g)	5.0 ±0.4	−3.5 ±0.6	SE, CT, GLC	119

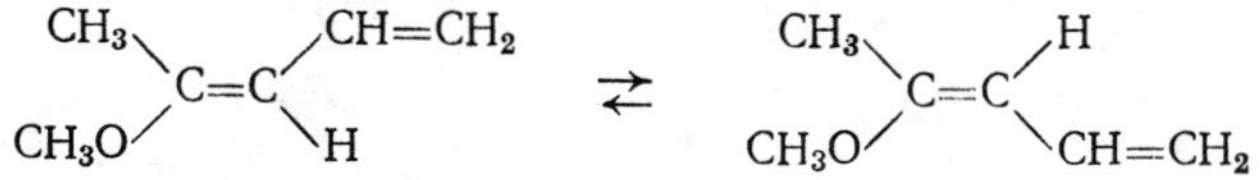

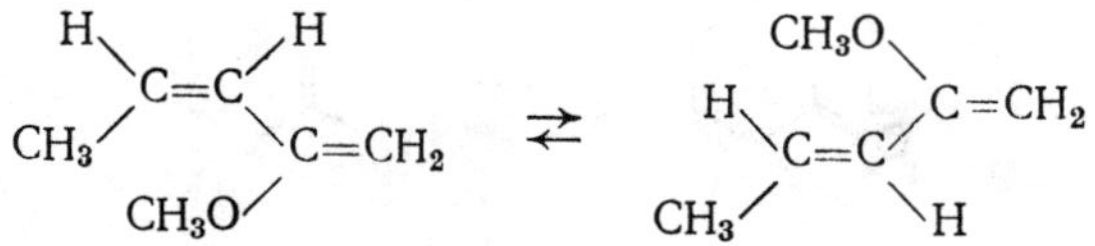

1.2
1.3

Table (Continued)

Formula	Type	Reaction	T, K (phase)	$\Delta_r H_T^\circ$ ($\Delta_r G_T^\circ$)	$\Delta_r S_T^\circ$	Technique	Reference
$C_6H_{10}O$	2.1	furan-CH₂-CH₃ ⇌ (endo-) methyl dihydrofuran; (exo-) ⇌ (endo-)	298 (g); 298 (g)	0.1 ±0.4; -4.8 ±0.4	-5.4 ±0.9; -1.9 ±0.9	SE, CT, GLC; SE, CT, GLC	119; 119
	1.1, 1.2	cyclohexenol ⇌ cyclohexanone	298 (l)	-25.35	—	RC	24
$C_6H_{10}O_2$	1.1, 1.2	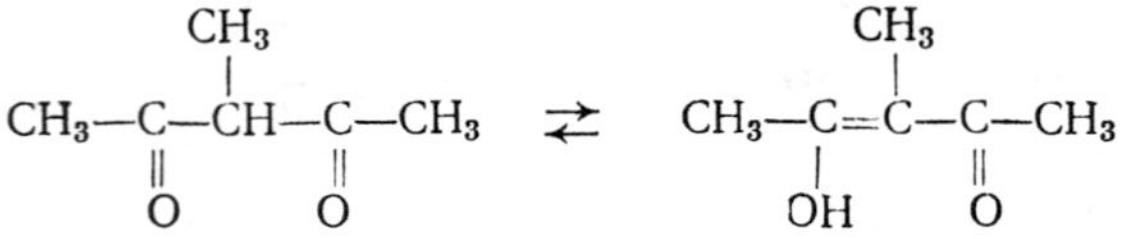 $CH_3-C(=O)-CH-C(=O)-CH_3$ with CH_3 ⇌ $CH_3-C(CH_3)=C(OH)-C(=O)-CH_3$	476 (g)	-12.6	—	SE, IR	77
			376 (g)	-5.4 ±0.4	—	SE, UV	79

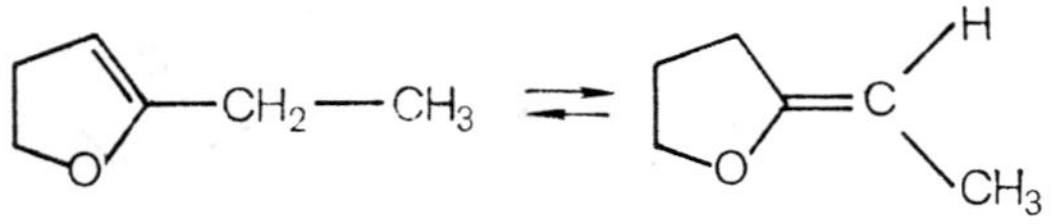

	Structure	T (K)	ΔH	ΔS	Method, Solvent	Ref.
1.3	$CH_3-CH_2-CH=CH-\overset{O}{\overset{\|}{C}}-OCH_3 \rightleftarrows$ $CH_3-CH=CH-CH_2-\overset{O}{\overset{\|}{C}}-OCH_3$ (cis+trans-) (cis+trans-)	390 (l)	(2.64 ±0.17)	—	SE, CT, GLC CH_3OH	56
	$\underset{CH_2=CH-CH}{\overset{CH_3}{\diagup}}\overset{O}{\underset{O-C-CH_3}{\diagdown}} \rightleftarrows \underset{CH_3}{\overset{H}{\diagup}}C=C\underset{H}{\overset{CH_2-O-\overset{O}{\overset{\|}{C}}-CH_3}{\diagdown}}$ (cis-)	335 (l)	−4.47	−2.03	SE, CT, GLC $CH_3\overset{O}{\overset{\|}{C}}$ OH, $HClO_3$	120
	$\underset{CH_2=CH-CH}{\overset{CH_3}{\diagup}}\overset{O}{\underset{O-C-CH_3}{\diagdown}} \rightleftarrows \underset{CH_3}{\overset{CH_2-O-\overset{O}{\overset{\|}{C}}-CH_3}{\diagup}}C=C\underset{H}{\overset{H}{\diagdown}}$ (trans-)	335 (l)	0.78	−0.23	SE, CT, GLC $CH_3\overset{O}{\overset{\|}{C}}$ OH, $HClO_3$	120
2.1	(trans-) ⇌ (cis-) lactones with H_3C and CH_3 substituents	298 (l)	(−0.59)	—	SE, CT, GLC $t-C_4H_9OH$	121

Table (Continued)

Formula	Type	Reaction	T, K (phase)	$\Delta_r H_T^\circ$ ($\Delta_r G_T^\circ$)	$\Delta_r S_T^\circ$	Technique	Reference
$C_6H_{10}O_3$	1.3	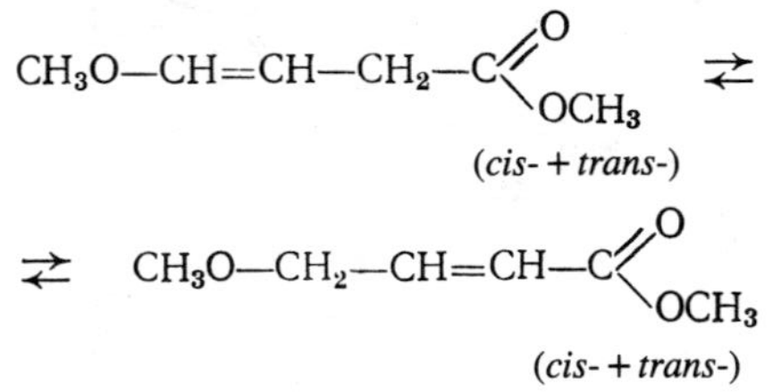 $CH_3O-CH=CH-CH_2-C(=O)OCH_3$ ⇌ $CH_3O-CH_2-CH=CH-C(=O)OCH_3$ (cis- + trans-) / (cis- + trans-)	365 (l)	(10.46 ±1.25)	—	SE, CT, GLC C_8H_{18}	56
	2.1	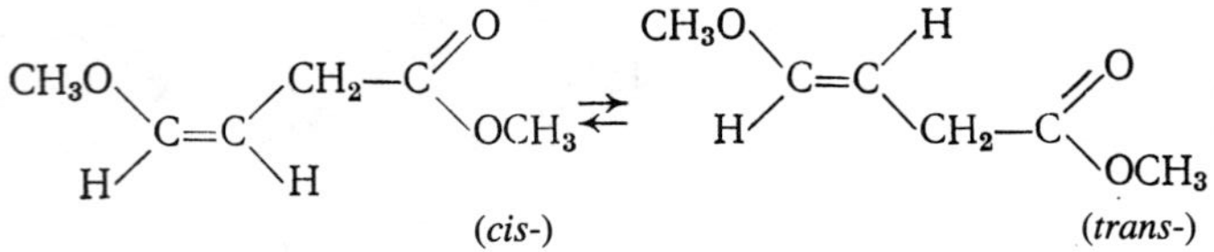 (cis-) ⇌ (trans-) methyl ester enol forms	298 (l)	(−0.33 ±0.08)	—	SE, CT, GLC	56
	2.1	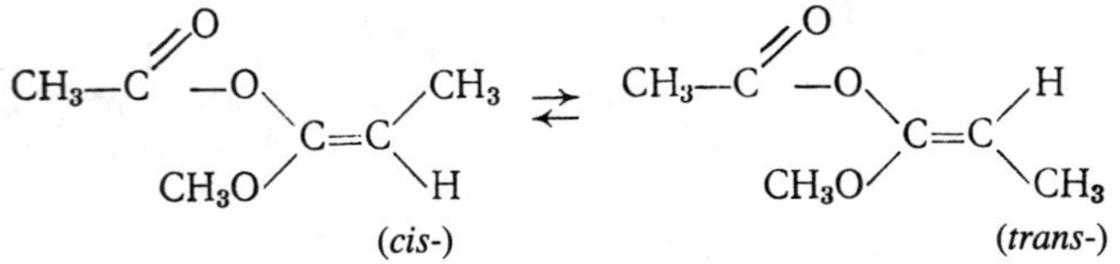 (cis-) ⇌ (trans-) enol ether	298 (l)	2.8 ±1.0	14.0 ±3.0	SE, CT, GLC, NMR CCl_4	118
$C_6H_{11}N$	1.1 1.2	 2-methylenepyridine ⇌ 2-methylpyridine	298 (l)	(−11.3)	—	SE, UV	67

Formula		Reaction	T (K), state	ΔH	ΔH (other)	Method	Ref.
$C_6H_{11}NO$	1.1 1.2	[2-methoxy-piperidine ⇌ 1-methyl-2-piperidone]	403 (l)	−72.78 ±2.09	—	RC	106
$C_6H_{11}NO_2$	1.3	$t\text{-}C_4H_9\text{-}C(O_2N)=CH_2 \rightleftharpoons t\text{-}C_4H_9\text{-}C(H)=C(H)NO_2$ (trans-)	344 (l)	−17.57 ±0.96	−23.26 ±1.09	SE, CT, GLC	122
$C_6H_{11}NS$	1.1 1.2	[2-methylthio-piperidine ⇌ 1-methyl-2-piperidinethione]	298 (g)	−8.8 ±13.4	—	RC	107
C_6H_{12}	1.3	$CH_2=CH\text{-}CH_2\text{-}C_3H_7 \rightleftharpoons CH_2=C(CH_3)\text{-}CH_2\text{-}C_2H_5$	298 (g)	−17.69 ±0.92	—	CC	123
		$CH_2=CH\text{-}CH_2\text{-}C_3H_7 \rightleftharpoons CH_2=C(C_2H_5)\text{-}CH_2\text{-}CH_3$	298 (g)	−14.31 ±1.17	—	CC	123
		$CH_2=CH\text{-}CH_2\text{-}C_3H_7 \rightleftharpoons CH_3\text{-}C(CH_3)=CH\text{-}C_2H_5$	298 (g)	−25.18 ±1.13	—	CC	123
		$CH_2=CH\text{-}CH_2\text{-}C_2H_5 \rightleftharpoons CH_2=CH\text{-}CH(CH_3)\text{-}C_2H_5$	298 (g)	9.91 ±1.05	—	CC	123

Table (Continued)

Formula	Type	Reaction	T, K (phase)	$\Delta_r H_T^\circ$ ($\Delta_r G_T^\circ$)	$\Delta_r S_T^\circ$	Technique	Reference
C_6H_{12}	1.3	$CH_2=CH-\underset{\underset{CH_3}{\|}}{CH}-C_2H_5 \rightleftarrows CH_2=CH-CH_2-\underset{\underset{CH_3}{\|}}{CH}-CH_3$	298 (g)	-1.76 ± 1.38	—	CC	123
		$CH_2=C-CH_2-CH_2-CH_3 \rightleftarrows CH_2=CH-CH_2-\underset{\underset{CH_3}{\|}}{CH}-CH_3$ with $\underset{CH_3}{\|}$	395 (g, l)	6.90 ± 0.42	-3.93 ± 1.25	SE, FE, CT, GLC	124
		$CH_2=C-\underset{\underset{CH_3}{\|}}{CH}-CH_3 \rightleftarrows CH_2=CH-\underset{\underset{CH_3}{\|}}{C}-CH_3$ with $\underset{CH_3}{\|}$ and $\underset{CH_3}{\|}$	298 (g)	4.81 ± 1.34	—	CC	123
		$CH_2=C-CH_2-CH_2-CH_3 \rightleftarrows CH_2=C-\underset{\underset{CH_3}{\|}}{CH}-CH_3$ with C_2H_5 and $CH_3\ CH_3$	298 (g)	-10.33 ± 1.30	—	CC	123
		$CH_2=C-CH_2-CH_2-CH_3 \rightleftarrows CH_3-C=CH-CH_2-CH_3$ with $\underset{CH_3}{\|}$ and $\underset{CH_3}{\|}$	548 (g)	-6.78	—	FE, CT, GLC	125
		$CH_2=C-CH_2-CH_2-CH_3 \rightleftarrows CH_3-C=CH-CH_2-CH_3$ with $\underset{CH_3}{\|}$ and $\underset{CH_3}{\|}$	503 (g)	-6.07 ± 0.88	-3.76 ± 1.67	FE, CT, GLC	126
		$CH_2=C-CH_2-CH_2-CH_3 \rightleftarrows CH_3-C=CH-CH_2-CH_3$ with $\underset{CH_3}{\|}$ and $\underset{CH_3}{\|}$	395 (g, l)	4.73 ± 0.62	-1.21 ± 1.46	SE, FE, CT, GLC	124
		$CH_2=C-\underset{\underset{CH_3}{\|}}{CH}-CH_3 \rightleftarrows CH_3-C=C-CH_3$ with $CH_3\ CH_3$ and $CH_3\ CH_3$	298 (g)	-3.47 ± 1.30	—	CC	123

Reaction	T (K)	ΔH	ΔS	Method	Ref.
$CH_2=CH-CH_2-C_3H_7 \rightleftharpoons \begin{matrix} C_2H_5 \\ \diagdown \\ C=C \\ \diagup \\ H \end{matrix} \begin{matrix} C_2H_5 \\ \diagup \\ \diagdown \\ H \end{matrix}$ (cis-)	518 (g)	−7.53	−6.27	FE, CT, GLC	127
$CH_2=CH-CH_2-C_3H_7 \rightleftharpoons \begin{matrix} C_2H_5 \\ \diagdown \\ C=C \\ \diagup \\ H \end{matrix} \begin{matrix} H \\ \diagup \\ \diagdown \\ C_2H_5 \end{matrix}$ (trans-)	486 (g)	−8.58 ±0.17	−11.92 ±0.38	SE, CT, GLC	128
$CH_2=CH-CH_2-C_3H_7 \rightleftharpoons \begin{matrix} CH_3 \\ \diagdown \\ C=C \\ \diagup \\ H \end{matrix} \begin{matrix} (i{-}C_3H_7) \\ \diagup \\ \diagdown \\ H \end{matrix}$ (cis-)	562 (g)	−7.61 ±0.50	−7.19 ±1.00	FE, CT, GLC	126
	298 (g)	−5.94 ±0.96	—	CC	123
$CH_2=CH-CH_2-C_3H_7 \rightleftharpoons \begin{matrix} CH_3 \\ \diagdown \\ C=C \\ \diagup \\ H \end{matrix} \begin{matrix} C_3H_7 \\ \diagup \\ \diagdown \\ H \end{matrix}$ (cis-)	561 (g)	−5.73 ±0.77	−11.71 ±1.67	FE, CT, GLC	129
	561 (g)	−9.35 ±1.13	−8.78 ±2.30	FE, CT, GLC	129
	298 (g)	−15.77 ±0.75	—	CC	123
	298 (g)	−10.67 ±1.00	—	CC	123
$CH_2=CH-CH_2-C_3H_7 \rightleftharpoons \begin{matrix} CH_3 \\ \diagdown \\ C=C \\ \diagup \\ H \end{matrix} \begin{matrix} C_3H_7 \\ \diagup \\ \diagdown \\ H \end{matrix}$ (cis-)	573 (g)	−8.78	—	FE, CT, GLC	125

Table (Continued)

Formula	Type	Reaction	T, K (phase)	$\Delta_r H^\circ_T$ ($\Delta_r G^\circ_{T_m}$)	$\Delta_r S^\circ_T$	Technique	Reference
C_6H_{12}	1.3	$CH_2=CH-CH_2-C_3H_7 \rightleftarrows \begin{array}{c} CH_3 \\\ C=C \\\ H \quad\quad\quad C_3H_7 \end{array}$ (trans-)	561 (g)	-6.86 ± 0.75	-3.35 ± 1.46	FE, CT, GLC	129
			298 (g)	-10.11 ± 1.14	—	RC, CT C_6H_{14}	130
		$CH_3-CH-CH_2-CH=CH_2 \rightleftarrows \begin{array}{c} CH_3 \quad (i-C_3H_7) \\\ C=C \\\ H \quad\quad\quad H \end{array}$ (cis-) CH_3	573 (g)	-11.71	—	FE, CT, GLC	125
		$CH_3-CH-CH_2-CH=CH_2 \rightleftarrows \begin{array}{c} H \quad\quad (i-C_3H_7) \\\ C=C \\\ CH_3 \quad\quad H \end{array}$ (trans-) CH_3	561 (g)	-10.29 ± 0.67	-3.76 ± 1.25	FE, CT, GLC	129
			548 (g)	-3.35	—	FE, CT, GLC	125
		$CH_2=C-CH_2-CH_2-CH_3 \rightleftarrows \begin{array}{c} (i-C_3H_7) \\\ C=C \\\ CH_3 \quad H \end{array}$ (cis-) CH_3	548 (g)	-5.73	—	FE, CT, GLC	125
			395 (g, l)	1.97 ± 0.25	-9.96 ± 0.67	SE, FE, CT, GLC	124

68

Reaction	T (K), phase	ΔH	ΔS	Method	Ref.
CH₂=C(CH₃)—CH₂—CH₂—CH₃ ⇌ CH₃—C(H)=C(H)—(i-C₃H₇) (trans-)	395 (g, l)	−1.57 ±0.17	−6.57 ±0.42	SE, FE, CT, GLC	124
CH₂=CH—CH(CH₃)—CH₂—CH₃ ⇌ CH₃—C(C₂H₅)=C(H)—CH₃ (trans-)	598 (g)	−0.71	—	FE, CT, GLC	125
CH₂=CH—CH(CH₃)—CH₂—CH₃ ⇌ CH₃—C(CH₃)=C(H)—C₂H₅ (cis-)	598 (g)	−14.85	—	FE, CT, GLC	125
CH₂=C(C₂H₅)—CH₂—CH₃ ⇌ CH₃—C(CH₃)=C(H)—C₂H₅ (cis-)	476 (g)	5.52 ±2.09	−2.18 ±4.39	FE, CT, GLC	126
CH₃—C(CH₃)=CH—CH₂—CH₃ ⇌ CH₃—C(H)=C(H)—(i-C₃H₇) (cis-)	548 (g)	−5.86	—	FE, CT, GLC	125
CH₃—C(CH₃)=CH—CH₂—CH₃ ⇌ CH₃—C(H)=C(H)—(i-C₃H₇) (trans-)	548 (g)	−2.51	—	FE, CT, GLC	125
CH₃—C(CH₃)=CH—CH₂—CH₃ ⇌ CH₃—C(CH₃)=C(H)—C₂H₅ (cis-)	298 (g)	4.68 ±1.17	—	CC	123

Table (Continued)

Formula	Type	Reaction	T, K (phase)	$\Delta_r H^\circ_T$ ($\Delta_r G^\circ_T$)	$\Delta_r S^\circ_T$	Technique	Reference
C_6H_{12}	1.3	$C_2H_5\text{-}C(C_2H_5)=CH_2 \rightleftarrows$ (cis-) CH$_3$(H)C=C(CH$_3$)(C$_2$H$_5$)	598 (g)	−7.95	—	FE, CT, GLC	125
	2.1	CH$_3$(H)C=C(C$_3$H$_7$)(H) (cis-) $\rightleftarrows$ C$_2$H$_5$(H)C=C(C$_2$H$_5$)(H) (cis-)	298 (g)	6.15 ±1.42	—	RC, CT C$_6$H$_{14}$	130
		C$_2$H$_5$(H)C=C(C$_2$H$_5$)(H) (cis-) $\rightleftarrows$ C$_2$H$_5$(H)C=C(H)(C$_2$H$_5$) (trans-)	298 (g)	−6.81 ±1.30	—	CC	123
		CH$_3$(H)C=C(C$_2$H$_5$)(C$_2$H$_5$) $\rightleftarrows$ CH$_3$(H)C=C(CH$_3$)(C$_2$H$_5$) (cis-)	298 (g)	−9.54 ±1.95	—	RC, CT C$_6$H$_{14}$	130
		CH$_3$(H)C=C(H)(C$_2$H$_5$) (trans-) $\rightleftarrows$ CH$_3$(H)C=C(CH$_3$)(C$_2$H$_5$)	476 (g)	1.37 ±0.23	−1.46 ±0.50	FE, CT, GLC	126
		CH$_3$(H)C=C(CH$_3$)(C$_2$H$_5$) (trans-) $\rightleftarrows$ CH$_3$(H)C=C(C$_2$H$_5$)(CH$_3$) (cis-)	298 (g)	0.92 ±1.05	—	CC	123

C$_6$H$_{12}$O						
	1.2	CH$_3$-C(CH$_3$)=CH-C$_3$H$_7$ (cis-) ⇌ CH$_3$-C(CH$_3$)=CH-C$_3$H$_7$ (trans-)	298 (g)	−1.55 ±1.17	—	CC	123
		CH$_3$-C(CH$_3$)=CH-(i-C$_3$H$_7$) (cis-) ⇌ CH$_3$-C(CH$_3$)=CH-(i-C$_3$H$_7$) (trans-)	298 (g)	−1.92 ±1.37	—	RC, CT	130
		cyclohexane ⇌ methylcyclopentane	298 (g)	−4.02 ±0.88	—	CC	123
	1.3	CH$_2$=C(OCH$_3$)−CH$_2$−C$_2$H$_5$ ⇌ CH$_3$O−C(C$_2$H$_5$)=CH−H (cis-)	490 (g)	−3.64 ±0.75	3.26 ±1.72	FE, CT, GLC	126
	2.1	C$_2$H$_5$−C(OC$_2$H$_5$)=CH−H (cis-) ⇌ C$_2$H$_5$−C(H)=CH−OC$_2$H$_5$ (trans-)	338 (l)	19.19	33.01	SE, CT, GLC	131
			319 (l)	18.07 ±1.17	43.34 ±3.76	SE, CT, GLC	132
		i−C$_3$H$_7$−C(OCH$_3$)=CH−H (cis-) ⇌ i−C$_3$H$_7$−C(H)=CH−OCH$_3$ (trans-)	298 (g)	0.78 ±0.39	−2.5 ±0.8	SE, GLC	91
			311 (l)	−2.34 ±0.33	−6.69 ±0.84	SE, CT, GLC	40
			298 (g)	−3.3 ±0.3	−4.4 ±0.6	SE, CT, GLC	74

Table (Continued)

Formula	Type	Reaction	T, K (phase)	$\Delta_r H_T^\circ$ ($\Delta_r G_T^\circ$)	$\Delta_r S_T^\circ$	Technique	Reference
$C_6H_{12}O$	2.1	$\mathrm{CH_3{-}C(O{-}i{-}C_3H_7){=}C(H)(H)}$ (cis-) ⇌ $\mathrm{CH_3{-}C(H){=}C(H)(O{-}i{-}C_3H_7)}$ (trans-)	311 (l)	2.38 ±0.20	−0.38 ±0.50	SE, CT, GLC	40
		$\mathrm{CH_3{-}C(C_2H_5){=}C(OCH_3)(H)}$ (cis-) ⇌ $\mathrm{CH_3{-}C(H){=}C(OCH_3)(C_2H_5)}$ (trans-)	298 (g)	10.88 ±0.41	−15.4 ±1.0	SE, CT, GLC	91
$C_6H_{12}OS$	1.3	$\mathrm{CH_2{=}CH{-}CH_2{-}S(O)(i{-}C_3H_7)}$ ⇌ $\mathrm{CH_3{-}C(H){=}C(S(O)(i{-}C_3H_7))(H)}$ (cis-)	293 (l)	(0.8)	—	SE, CT, GLC $t{-}C_4H_9OH$	57
		$\mathrm{CH_2{=}CH{-}CH_2{-}S(O)(i{-}C_3H_7)}$ ⇌ $\mathrm{CH_3{-}C(H){=}C(H)(S(O)(i{-}C_3H_7))}$ (trans-)	293 (l)	(−6.2)	—	SE, CT, GLC $t{-}C_4H_9OH$	57
	2.1	$\mathrm{CH_3{-}C(S(O)(i{-}C_3H_7)){=}C(H)(H)}$ (cis-) ⇌ $\mathrm{CH_3{-}C(H){=}C(H)(S(O)(i{-}C_3H_7))}$ (trans-)	293 (l)	(−7.0)	—	SE, CT, GLC $t{-}C_4H_9OH$	57

72

C$_6$H$_{12}$O$_2$	2.1	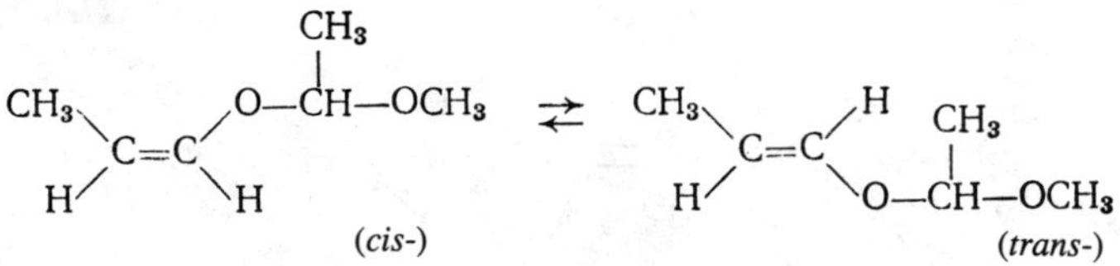 CH$_3$\O—CH—OCH$_3$/C=C\H/H (cis-) ⇌ CH$_3$\C=C/H—CH\H/CH$_3$—O—CH—OCH$_3$ (trans-)	298 (l)	2.9 ±0.2	0.3 ±0.5	SE, CT, GLC	92
		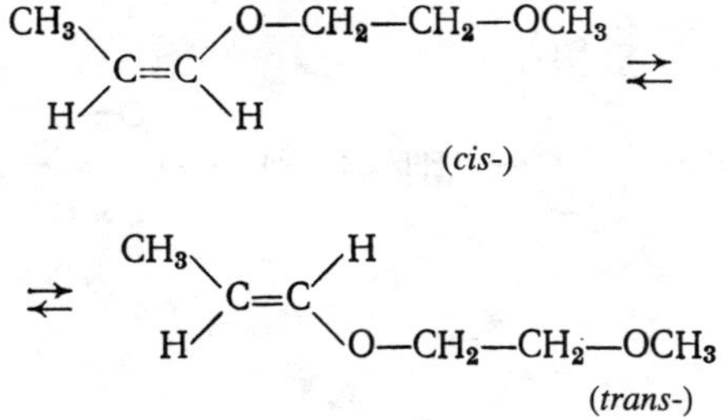 CH$_3$\C=C/O—CH$_2$—CH$_2$—OCH$_3$/H—H (cis-) ⇌ CH$_3$\C=C/H—H/O—CH$_2$—CH$_2$—OCH$_3$ (trans-)	298 (l)	0.2 ±0.2	−2.9 ±0.6	SE, CT, GLC	92
	1.3	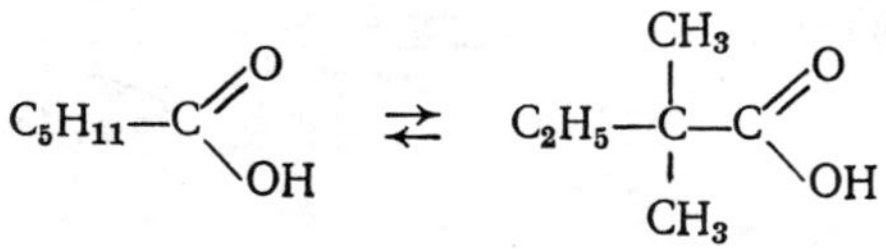 CH$_3$—C(C$_2$H$_5$)(C$_5$H$_{11}$)—COOH ⇌ CH$_3$—C(C$_2$H$_5$)(C$_5$H$_{11}$)—COOH	298 (l)	−14.81	—	cc	58
	2.1	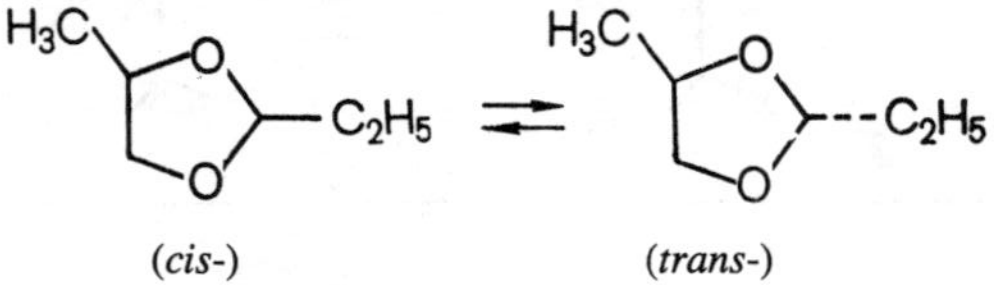 H$_3$C—(trans dioxolane with C$_2$H$_5$) ⇌ H$_3$C—(cis dioxolane with C$_2$H$_5$)	298 (l)	(1.09 ±0.02)	—	SE, CT, GLC (C$_2$H$_5$)$_2$O	93
			341 (l)	1.37	0.84	SE, CT, GLC CCl$_4$	84

Table (Continued)

Formula	Type	Reaction	T, K (phase)	$\Delta_r H°_T$ ($\Delta_r G°_T$)	$\Delta_r S°_T$	Technique	Reference
$C_6H_{12}O_2$	2.1	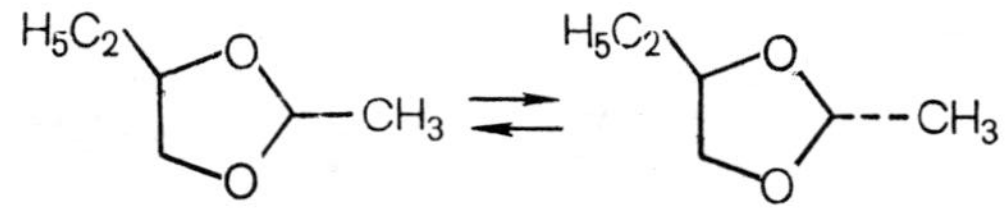 (cis-) ⇌ (trans-) [1,3-dioxolane with CH₃ and C₂H₅ substituents]	298 (l)	(1.17 ±0.04)	—	SE, CT, GLC $(C_2H_5)_2O$	93
		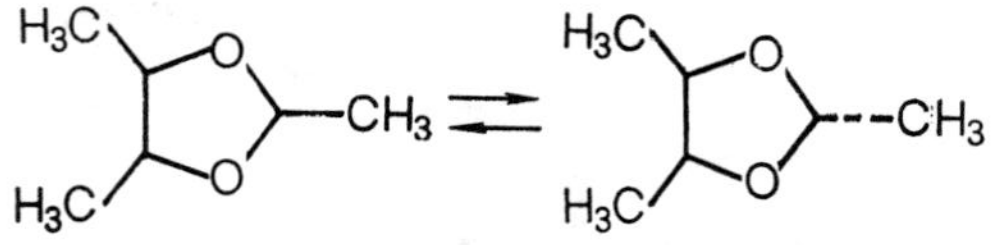 (syn-) ⇌ (anti-) [1,3-dioxolane with CH₃ groups]	298 (l)	(2.94 ±0.13)	—	SE, CT, GLC $(C_2H_5)_2O$	93
		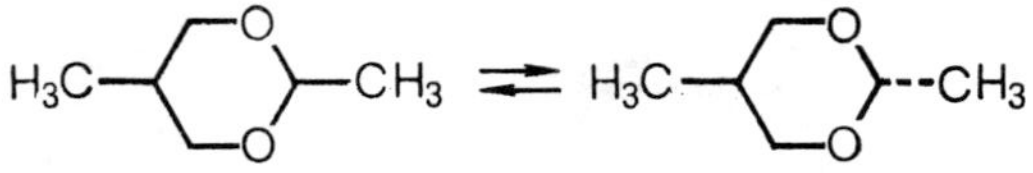 (cis-) ⇌ (trans-) [1,3-dioxane with CH₃ groups]	333 (l)	(−4.60 ±0.04)	—	SE, CT, GLC	133
$C_6H_{12}O_2S$	1.3	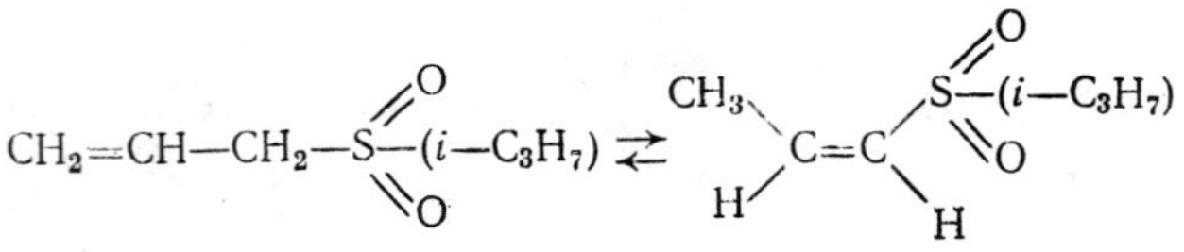 $CH_2=CH-CH_2-S-(i-C_3H_7)$ ⇌ [cis alkene with S-(i-C₃H₇)]	293 (l)	(8.2)	—	SE, CT, GLC $t-C_4H_9OH$	57

Formula		Structure	T, K (phase)	Value	±	Method / Solvent	Ref.
		CH$_2$=CH—CH$_2$—S—(i-C$_3$H$_7$) ⇌ CH$_3$\C=C/H, S(i-C$_3$H$_7$)(O)(O) (*trans*-)	293 (l)	(−5.8)	—	SE, CT, GLC t—C$_4$H$_9$OH	57
C$_6$H$_{12}$O$_3$	2.1	CH$_3$\C=C/H (*cis*-) ⇌ CH$_3$\C=C/H (*trans*-) S(i-C$_3$H$_7$)(O)(O)	293 (l)	(−13.9)	—	SE, CT, GLC t—C$_4$H$_9$OH	57
	2.1	CH$_3$\C=C/H—O—CH(OCH$_3$)$_2$ (*cis*-) ⇌ CH$_3$\C=C/H—O—CH(OCH$_3$)$_2$ (*trans*-)	298 (l)	3.5 ±1.0	3.5 ±3.4	SE, CT, GLC	134
		CH$_3$ 1,3-dioxane-OCH$_3$ (*cis*-) ⇌ CH$_3$ 1,3-dioxane-OCH$_3$ (*trans*-)	298 (l)	(−1.51)	—	SE, CT, GLC, NMR CCl$_4$	135
C$_6$H$_{12}$S	1.3	CH$_2$=CH—CH$_2$—S—(i-C$_3$H$_7$) ⇌ CH$_3$\C=C/H, S—(i-C$_3$H$_7$) (*cis*-)	293 (l)	(−10.6)	—	SE, CT, GLC t—C$_4$H$_9$OH	57
		CH$_2$=CH—CH$_2$—S—(i-C$_3$H$_7$) ⇌ CH$_3$\C=C/H, S—(i-C$_3$H$_7$) (*cis*-)	293 (l)	(−11.6)	—	SE, CT, GLC t—C$_4$H$_9$OH	57

Table (Continued)

Formula	Type	Reaction	T, K (phase)	$\Delta_r H^\circ_T$ ($\Delta_r G^\circ_T$)	$\Delta_r S^\circ_T$	Technique	Reference
$C_6H_{12}S$	2.1	cis ⇌ trans (CH₃−CH=CH−S−(i−C₃H₇))	293 (l)	(−1.0)	—	SE, CT, GLC t−C_4H_9OH	57
$C_6H_{12}S_2$	2.1	cis ⇌ trans (2-methyl-6-methyl-1,3-dithiane)	371 (l)	7.01 ±0.10	5.33 ±0.26	SE, CT, GLC CCl_4	136
$C_6H_{13}Cl$	1.3	CH_3−CHCl−CH_2−C_3H_7 ⇌ CH_3−CH_2−CHCl−C_3H_7	298 (l)	(5.27 ±0.08)	—	SE, CT, GLC $CHCl_3$	137
		CH_3−CCl(CH₃)−CH_2−C_2H_5 ⇌ CH_3−CH_2−CCl(CH₃)−C_2H_5	330 (l)	−0.61 ±0.42	0.38 ±1.25	SE, CT, GLC	96
		CH_3−CCl(CH₃)−CH_2−C_2H_5 ⇌ CH_3−CH_2−CCl(CH₃)−C_2H_5	348 (l)	2.80 ±0.29	4.94 ±0.75	SE, CT, GLC (benzene)	138
		CH_3−CCl(CH₃)−CH_2−C_2H_5 ⇌ CH_3−CH(CH₃)−CHCl−C_2H_5	348 (l)	7.74 ±1.25	8.95 ±4.18	SE, CT, GLC (benzene)	138

Reactant	Product	T (K)	ΔH	ΔS	Method	Ref.
CH₃—CH(Cl)—CH(C₂H₅)—CH₃	CH₃—CH(CH₃)—CH(Cl)—C₂H₅ (l.b.+h.b.)	348 (l)	3.64 ±1.25	10.21 ±2.10	SE, CT, GLC	138
CH₃—CH(Cl)—CH(C₂H₅)—CH₃	CH₃—CH(CH₃)—CH(Cl)—C₂H₅ (l.b.)	348 (l)	2.38 ±1.25	1.38 ±2.10	SE, CT, GLC	138
CH₃—CH(Cl)—CH(C₂H₅)—CH₃	CH₃—CH(CH₃)—CH(Cl)—C₂H₅ (h.b.)	348 (l)	4.89 ±1.25	7.49 ±2.10	SE, CT, GLC	138
CH₃—CH₂—C(Cl)(CH₃)—C₂H₅	CH₃—CH(CH₃)—CH(Cl)—C₂H₅ (l.b.+h.b.)	348 (l)	8.58 ±1.25	14.22 ±3.76	SE, CT, GLC	138
CH₃—CH₂—C(Cl)(CH₃)—C₂H₅	CH₃—CH(CH₃)—CH(Cl)—C₂H₅ (l.b.)	348 (l)	7.32 ±1.25	−0.38 ±2.10	SE, CT, GLC	138
CH₃—CH₂—C(Cl)(CH₃)—C₂H₅	CH₃—CH(CH₃)—CH(Cl)—C₂H₅ (h.b.)	348 (l)	9.83 ±1.25	5.73 ±2.10	SE, CT, GLC	138

Table (Continued)

Formula	Type	Reaction	T, K (phase)	$\Delta_r H_T^\circ$ ($\Delta_r G_T^\circ$)	$\Delta_r S_T^\circ$	Technique	Reference
$C_6H_{13}Cl$	2.1	$CH_3-CH-CH-C_2H_5$ (l.b.) $\rightleftarrows$ $CH_3-CH-CH-C_2H_5$ (h.b.) with Cl and CH_3 substituents	348 (l)	2.38 ±1.17	6.11 ±3.64	SE, CT, GLC	138
C_6H_{14}	1.3	$CH_3-CH-CH_2-C_2H_5$ $\rightleftarrows$ $CH_3-CH_2-CH-C_2H_5$ with CH_3 substituents	368 (g, l)	0.92 ±0.46	4.05 ±1.25	SE, FE, CT, GLC	139
$C_6H_{14}N_2S$	1.1 1.2	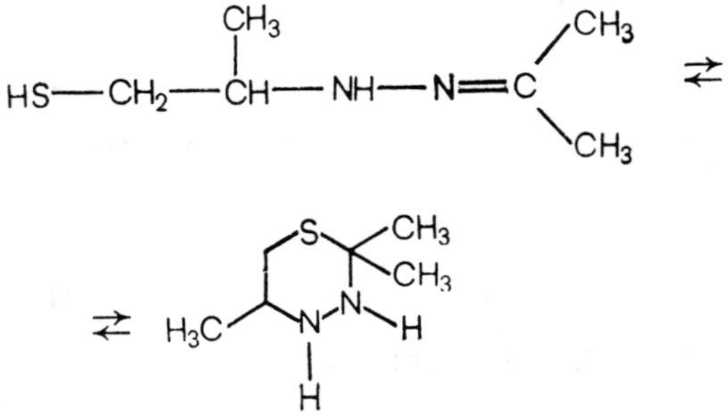 $HS-CH_2-CH-NH-N=C$ structure $\rightleftarrows$ cyclic thiazine structure	333 (l)	21.2 ±0.8	72.0 ±3.0	SE, NMR 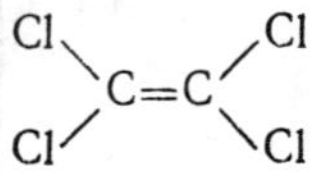 ($CCl_2=CCl_2$)	140
$C_6H_{14}O$	2.1	$CH_3-CH-CH-C_2H_5$ (threo-) $\rightleftarrows$ $CH_3-CH-CH-C_2H_5$ (erythro-) with OH and CH_3 substituents	516 (g)	0.3 ±0.2	0.9 ±0.5	FE, CT, GLC	142

Formula	Ratio	Structures	T (K)	ΔH	ΔG	Method, Solvent	Ref.
$C_6H_{11}S$	1.3	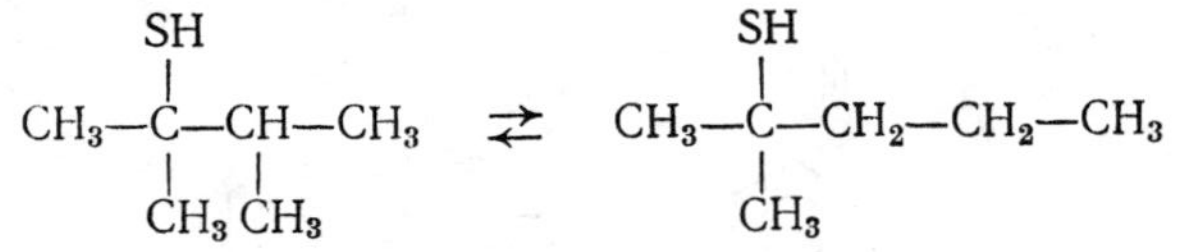 $CH_3-C(SH)(CH_3)-CH-CH_3 \rightleftarrows CH_3-C(SH)(CH_3)-CH_2-CH_2-CH_3$	298 (g)	-0.63 ± 1.23	—	CC	94
$C_7H_4N_4S$	1.1 / 1.2	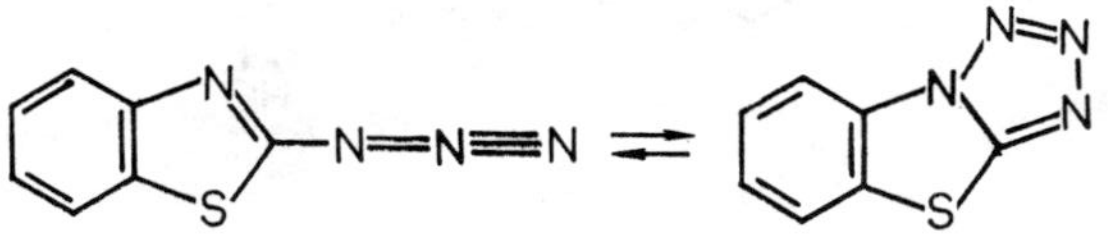 benzothiazole azide ⇌ tetrazole tautomers	331 (l)	4.12	—	SE, IR (dioxane)	141
$C_7H_5ClN_2O_4$	1.1 / 1.2	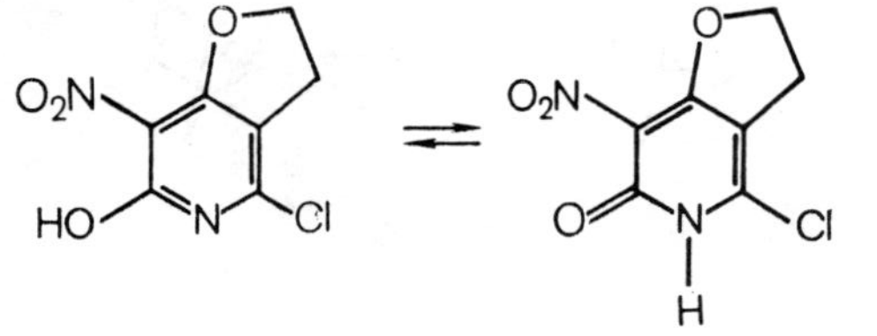 4-chloro-7-nitro-2,3-dihydrofuro[3,2-c]pyridin-6-ol ⇌ 6(7H)-one	323 (l)	-4.12	-5.94	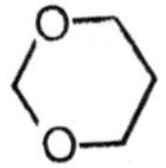 SE, UV; $C_2H_5OH-H_2O$	65
$C_7H_5Cl_3N_2O$	1.1 / 1.2	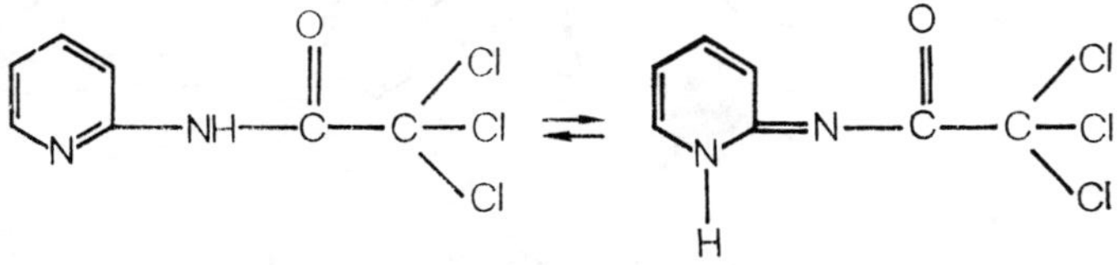 $O=C(CCl_3)-NH$-pyridin-2-yl ⇌ $O=C(CCl_3)-N=$ pyridin-2(1H)-ylidene	321 (l)	4.45	-3.14	SE, UV; H_2O	69
$C_7H_6ClNO_2$	1.1 / 1.2	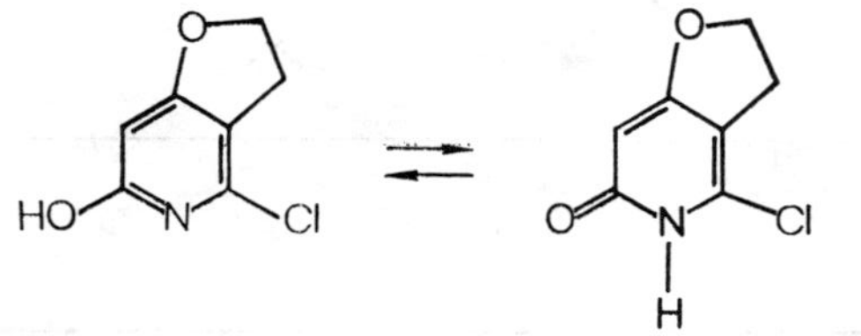 4-chloro-2,3-dihydrofuro[3,2-c]pyridin-6-ol ⇌ 6(7H)-one	323 (l)	-27.17	-57.35	SE, UV; $C_2H_5OH-H_2O$	65

Table (Continued)

Formula	Type	Reaction	T, K (phase)	$\Delta_r H^\circ_T$ ($\Delta_r G^\circ_T$)	$\Delta_r S^\circ_T$	Technique	Reference
$C_7H_6Cl_2N_2O$	1.1, 1.2	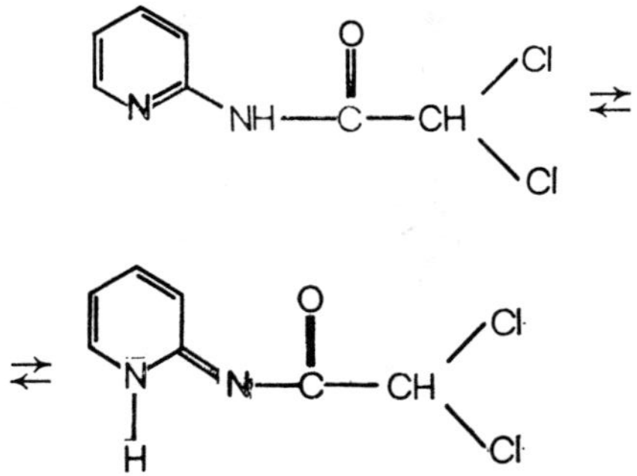	321 (l)	8.98	−5.44	SE, UV H_2O	69
C_7H_8	1.2		298 (g)	113.48 ±1.00	—	SE, CT, GLC HCl	143
			298 (g)	91.5 ±4.9	—	CC	144
$C_7H_8O_2$	1.3	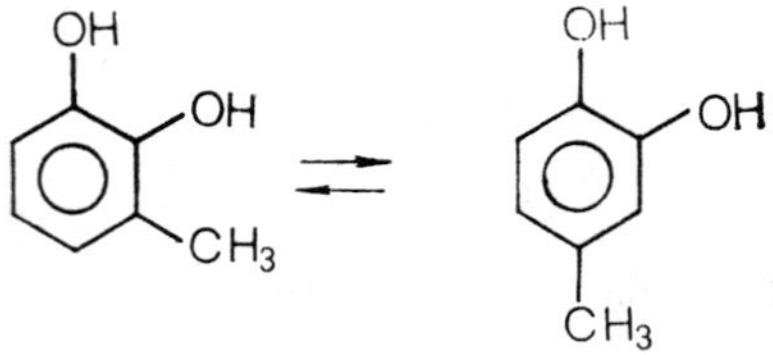	298 (g)	0.9 ±2.1	—	CC	256

Formula		Structures	T (K)	value 1	value 2	Method	Ref.
$C_7H_8O_2S$	1.1, 1.2	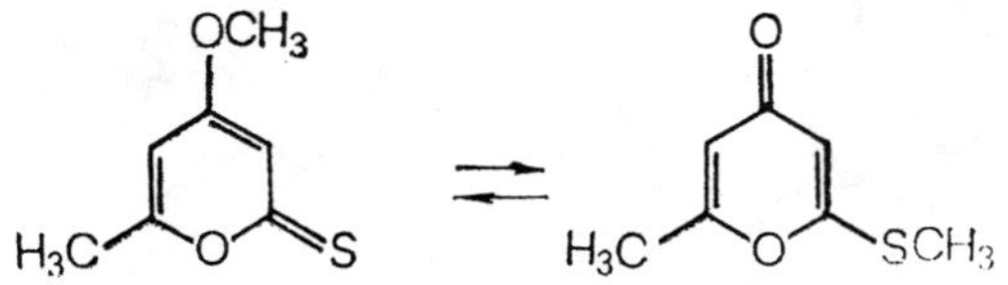 OCH₃/H₃C pyranthione ⇌ SCH₃/H₃C pyranone	298 (g)	-79.5 ± 15.1	—	RC	107
$C_7H_8O_3$	1.1, 1.2	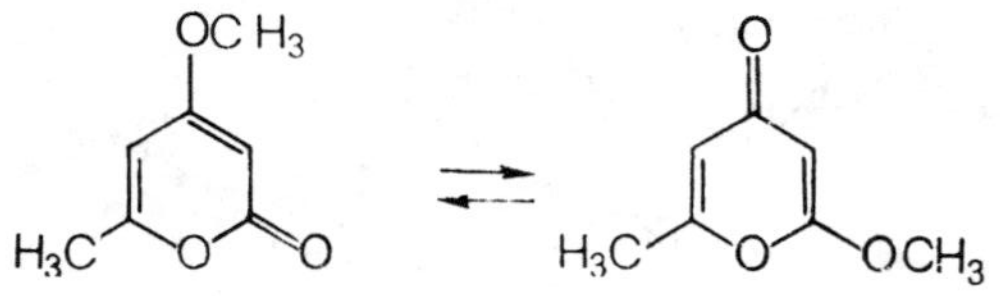 OCH₃/H₃C pyranone ⇌ OCH₃/H₃C pyranone	298 (g)	36.8 ± 8.8	—	RC	107
C_7H_9N	1.3	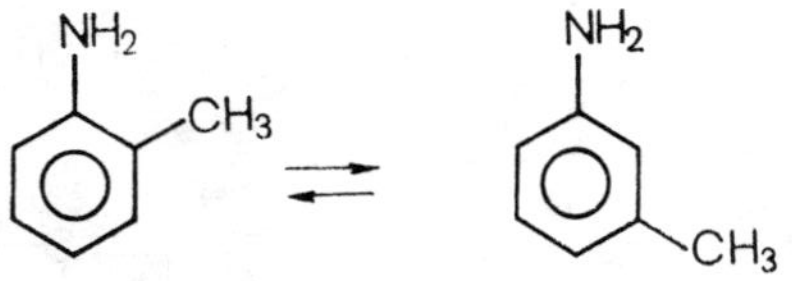 o-toluidine ⇌ m-toluidine	298 (g)	6.3	0.50	RC, ST	257
		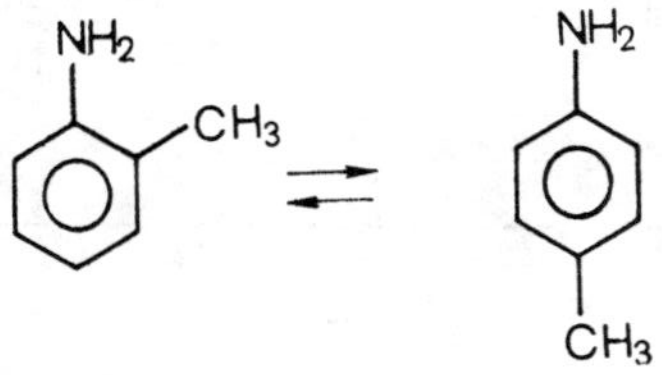 o-toluidine ⇌ p-toluidine	298 (g)	-13.0	-2.83	RC, ST	257

Table (Continued)

Formula	Type	Reaction	T, K (phase)	$\Delta_r H_T^\circ$ ($\Delta_r G_T^\circ$)	$\Delta_r S_T^\circ$	Technique	Reference
$C_7H_9N_5O$	1.1, 1.2	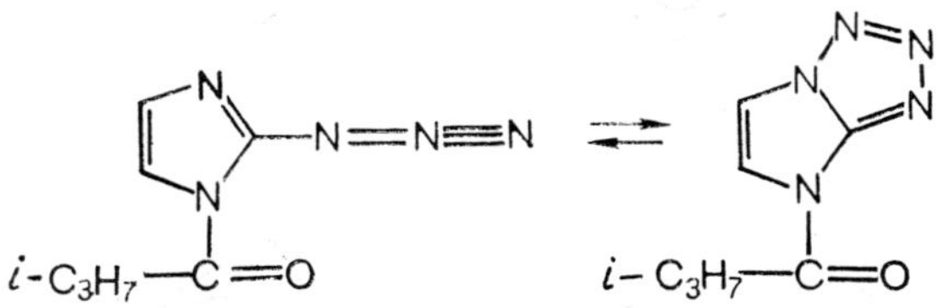 (azole-N=N-azole-i-C₃H₇ isomerization, with CH₃–C(=O)–CH₃)	293 (l)	−4.6 ±1.3	−26.8 ±4.2	SE, NMR	70
C_7H_{10}	2.1	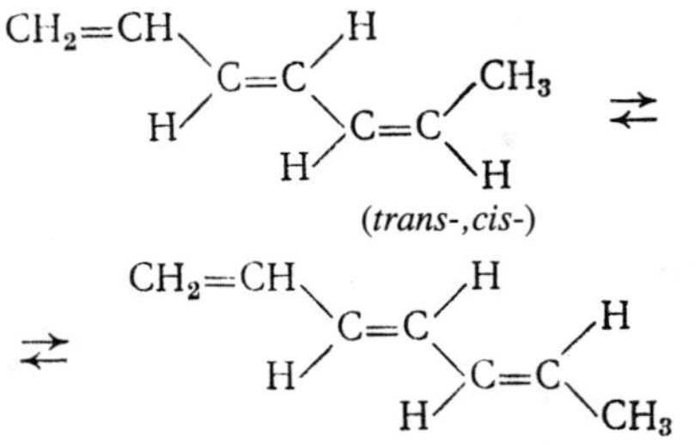 (trans-,cis-) ⇌ (trans-,trans-) pentadiene isomers	493 (g)	−3.51 ±0.17	0.50 ±0.38	SE, CT, GLC	145
	1.2	methylenecyclohexane ⇌ bicyclic isomer	298 (l)	−125.45 ±0.33	—	SE, CT, GLC	143

Formula		Reaction	T (state)	ΔH		Method	Ref.
C₇H₁₀O₂	1.3	(norbornene ⇌ norbornene isomer)	298 (g)	−9.1 ±4.1	—	CC	144
			298 (g)	−9.96 ±3.18	—	CC	147
		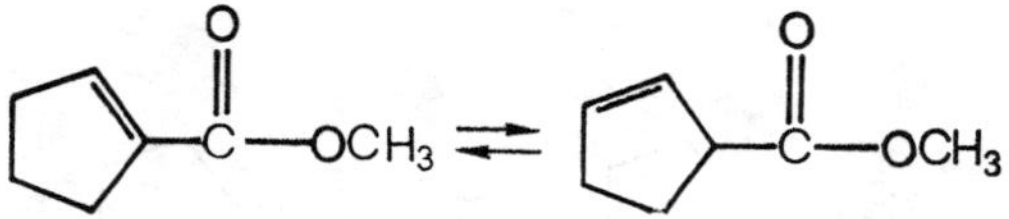 methyl cyclopent-1-enecarboxylate ⇌ methyl cyclopent-2-enecarboxylate	373 (l)	(8.78 ±0.84)	—	SE, CT, GLC CH₃OH	56
C₇H₁₁ClO	1.3	 1-chloro-6-methoxycyclohexene ⇌ 2-chloro-1-methoxycyclohexene	298 (l)	14.1 ±0.7	20.8 ±2.0	SE, CT, GLC ⬡	35
C₇H₁₂	1.3	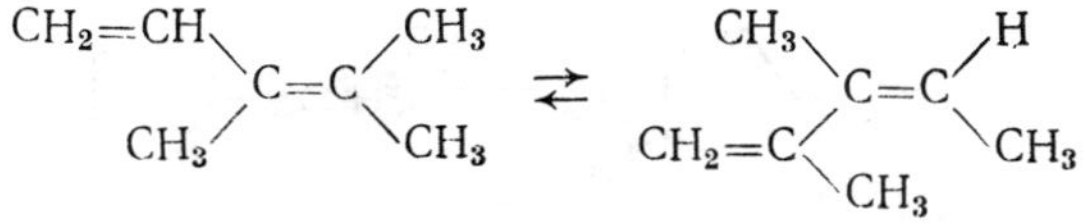 CH₂=CH−C(CH₃)₂−CH₃ ⇌ (trans-) CH₃−C(=CH−CH₃)−CH₂−CH=CH₂	549 (g)	2.44 ±0.20	3.0 ±1.5	SE, GLC	148
	2.1	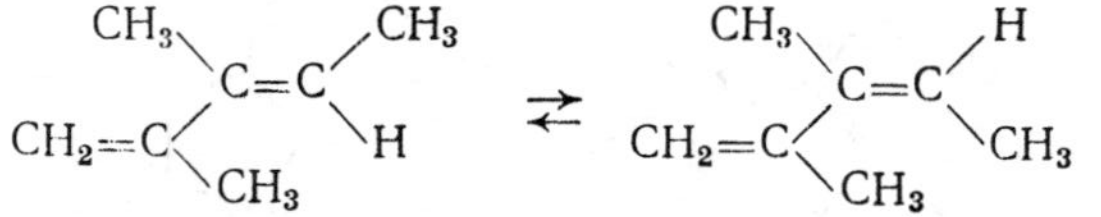 (cis- ⇌ trans-)	658 (g)	5.60 ±0.25	2.0 ±1.5	SE, GLC	148
	1.3	CH≡C−CH₂−CH₂−C₃H₇ ⇌ CH₃−C≡C−CH₂−CH₂−C₃H₇	298 (l)	−18.99 ±3.18	—	RC, CT C₆H₁₄	112
	1.3	CH₃−C≡C−CH₂−C₃H₇ ⇌ CH₃−CH₂−C≡C−C₃H₇	298 (l)	−2.01 ±3.10	—	RC, CT C₆H₁₄	112

83

Table (Continued)

Formula	Type	Reaction	T, K (phase)	$\Delta_r H_T^\circ$ ($\Delta_r G_T^\circ$)	$\Delta_r S_T^\circ$	Technique	Reference
C_7H_{12}		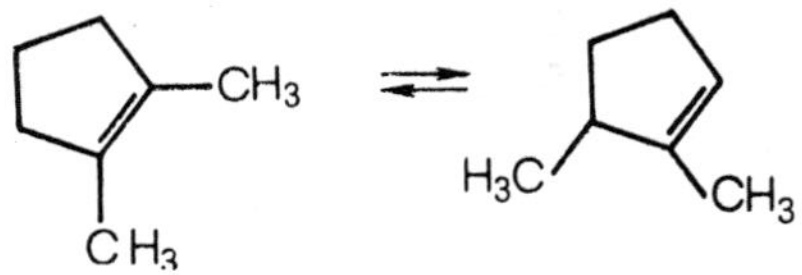 1,2-dimethylcyclopentene ⇌ 2,3-dimethylcyclopentene	583 (g)	9.20	6.48	FE, CT, GLC	149
	1.3	 1-ethylcyclopentene ⇌ 3-ethylcyclopentene	523 (g)	(10.83 ±0.17)	—	FE, CT, GLC	73
		 1-ethylcyclopentene ⇌ 4-ethylcyclopentene	523 (g)	(10.62 ±0.29)	—	FE, CT, GLC	73
		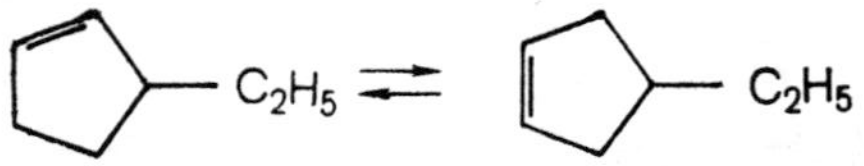 1-ethylcyclopentene ⇌ 3-ethylcyclopentene	411 (g, l)	4.18	—	SE, FE, CT, GLC	116
	1.2	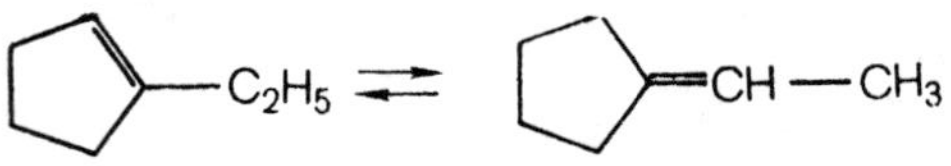 1-ethylcyclopentene ⇌ ethylidenecyclopentane	298 (l)	1.55 ±1.17	—	CC	114
	1.3	1-ethylcyclopentene ⇌ ethylidenecyclopentane	298 (g)	1.63 ±1.28	—	CC	115
		1-ethylcyclopentene ⇌ ethylidenecyclopentane	523 (g)	(8.16 ±0.13)	—	FE, CT, GLC	73

	Reaction	T (K) (phase)	ΔH		Method	Ref.
1.2 1.3	![structure] -CH=CH₂ ⇌ ![structure] =CH-CH₃					
1.2	![cyclopentane with C₂H₅] ⇌ ![cyclopentene with CH₂]	298 (l)	−23.34 ±1.31	—	CC	114
	![cyclopentene with C₂H₅] ⇌ ![cyclopentene with CH₃]	 298 (l)	−5.65 ±3.15	—	CC	115
1.3	![cyclohexene CH₃] ⇌ ![cyclohexene CH₃]	 298 (l)	−22.88 ±1.12	—	CC	114
		463 (g, l)	8.07 ±0.21	1.30 ±0.71	SE, FE, CT, GLC	150
		523 (g)	(8.83 ±0.17)	—	FE, CT, GLC	73
1.3	 ![cyclohexene CH₃] ⇌ ![cyclohexene CH₃]	463 (g, l)	5.81 ±0.29	−0.29 ±0.71	SE, FE, CT, GLC	150
		523 (g)	(8.24 ±0.17)	—	FE, CT, GLC	73
	![cyclohexene CH₃] ⇌ ![cyclohexene CH₃]	 411 (g, l)	−2.38	—	SE, FE, CT, GLC	116

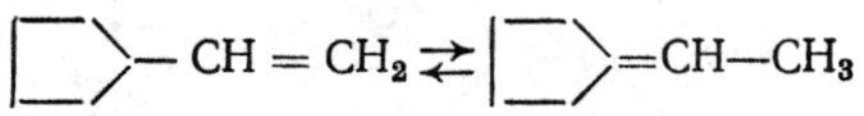

85

Table (Continued)

Formula	Type	Reaction	T, K (phase)	$\Delta_r H_T^\circ$ ($\Delta_r G_T^\circ$)	$\Delta_r S_T^\circ$	Technique	Reference
C_7H_{12}	1.2, 1.3	 1-methylcyclopentene ⇌ methylenecyclopentane	523 (g)	(15.44 ±0.17)	—	SE, FE, CT, GLC	73
			298 (g)	10.04	—	CC	32
$C_7H_{12}O$	1.2, 1.3	$\begin{array}{c}CH_3\\CH_3\end{array}$C=CH—CH$_2$—C(=O)CH$_3$ ⇌ $\begin{array}{c}CH_3\\CH_3\end{array}$CH—CH=CH—C(=O)CH$_3$ (cis- + trans-)	463 (g, l)	7.15 ±0.29	−15.85 ±0.71	SE, FE, CT, GLC	150
		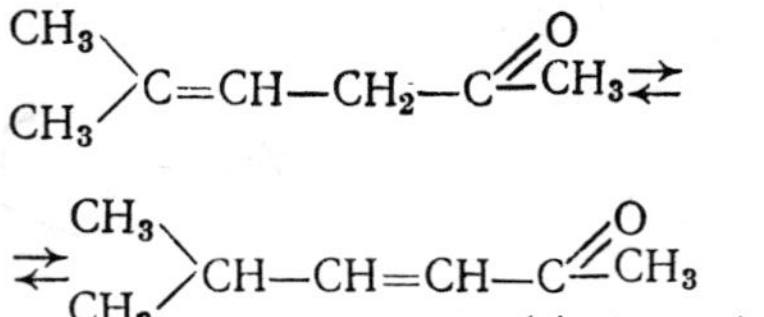 i-C_3H_7, H \ C=C / H, H ⇌ \ C=C / O, H	328 (g)	1.25 ±2.09	−2.9 ±5.9	SE, CT, NMR t-C_4H_9OH	151
	1.3	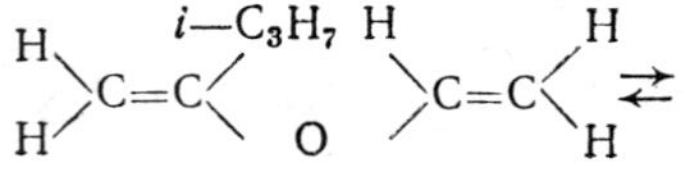 H, H C=C / \ C=C H, H	298 (g)	−1.4 ±0.4	18.9 ±1.1	SE, CT, GLC	74

Structure	T (K)	ΔH	ΔG	Method	Ref
CH₃₂C=C(H)(CH₃)/CH₃ ⇌ (H)(CH₃)C=C(H)(H)			2.1		
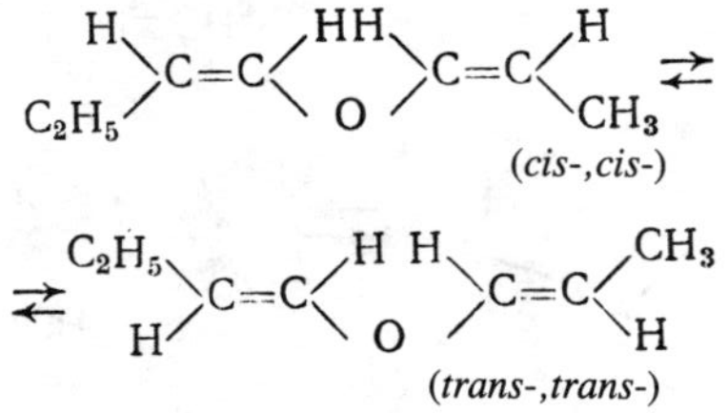 C₂H₅(H)C=C(HH)/O(CH₃)(H) ⇌ (cis-,cis-) C₂H₅(H)C=C(H H)/O(CH₃)(H) (trans-,trans-)	298 (g)	5.5 ±0.5	5.1 ±1.1	SE, CT, GLC	74
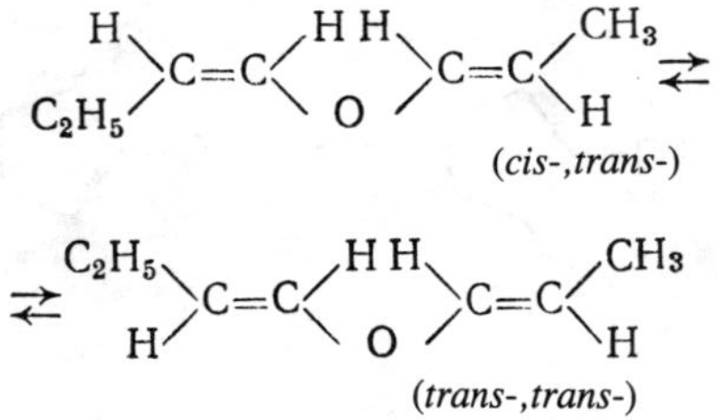 H(C₂H₅)C=C(H H)/O(CH₃)(H) (cis-,trans-) C₂H₅(H)C=C(HH)/O(CH₃)(H) (trans-,trans-)	298 (g)	2.2 ±0.6	3.1 ±1.6	SE, CT, GLC	74
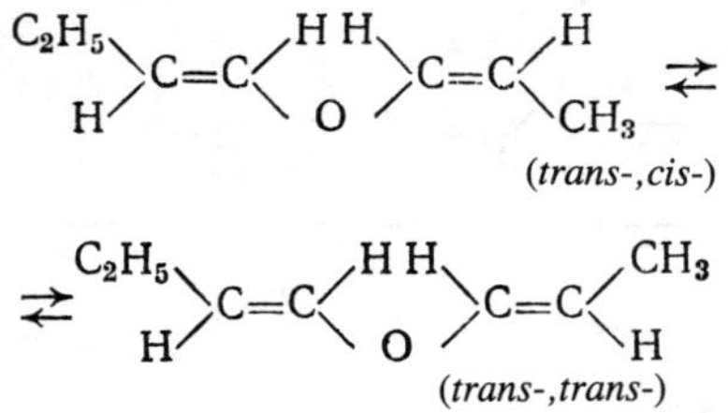 C₂H₅(H)C=C(H)(H)/O(CH₃)(H) (trans-,cis-) C₂H₅(H)C=C(H H)/O(CH₃)(H) (trans-,trans-)	298 (g)	3.3 ±0.7	2.9 ±1.7	SE, CT, GLC	74

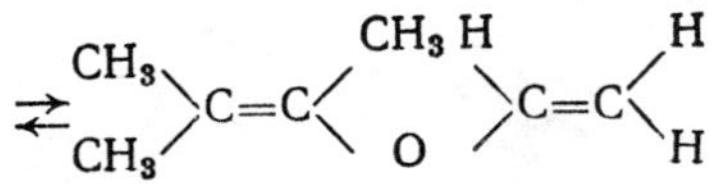

Table (*Continued*)

Formula	Type	Reaction	T, K (phase)	$\Delta_r H_T^\circ$ ($\Delta_r G_T^\circ$)	$\Delta_r S_T^\circ$	Technique	Reference
$C_7H_{12}O$	2.1	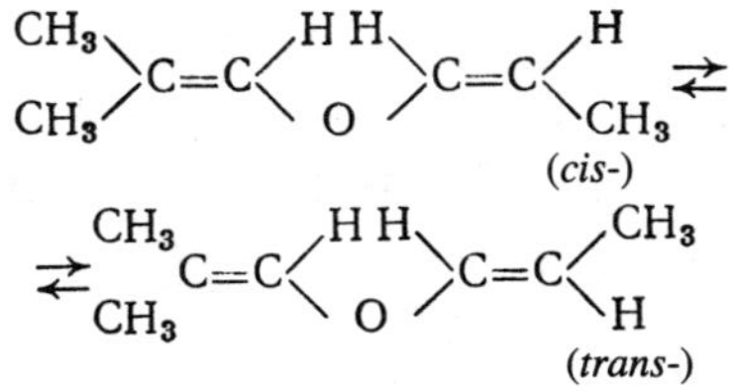 $CH_3\!\!>\!\!C=C\!\!<\!\!_O^{HH}\!\!>\!\!C=C\!\!<\!\!_{CH_3}^{H}$ (cis-) $\rightleftarrows$ $CH_3\!\!>\!\!C=C\!\!<\!\!_O^{HH}\!\!>\!\!C=C\!\!<\!\!_H^{CH_3}$ (trans-)	298 (g)	2.7 ±0.4	1.5 ±0.9	SE, CT, GLC	74
	1.3	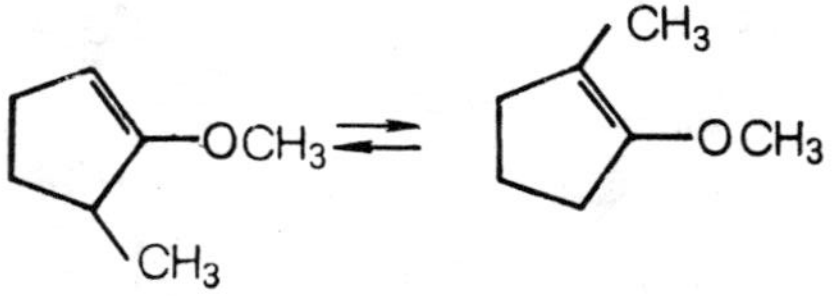 (cyclopentene with OCH₃ and CH₃) $\rightleftarrows$ (cyclopentene with OCH₃ and CH₃)	298	(1.72)	—	SE, CT, GLC	56
	1.2, 1.3	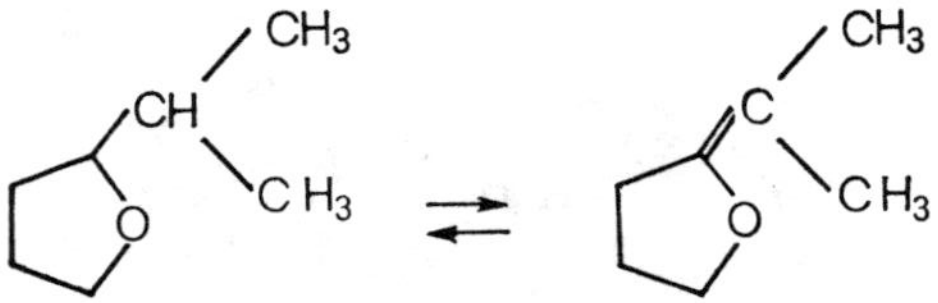 (tetrahydrofuran derivative) $\rightleftarrows$ (dihydrofuran derivative)	298 (g)	−2.6 ±0.4	−1.7 ±0.9	SE, CT, GLC	119
$C_7H_{12}O_2$	1.2, 1.3	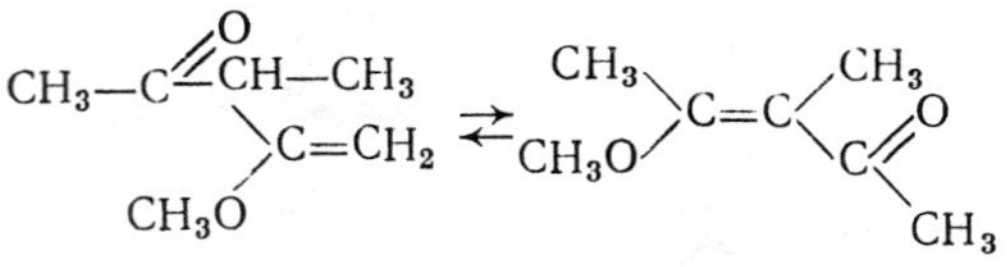 $CH_3-C(=O)-CH-CH_3 \atop CH_3O-C=CH_2$ $\rightleftarrows$ $CH_3\!\!>\!\!C=C\!\!<\!\!_{CH_3O}^{CH_3}\!\!C(=O)CH_3$ (cis-)	298 (g)	−4.0 ±0.3	−20.9 ±1.0	SE, CT, GLC	108

1.2 1.3	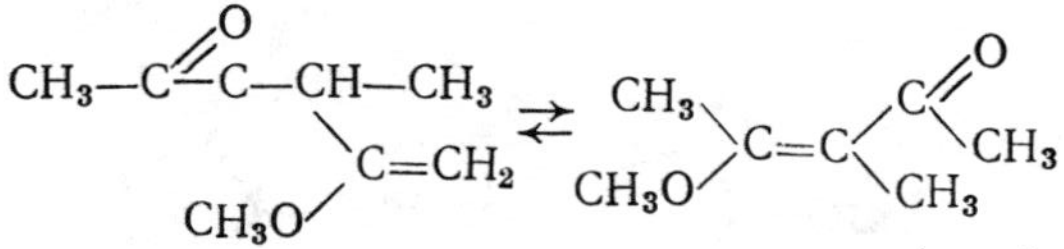 CH₃–C(=O)–C–CH–CH₃ ⇌ CH₃O–C=C(CH₃)–C(=O)–CH₃ with C=CH₂/CH₃O (trans-)	298 (g)	−3.6 ±0.4	−7.2 ±1.2	SE, CT, GLC  ⬡	108
2.1	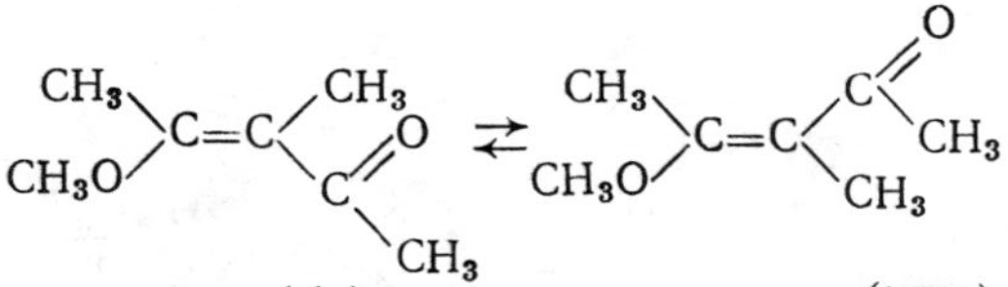 CH₃O,CH₃ C=C ⇌ CH₃O,CH₃ C=C(CH₃)–C(=O)–CH₃ (cis-) ... (trans-)	298 (g)	0.5 ±0.2	13.6 ±0.6	SE, CT, GLC ⬡	108
	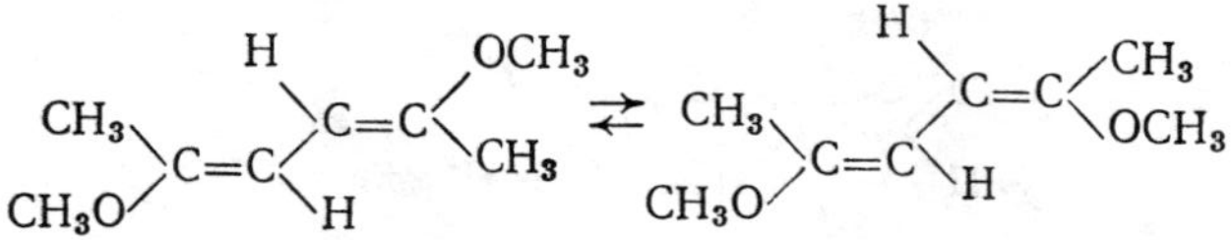 H,OCH₃ C=C CH₃,H ⇌ CH₃O,CH₃ C=C OCH₃,H (cis-,cis-) ... (cis-,trans-)	298 (l)	3.9 ±0.6	4.5 ±1.4	SE, CT, GLC ⬡	152
	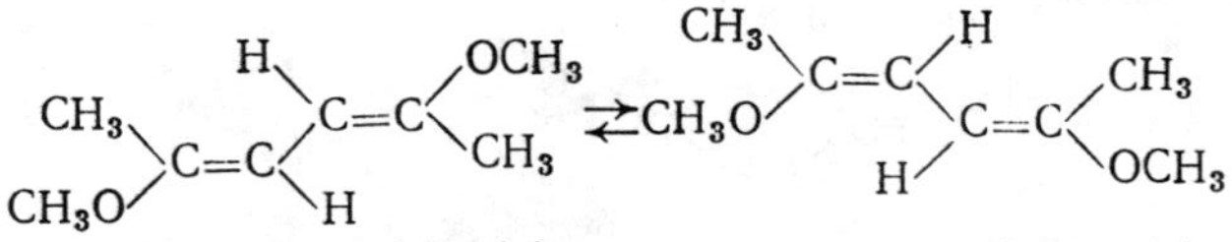 H,OCH₃ C=C CH₃,H ⇌ CH₃O,CH₃ C=C H,OCH₃ (cis-,cis-) ... (trans-,cis-)	298 (l)	12.2 ±0.7	4.2 ±2.0	SE, CT, GLC  ⬡	152
	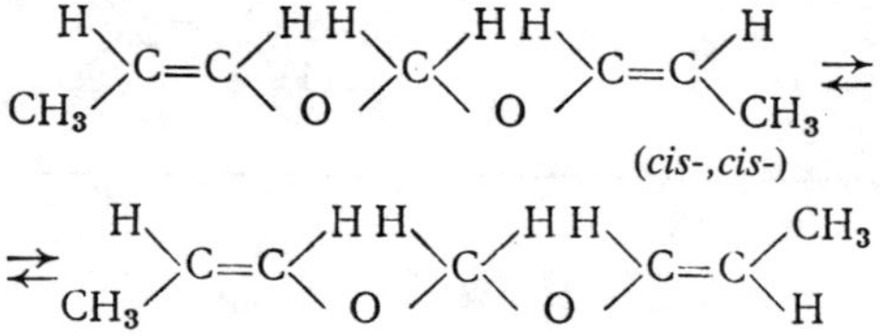 H,H C=C H,H / O / H,CH₃ ⇌ H,H C=C CH₃,H / O / H,CH₃ (cis-,cis-) ... (cis-,trans-)	298 (l)	2.0 ±0.3	5.3 ±0.8	SE, CT, GLC ⬡	92

Table (Continued)

Formula	Type	Reaction	T, K (phase)	$\Delta_r H_T^\circ$ ($\Delta_r G_T^\circ$)	$\Delta_r S_T^\circ$	Technique	Reference
$C_7H_{12}O_2$		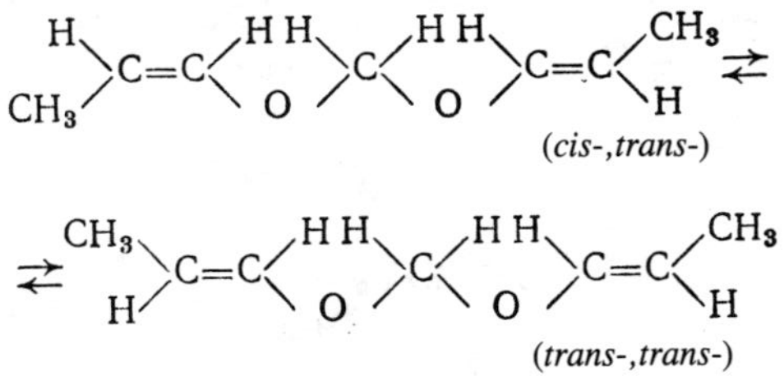 (cis-,trans-) ⇌ (trans-,trans-)	298 (l)	2.0 ±0.2	5.6 ±0.5	SE, CT, GLC	92
		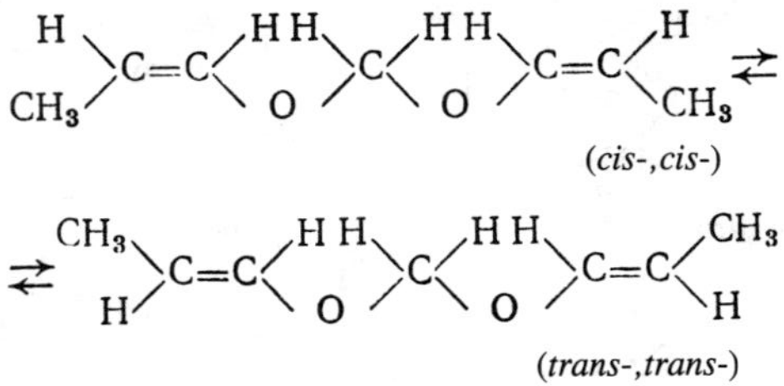 (cis-,cis-) ⇌ (trans-,trans-)	298 (l)	4.0 ±0.2	0.3 ±0.4	SE, CT, GLC	92
	1.3	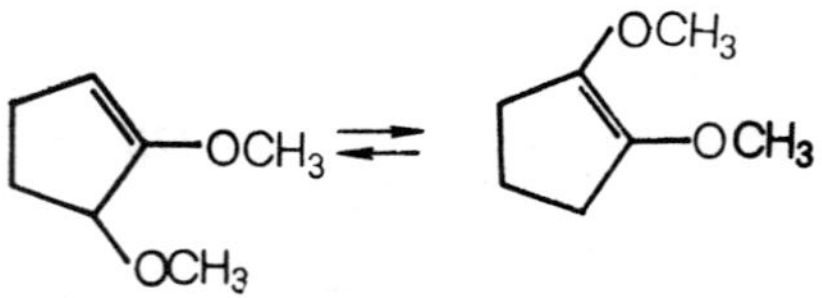	298 (l)	11,9 ±0,2	6,9 ±0,6	SE, CT, GLC	81

Formula	Ratio	Structure	T (K), phase	ΔH	—	Method, solvent	Ref.
$C_7H_{12}O_3$	2.1	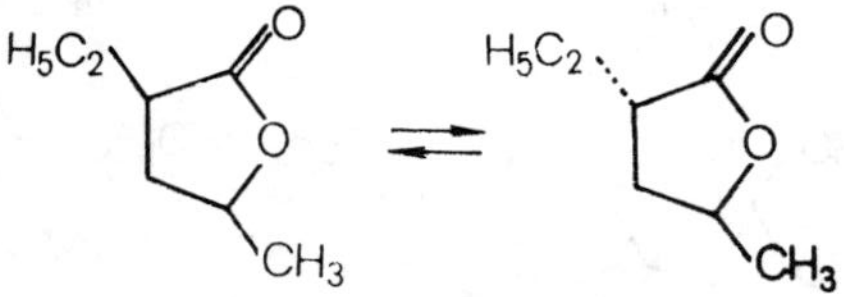 trans/cis lactone (H₅C₂, CH₃ substituted γ-butyrolactone)	298 (l)	(0.01)	—	SE, CT, GLC; t-C_4H_9OH	121
	1.3	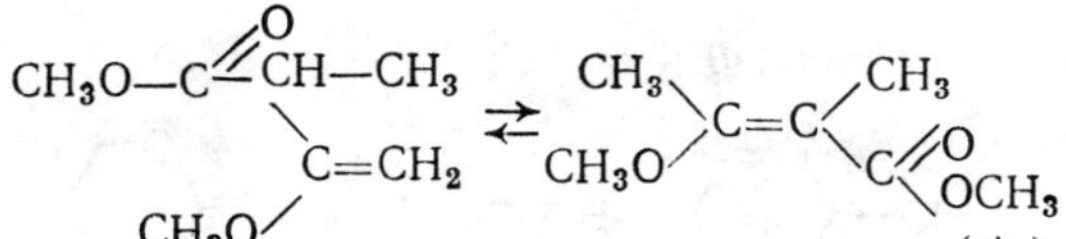 $CH_3O-CO-CH(CH_3)-C(OCH_3)=CH_2$ ⇌ $CH_3-C(=C(OCH_3)(COOCH_3))-CH_3$ (cis-)	298 (g)	−1.9 ±1.1	−11.2 ±2.9	SE, CT, GLC; C₆H₆	108
	1.2, 1.3	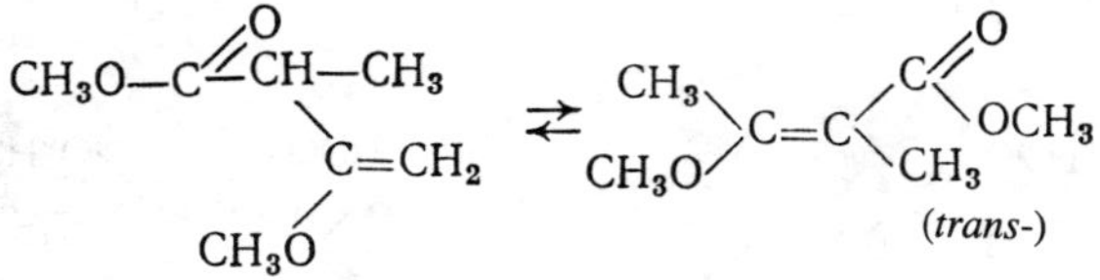 analogous enol ether equilibrium (trans-)	298 (g)	−9.1 ±0.7	−7.3 ±1.8	SE, CT, GLC; C₆H₆	108
$C_7H_{12}O_4$	2.1	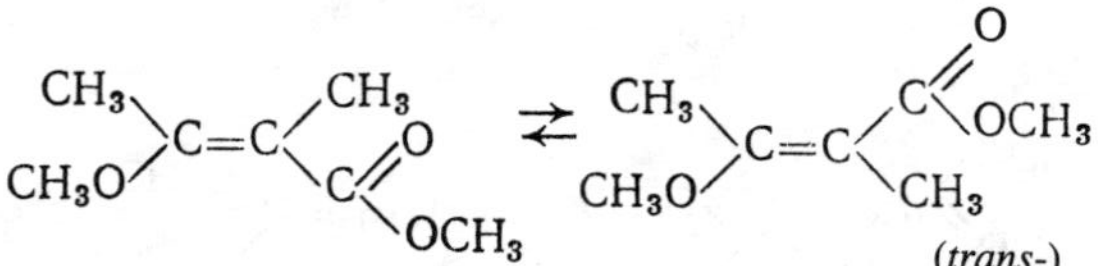 (cis-/trans- methoxy ester equilibrium)	298 (g)	−7.2 ±0.3	3.9 ±0.7	SE, CT, GLC; C₆H₆	108
	1.2, 1.3	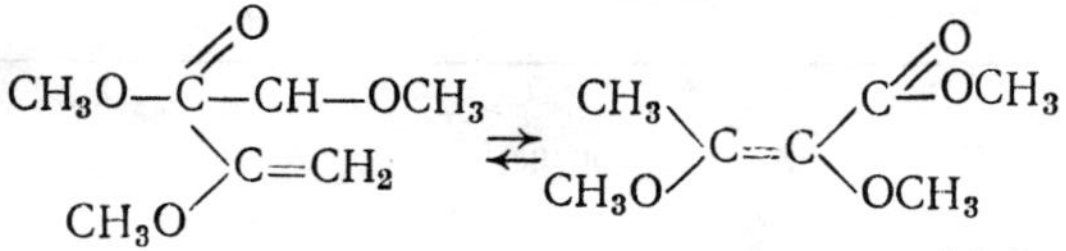 $CH_3O-CO-CH(OCH_3)-C(OCH_3)=CH_2$ ⇌ cis isomer	298 (g)	−5.3 ±0.1	1.0 ±0.1	SE, CT, GLC; CCl₄	108

Table (Continued)

Formula	Type	Reaction	T, K (phase)	$\Delta_r H_T^\circ$ ($\Delta_r G_T^\circ$)	$\Delta_r S_T^\circ$	Technique	Reference
$C_7H_{12}O_4$	1.3	$CH_3O-\underset{\underset{CH_3O}{\|}}{\overset{\overset{O}{\|}}{C}}-CH-OCH_3 \rightleftharpoons \underset{C=CH_2}{} \quad \underset{CH_3O}{CH_3}>C=C<\underset{C\overset{\|}{\underset{O}{}}OCH_3}{OCH_3}$ (trans-)	298 (g)	1.6 ±0.1	5,8 ±0.2	SE, CT, GLC CCl_4	108
	2.1	$\underset{CH_3O}{\overset{CH_3}{}}C=C\underset{OCH_3}{\overset{\overset{O}{\|}}{OCH_3}}$ (cis-) $\rightleftharpoons \underset{CH_3O}{\overset{CH_3}{}}C=C\underset{C\overset{\|}{\underset{O}{}}OCH_3}{OCH_3}$ (trans-)	298 (g)	7,2 ±0.2	5.5 ±0.4	SE, CT, GLC CCl_4	108
$C_7H_{13}BrO_2$	2.1	(cis-) $\rightleftharpoons$ (trans-) [1,3-dioxane with CH(CH_3)_2 and Br]	298 (l)	(−6.02)	—	SE, CT, GLC $(C_2H_5)_2O$	153
		(cis-, trans-) $\rightleftharpoons$ (trans-, cis-) [dioxane with CH_2Br, CH_3]	298 (l)	(0.46 ±0.02)	—	SE, CT, GLC (C_6H_{12})	154

$C_7H_{13}ClO_2$	2.1	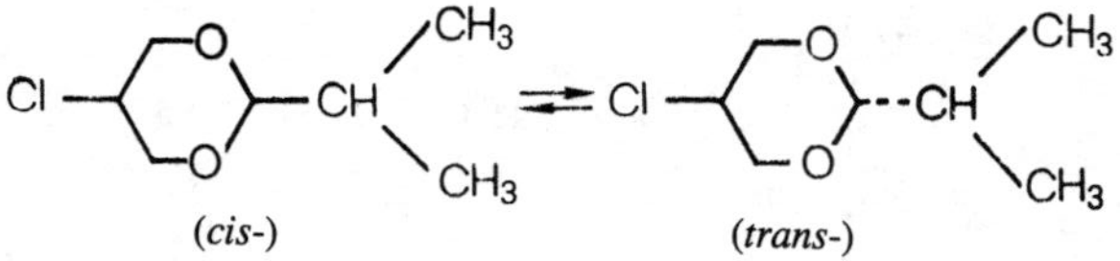 (trans-) ⇌ (cis-)	298 (l)	(−5.02)	—	SE, CT, GLC $(C_2H_5)_2O$	153
		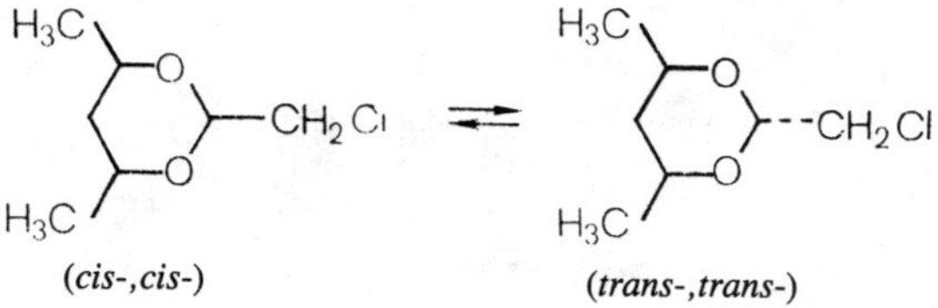 (trans-,trans-) ⇌ (cis-,cis-)	298 (l)	(17.52 ±0.25)	—	SE, CT, GLC $(CH_3)_2O$	151
		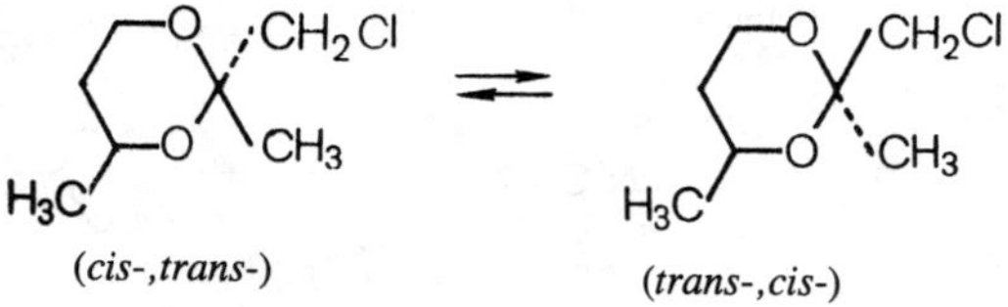 (trans-,cis-) ⇌ (cis-,trans-)	298 (l)	(−0.04 ±0.02)	—	SE, CT, GLC  ⬡	154
$C_7H_{13}FO_2$	2.1	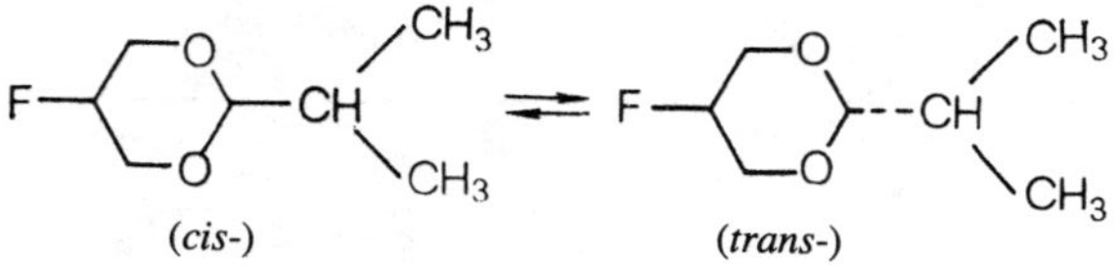 (trans-) ⇌ (cis-)	298 (l)	(2.59)	—	SE, CT, GLC $(C_2H_5)_2O$	153

Table (Continued)

Formula	Type	Reaction	T, K (phase)	$\Delta_r H_T^\circ$ ($\Delta_r G_T^\circ$)	$\Delta_r S_T^\circ$	Technique	Reference
$C_7H_{13}NO_2$	1.1, 1.2	$\begin{array}{c}\text{O} \quad C_2H_5 \\ \parallel \quad / \\ C-N \\ \quad \backslash C_2H_5 \\ \mid \\ HO-C=CH_2 \end{array} \rightleftharpoons \begin{array}{c} \text{O} \quad C_2H_5 \\ \parallel \quad / \\ C-N \\ \quad \backslash C_2H_5 \\ \mid \\ O=C-CH_3 \end{array}$	318 (l)	6.94 ±0.21	29.20 ±1.05	SE, NMR CCl_4	155
$C_7H_{13}NO_4$	2.1	(cis-) ⇌ (trans-) 1,3-dioxane with O_2N and CH-CH$_3$/CH$_3$ groups	298 (l)	(1.59)	—	SE, CT, NMR CCl_4	153
C_7H_{14}	1.3	$CH_2=CH-CH-CH_2-CH_2-CH_2-C_2H_5 \rightleftharpoons CH_2=C-CH_2-CH_2-CH-CH-CH_3$ (with CH_3 groups)	298 (g)	−22.50 ±1.67	—	CC	156
		$CH_2=C-CH_2-CH-CH_3 \rightleftharpoons CH_2=CH-CH_2-CH-CH_3$ (with CH_3 groups)	298 (g)	4.48 ±1.67	—	CC	156
		$CH_2=CH-CH-CH_2-CH_2-CH_2-C_2H_5 \rightleftharpoons CH_2=C-CH-CH_3$ (with C_2H_5, CH_3 groups)	298 (g)	−18.24 ±1.34	—	CC	156
		$CH_2=C-CH-CH_3 \rightleftharpoons CH_2=C-C-CH_3$ (with C_2H_5, CH_3, CH_3 groups)	298 (g)	−5.94 ±1.25	—	CC	156

Reaction	T (K), state	ΔH	ΔS	Method	Ref
$CH_2=CH-CH_2-C_4H_9 \rightleftarrows \begin{array}{c} CH_3 \\ H \end{array} C=C \begin{array}{c} C_4H_9 \\ H \end{array}$ (cis-)	596 (g)	−7.61 ±0.10	−2.89 ±0.21	FEC CT, GLC	157
$CH_2=CH-CH_2-C_4H_9 \rightleftarrows \begin{array}{c} CH_3 \\ H \end{array} C=C \begin{array}{c} H \\ C_4H_9 \end{array}$ (trans-)	298 (l)	−6.78 ±1.20	—	CC	158
$CH_2=CH-CH_2-CH_2-C_3H_7 \rightleftarrows \begin{array}{c} C_2H_5 \\ H \end{array} C=C \begin{array}{c} C_3H_7 \\ H \end{array}$ (cis-)	596 (g)	−10.00 ±0.21	−2.01 ±0.42	FE, CT, GLC	157
$CH_2=CH-CH_2-CH_2-C_3H_7 \rightleftarrows \begin{array}{c} C_2H_5 \\ H \end{array} C=C \begin{array}{c} H \\ C_3H_7 \end{array}$ (trans-)	596 (g)	−7.61 ±1.09	−3.93 ±2.09	FE, CT, GLC	157
$CH_2=CH-CH_2-C_4H_9 \rightleftarrows \begin{array}{c} C_2H_5 \\ CH_3 \end{array} C=C \begin{array}{c} C_2H_5 \\ H \end{array}$ (cis-)	596 (g)	−10.88 ±0.38	−2.43 ±0.63	FE, CT, GLC	157
$CH_2=CH-CH_2-C_4H_9 \rightleftarrows CH_3-\underset{CH_3}{\overset{CH_3}{C}}=CH-CH-CH_3$	298 (g)	−18.11 ±0.92	—	CC	156
$CH_3-\underset{CH_3}{\overset{CH_3}{C}}=CH-CH-CH_3$	298 (g)	−27.40 ±0.84	—	CC	156
$\begin{array}{c} CH_3 \\ H \end{array} C=C \begin{array}{c} (t-C_4H_9) \\ H \end{array}$ (cis-)	298 (g)	16.06 ±1.05	—	CC	156

Table (Continued)

Formula	Type	Reaction	T, K (phase)	$\Delta_r H_T^\circ$ ($\Delta_r G_T^\circ$)	$\Delta_r S_T^\circ$	Technique	Reference
C_7H_{14}		$\underset{(cis\text{-})}{\overset{C_2H_5}{\underset{H}{>}}C=C\overset{C_3H_7}{\underset{H}{<}}} \rightleftarrows \underset{(trans\text{-})}{\overset{CH_3}{\underset{H}{>}}C=C\overset{H}{\underset{C_4H_9}{<}}}$	298 (l)	−5.19 ±1.09	—	CC	158
		$\underset{(cis\text{-})}{\overset{CH_3}{\underset{H}{>}}C=C\overset{C_4H_9}{\underset{H}{<}}} \rightleftarrows \underset{(trans\text{-})}{\overset{CH_3}{\underset{H}{>}}C=C\overset{H}{\underset{C_4H_9}{<}}}$	298 (l)	−4.39 ±1.17	—	CC	158
		$\underset{(cis\text{-})}{\overset{CH_3}{\underset{H}{>}}C=C\overset{(t\text{-}C_4H_9)}{\underset{H}{<}}} \rightleftarrows \underset{(trans\text{-})}{\overset{CH_3}{\underset{H}{>}}C=C\overset{H}{\underset{(t\text{-}C_4H_9)}{<}}}$	298 (g)	−16.15 ±1.05	—	CC	156
		$\underset{(cis\text{-})}{\overset{C_2H_5}{\underset{H}{>}}C=C\overset{C_3H_7}{\underset{H}{<}}} \rightleftarrows \underset{(trans\text{-})}{\overset{C_2H_5}{\underset{H}{>}}C=C\overset{H}{\underset{C_3H_7}{<}}}$	298 (g)	−4.98 ±1.20	—	CC	158
		$\underset{(cis\text{-})}{\overset{C_2H_5}{\underset{CH_3}{>}}C=C\overset{C_2H_5}{\underset{H}{<}}} \rightleftarrows \underset{(trans\text{-})}{\overset{C_2H_5}{\underset{CH_3}{>}}C=C\overset{H}{\underset{C_2H_5}{<}}}$	298 (g)	2.59 ±0.88	—	CC	156

$C_7H_{14}O$	1.3		500 (g)	(0.0)	—	FE, CT, GLC	159
		cis- ⇌ trans- (1,3-dimethylcyclopentane)					
		$CH_2=C(OCH_3)-CH_2-CH(CH_3)_2$ ⇌ cis-$CH_3O,CH_3/C=C/i-C_3H_7,H$	298 (g)	0.60 ±0.37	−6.4 ±0.8	SE, CT, GLC	91
		$CH_2=C(OCH_3)-CH_2-CH(CH_3)_2$ ⇌ trans-$CH_3O,H/C=C/i-C_3H_7,CH_3$ 	298 (g)	12.16 ±0.40	9.1 ±0.9	SE, CT, GLC	91
		$CH_3-C(OCH_3)=C(CH_3)-CH_2-CH_3$ ⇌ cis-$CH_3,H/C=C/i-C_3H_7,OCH_3$	298 (g)	−9.03 ±0.34	−22.1 ±0.7	SE, CT, GLC	160
		$CH_3-C(OCH_3)=C(CH_3)-CH_2-CH_3$ ⇌ trans-$CH_3,CH_3/C=C/OCH_3,H$ (i-C_3H_7)	298 (g)	2.58 ±0.42	−2.7 ±1.0	SE, CT, GLC	160
	2.1	$CH_3,CH_3O/C=C/i-C_3H_7,H$ (cis) ⇌ $CH_3O,H/C=C/i-C_3H_7,CH_3$ (trans) 	298 (g)	11.46 ±0.35	15.2 ±0.7	SE, CT, GLC	91

Table (Continued)

Formula	Type	Reaction	T, K (phase)	$\Delta_r H_T^\circ$ ($\Delta_r G_T^\circ$)	$\Delta_r S_T^\circ$	Technique	Reference
$C_7H_{14}O$	2.1	$\begin{array}{c}CH_3\\ \diagdown\\ H\end{array}C=C\begin{array}{c}(i-C_3H_7)\\ \diagup\\ OCH_3\end{array}$ (cis-) $\rightleftarrows$ $\begin{array}{c}CH_3\\ \diagdown\\ H\end{array}C=C\begin{array}{c}OCH_3\\ \diagup\\ (i-C_3H_7)\end{array}$ (trans-)	298 (g)	11.60 ±0.39	19.5 ±0.9	SE, CT, GLC	160
		$\begin{array}{c}C_2H_5O\\ \diagdown\\ H\end{array}C=C\begin{array}{c}(i-C_3H_7)\\ \diagup\\ H\end{array}$ (cis-) $\rightleftarrows$ $\begin{array}{c}C_2H_5O\\ \diagdown\\ H\end{array}C=C\begin{array}{c}H\\ \diagup\\ (i-C_3H_7)\end{array}$ (trans-)	311 (l)	−3.26 ±0.25	−6.27 ±0.84	SE, CT, GLC	40
		$\begin{array}{c}CH_3\\ \diagdown\\ H\end{array}C=C\begin{array}{c}O-(i-C_4H_9)\\ \diagup\\ H\end{array}$ (cis-) $\rightleftarrows$ $\begin{array}{c}CH_3\\ \diagdown\\ H\end{array}C=C\begin{array}{c}H\\ \diagup\\ O-(i-C_4H_9)\end{array}$ (trans-)	311 (l)	−1.80 ±0.23	−9.04 ±0.75	SE, CT, GLC	40
		$\begin{array}{c}CH_3\\ \diagdown\\ H\end{array}C=C\begin{array}{c}O-(t-C_4H_9)\\ \diagup\\ H\end{array}$ (cis-) $\rightleftarrows$ $\begin{array}{c}CH_3\\ \diagdown\\ H\end{array}C=C\begin{array}{c}H\\ \diagup\\ O-(t-C_4H_9)\end{array}$ (trans-)	311 (l)	2.85 ±0.56	−0.58 ±1.80	SE, CT, GLC	40
		(CH₃, OH cis-cyclohexanol) $\rightleftarrows$ (CH₃, OH trans-cyclohexanol)	393 (l)	(−2.02 ±0.08)	—	SE, CT, IR	161
		(methylcyclohexanol isomerization)	353 (l)	(−3.55)	—	SE, CT, GLC	162

	Structure	T (K)	Value	±	Method	Ref
	cis-3-methylcyclohexanol ⇌ trans-3-methylcyclohexanol	363 (g)	−3.9 ±0.6	−0.8 ±1.1	FE, CT, GLC	163
		373 (l)	(2.43 ±0.04)	—	SE, CT, GLC	161
		353 (l)	(2.84)	—	SE, CT, GLC ⬡	162
	cis-4-methylcyclohexanol ⇌ trans-4-methylcyclohexanol	523 (g)	3.8 ±0.8	3.6 ±1.6	FE, CT, GLC	163
		418 (l)	(−2.02 ±0.08)	—	SE, CT, IR	161
		353 (l)	(−2.47)	—	SE, CT, GLC ⬡	162
		523 (g)	−1.7 ±0.5	−0.2 ±0.9	FE, CT, GLC	163
$C_7H_{14}OS$ 1.2 1.3	$CH_2=CH-CH_2-S(=O)(t-C_4H_9)$ ⇌ $(CH_3)(H)C=C(H)(S(=O)(t-C_4H_9))$ (cis-)	293 (l)	(−1.0)	—	SE, CT, GLC $t-C_4H_9OH$	57

Table (Continued)

Formula	Type	Reaction	T, K (phase)	$\Delta_r H_T^\circ$ ($\Delta_r G_T^\circ$)	$\Delta_r S_T^\circ$	Technique	Reference
$C_7H_{14}OS$	1.2 1.3	$CH_2=CH-CH_2-S(=O)(t-C_4H_9) \rightleftarrows (CH_3)(H)C=C(H)S(=O)(t-C_4H_9)$ (trans-/cis-)	293 (l)	(−7.6)	—	SE, CT, GLC $t-C_4H_9OH$	57
	2.1	$(CH_3)(H)C=C(H)S(=O)(t-C_4H_9)$ cis ⇌ trans	293 (l)	(−6.6)	—	SE, CT, GLC $t-C_4H_9OH$	57
$C_7H_{14}O_2$	2.1	$(CH_3)(H)C=C(H)O-C(CH_3)_2-OCH_3$ cis ⇌ trans	298 (l)	(3.2 ±0.2)	—	SE, CT, GLC	92
	1.3	$CH_2-CH-CH_2-O-C_4H_9 \rightleftarrows CH_2-CH-CH_2-O-(t-C_4H_9)$ (epoxide rearrangement)	298 (l)	−23.8 ±2.9	—	CC	164
	2.1	2-isopropyl-4-methyl-1,3-dioxolane (cis-) ⇌ (trans-)	341 (l)	0.15	−2.93	SE, CT, NMR CCl_4	84

Compound	T (K)	ΔG (kJ/mol)		Method	Ref
2-methyl-4-isopropyl-1,3-dioxolane (cis ⇌ trans)	298 (l)	(1.00 ±0.04)	—	SE, CT, GLC (C$_2$H$_5$)$_2$O	93
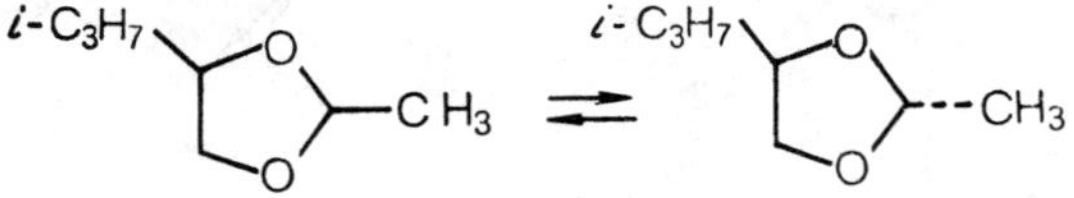 2-ethyl-4,5-dimethyl-1,3-dioxolane (syn ⇌ ant)	298 (l)	(1.35 ±0.03)	—	SE, CT, GLC (C$_2$H$_5$)$_2$O	93
2-methyl-4,5-dimethyl-1,3-dioxolane (cis ⇌ ant)	298 (l)	(2.60 ±0.11)	—	SE, CT, GLC (C$_2$H$_5$)$_2$O	93
2-methyl-5-ethyl-1,3-dioxane (cis ⇌ trans)	333 (l)	(−3.72 ±0.04)	—	SE, CT, GLC	165
 2-ethyl-5-methyl-1,3-dioxane (cis ⇌ trans)	333 (l)	(−4.10 ±0.04)	—	SE, CT, GLC	133

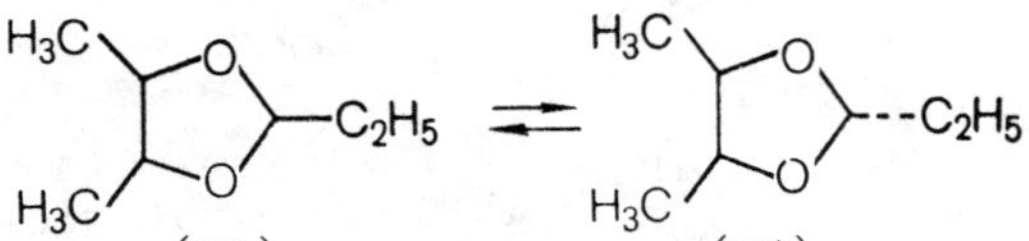

Table (Continued)

Formula	Type	Reaction	T, K (phase)	$\Delta_r H_T^\circ$ ($\Delta_r G_T^\circ$)	$\Delta_r S_T^\circ$	Technique	Reference
$C_7H_{14}O_2$	2.1	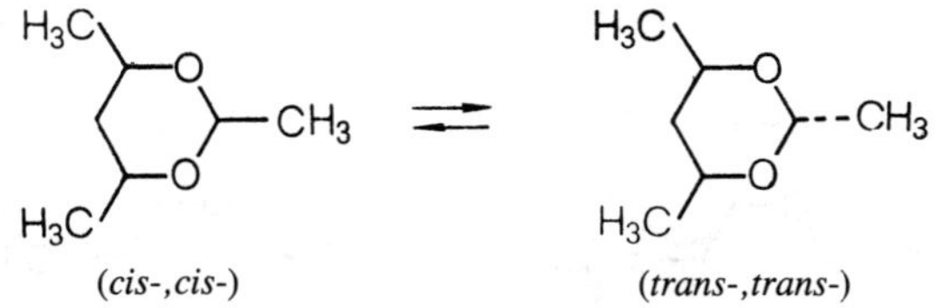 (cis-, cis-) ⇌ (trans-, trans-) 2,4,6-trimethyl-1,3-dioxane	298 (l)	(16.61)	—	SE, CT, GLC $(C_2H_5)_2O$	166
$C_7H_{14}O_2S$	1.2 1.3	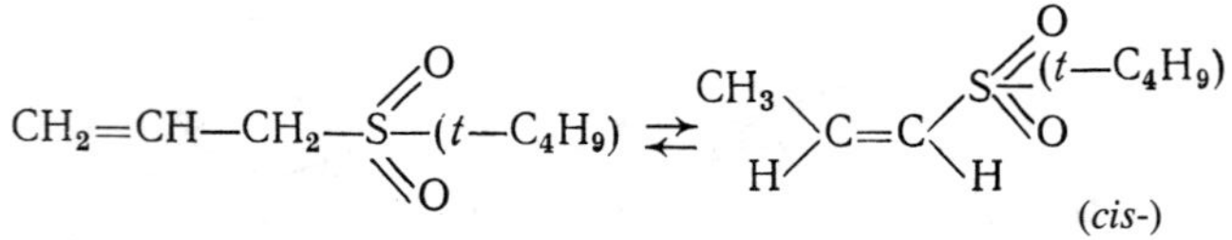 $CH_2=CH-CH_2-S(=O)_2-(t-C_4H_9)$ ⇌ CH_3, H C=C $(t-C_4H_9)SO_2$, H (cis-)	293 (l)	(4.9)	—	SE, CT, GLC $t-C_4H_9OH$	57
		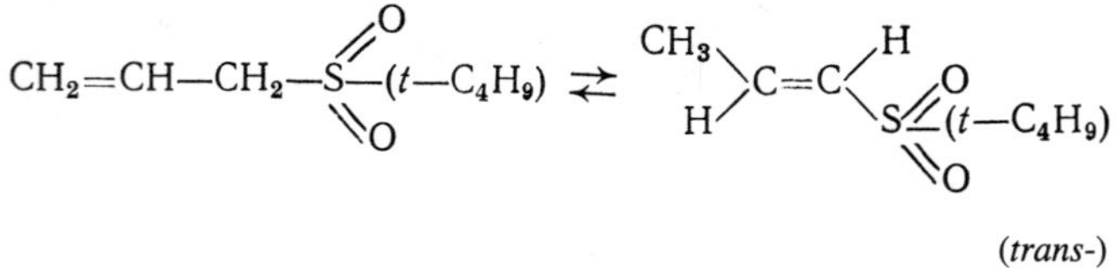 $CH_2=CH-CH_2-S(=O)_2-(t-C_4H_9)$ ⇌ CH_3, H C=C H, $(t-C_4H_9)SO_2$ (trans-)	293 (l)	(−9.1)	—	SE, CT, GLC $t-C_4H_9OH$	57
	2.1	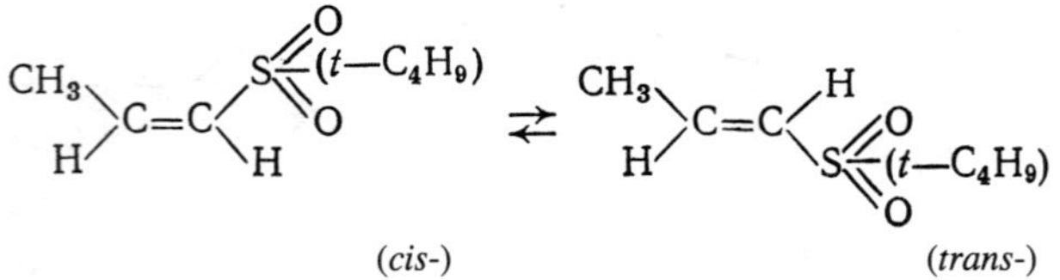 (cis-) ⇌ (trans-)	293 (l)	(14.1)	—	SE, CT, GLC $t-C_4H_9OH$	57

$C_7H_{14}O_3$	2.1	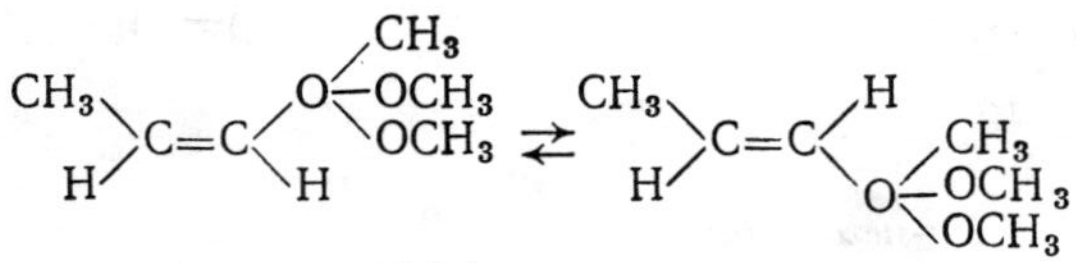 CH₃\C=C/H ⇌ CH₃\C=C/H (cis-) (trans-) with OCH₃/OCH₃/CH₃ groups	298 (l)	3.0 ±0.6	0.0 ±2.0	SE, CT, GLC 134
		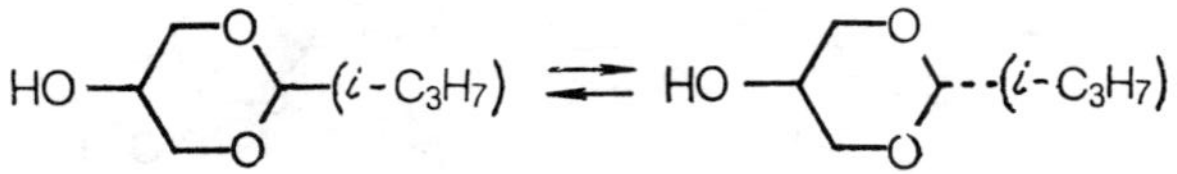 (i-C₃H₇) dioxane HO (cis-) ⇌ (trans-)	353 (l)	(−3.83)	—	SE, CT, GLC 153
		2-OCH₃ dimethyl dioxane (cis-,cis-) ⇌ (trans-,trans-)	298 (l)	(−1.72) ±0.13	—	SE, CT, GLC ($C_2H_5)_2O$ 167
		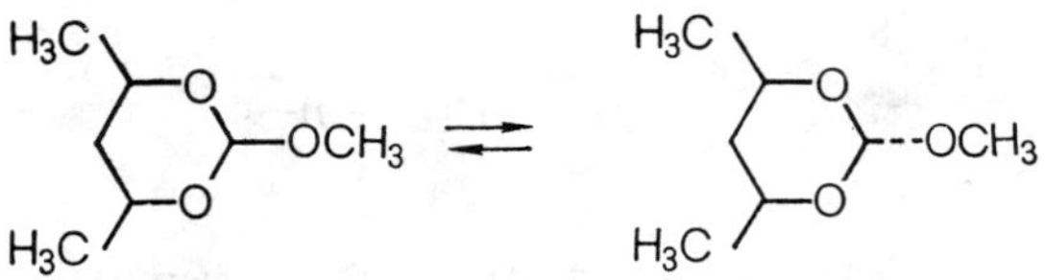	298 (l)	(−1.72)	—	SE, CT, NMR ($C_2H_5)_2O$ 166
$C_7H_{14}S$	1.3	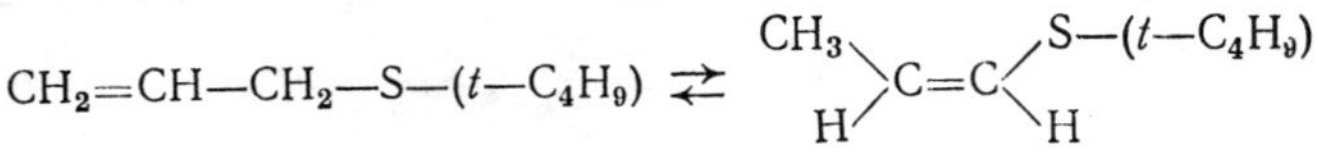 $CH_2=CH-CH_2-S-(t-C_4H_9)$ ⇌ CH₃\C=C/H with S-(t-C₄H₉) (cis-)	293 (l)	(−8.4)	—	SE, CT, GLC $t-C_4H_9OH$ 57

Table (*Continued*)

Formula	Type	Reaction	T, K (phase)	$\Delta_r H^\circ_T$ ($\Delta_r G^\circ_T$)	$\Delta_r S^\circ_T$	Technique	Reference
$C_7H_{14}S$	1.3	$CH_2=CH-CH_2-S-(t-C_4H_9) \rightleftarrows \underset{H}{\overset{CH_3}{}}C=C\underset{S-(t-C_4H_9)}{\overset{H}{}}$ (*trans*-)	293 (l)	(−9.6)	—	SE, CT, GLC $t-C_4H_9OH$	57
	2.1	$\underset{H}{\overset{CH_3}{}}C=C\underset{H}{\overset{S-(t-C_4H_9)}{}}$ (*cis*-) $\rightleftarrows \underset{H}{\overset{CH_3}{}}C=C\underset{S-(t-C_4H_9)}{\overset{H}{}}$ (*trans*-)	293 (l)	(−1.15)	—	SE, CT, GLC $t-C_4H_9OH$	57
$C_7H_{14}S_2$	2.1	(*cis*-) ⇌ (*trans*-) [1,3-dithiane structures with H_3C, C_2H_5 substituents]	298 (l)	(4.81) ±0.08	—	SE, CT, GLC $CHCl_3$	137
		(*cis*,*cis*-) ⇌ (*trans*,*trans*-) [1,3-dithiane structures with H_3C, CH_3 substituents]	342 (l)	(7.40) ±0.04	—	SE, CT, GLC $CHCl_3$	137
$C_7H_{15}Br$	1.3	$CH_3-CH(Br)-CH_2-CH_2-CH_2-C_3H_7 \rightleftarrows CH_3-CH_2-CH_2-CH_2-CH(Br)-CH-C_3H_7$	332 (l)	−1.13 ±1.05	−11.8 ±3.35	SE, CT, GLC	95

$C_7H_{15}Cl$	1.3	CH₃—CH—CH₂—C₄H₉ ⇌ CH₃—CH₂—CH—C₄H₉ (Br, Br)	332 (l)	−0.33 ±0.56	−1.80 ±2.1	SE, CT, GLC	95
		CH₃—CH—CH₂—C₃H₇ ⇌ CH₃—CH₂—CH₂—CH—C₃H₇ (Cl, Cl)	400 (g, l)	0.38 ±0.42	−8.03 ±1.25	SE, FE, CT, GLC	96
		CH₃—CH—CH₂—C₄H₉ ⇌ CH₃—CH₂—CH—C₄H₉ (Cl, Cl)	400 (g, l)	0.15 ±0.42	−0.67 ±1.25	SE, FE, CT, GLC	96
		CH₃—C(CH₃)(CH₃)—CH—C₂H₅ ⇌ CH₃—CH—C(CH₃)(CH₃)—C₂H₅ (Cl, Cl)	400 (l)	−0.42 ±0.84	2.09 ±1.67	SE, CT, GLC ⬡	168
		C₂H₅—CH₂—CH—C₃H₇ ⇌ C₂H₅—CH—CH₂—C₃H₇ (Cl, Cl)	400 (l)	−0.23 ±0.10	6.90 ±0.25	SE, FE, CT, GLC	157
C_7H_{16}	1.3	CH₃—CH(CH₃)—CH₂—CH—C₃H₇(CH₃) ⇌ CH₃—CH₂—CH—CH—C₃H₇ (CH₃)(CH₃)	368 (g, l)	−0.10 ±0.21	−1.97 ±0.42	SE, FE, CT, GLC	169
		CH₃—CH(CH₃)—CH—CH₃(CH₃) ⇌ CH₃—(CH₂)₅—CH₃	310 (l)	(4.40 ±0.60)	—	SE, CT, GLC H₃PO₄	170
		CH₃—CH(CH₃)—CH₂—CH(CH₃)—CH₃⇌CH₃—CH₂—CH₂—CH₂—CH—CH₃ (CH₃)	310 (l)	(0.82 ±0.6)	—	SE, CT, GLC H₃PO₄	170
		CH₃—CH(CH₃)—CH₂—CH(CH₃)—CH₃⇌CH₃—CH₂—CH₂—CH₂—CH—CH₃ (CH₃)	310 (l)	(0.00 ±0.60)	—	SE, CT, GLC H₃PO₄	170
		CH₃—CH(CH₃)—CH₂—CH(CH₃)—CH₃ ⇌ CH₃—CH₂—CH₂—CH—CH₂—CH₃ (C₂H₅)	310 (l)	(8.00 ±0.60)	—	SE, CT, GLC H₃PO₄	170

105

Table (Continued)

Formula	Type	Reaction	T, K (phase)	$\Delta_r H_T^\circ$ ($\Delta_r G_T^\circ$)	$\Delta_r S_T^\circ$	Technique	Reference
C_7H_{16}	1.3	$CH_3-CH-CH_2-CH-CH_3 \rightleftarrows CH_3-CH-CH-CH_2-CH_3$ with CH_3 groups / CH_3 CH_3	310 (l)	(2.62 ±0.60)	—	SE, CT, GLC H_3PO_4	170
		$CH_3-CH-CH_2-CH-CH_3 \rightleftarrows CH_3-CH-C-CH-CH_3$ with CH_3 / CH_3 CH_3 CH_3	310 (l)	(2.04 ±0.60)	—	SE, CT, GLC H_3PO_4	170
$C_7H_{16}N_2S$	1.1, 1.2	(cyclic thiadiazine ⇌ open-chain thiol-hydrazone structures)	333 (l)	17.3 ±1.2	68.0 ±4.0	SE, NMR $Cl_2C=CCl_2$	140
$C_8H_6F_6O_4$	2.1	(cis-/trans- fluorinated diester isomerization)	298 (l)	−21.2	2.82	SE, CT, GLC CH_3-S-CH_3, $=O$	171

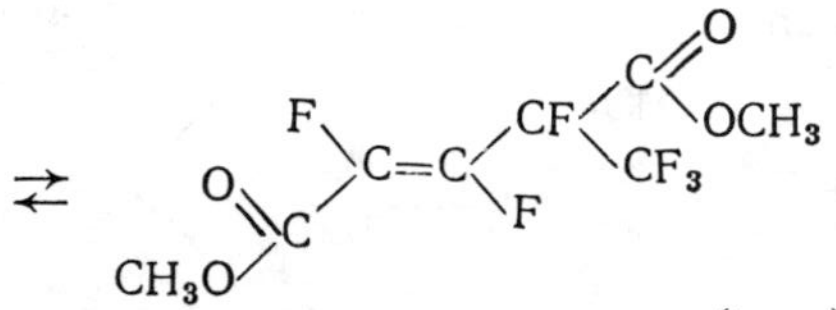

C_8H_9Br	1.3	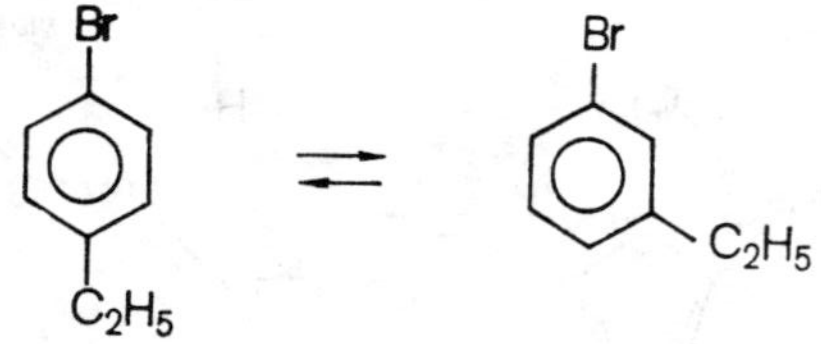 (structure: trans- F,CF₃,CF,F vinyl diester with OCH₃ and CH₃O groups) ⇌	330 (l)	−1.20 ±0.33	3.00 ±1.00	SE, CT, GLC	172
		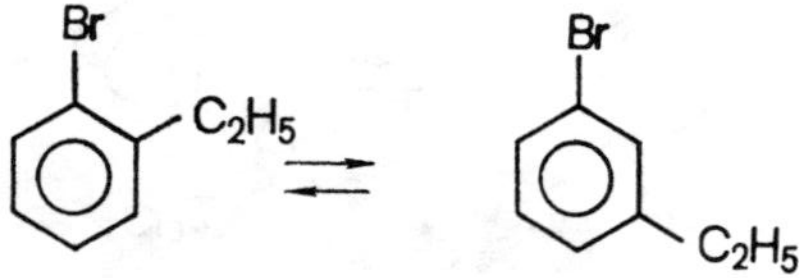 4-bromoethylbenzene ⇌ 3-bromoethylbenzene	330 (l)	−0.80 ±0.13	4.70 ±0.50	SE, CT, GLC	172
		2-bromoethylbenzene ⇌ 3-bromoethylbenzene					
$C_8H_9NO_2$	1.1 1.2	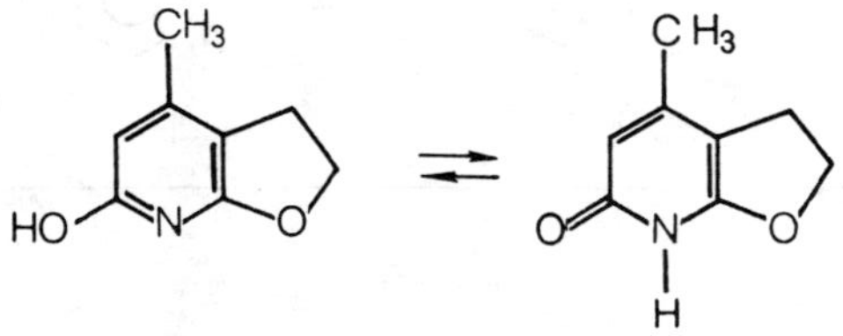 (furopyridinone tautomers: 4-methyl-2,3-dihydrofuro[3,2-b]pyridin-6(5H)-one ⇌ 6-hydroxy-4-methyl-2,3-dihydrofuro[3,2-b]pyridine)	323 (l)	−8.58	−23.09	SE, UV $C_2H_5OH-H_2O$	65

107

Table (*Continued*)

Formula	Type	Reaction	T, K (phase)	$\Delta_r H_T^\circ$ ($\Delta_r G_T^\circ$)	$\Delta_r S_T^\circ$	Technique	Reference
C_8H_{10}	1.3	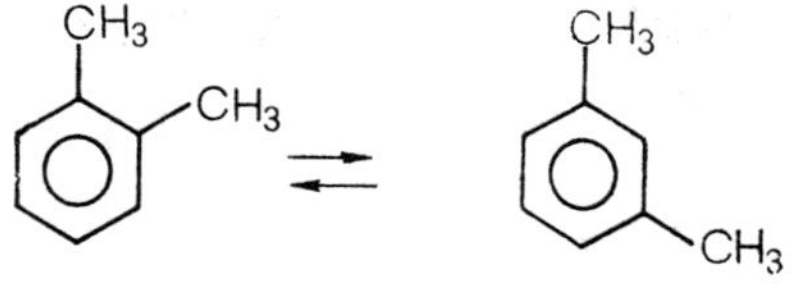 o-xylene ⇌ m-xylene	298 (g)	−1.76 ±1.20	−1.43 ±1.00	CC, ST	258
C_8H_{10}	1.3	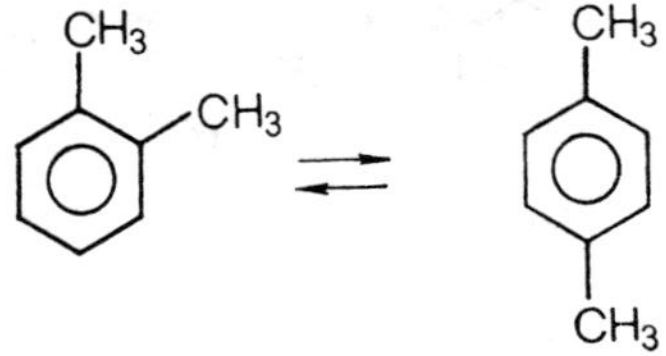 o-xylene ⇌ p-xylene	298 (g)	−1.05 ±1.38	−1.43 ±1.00	CC, ST	258
$C_8H_{10}OS$	2.1	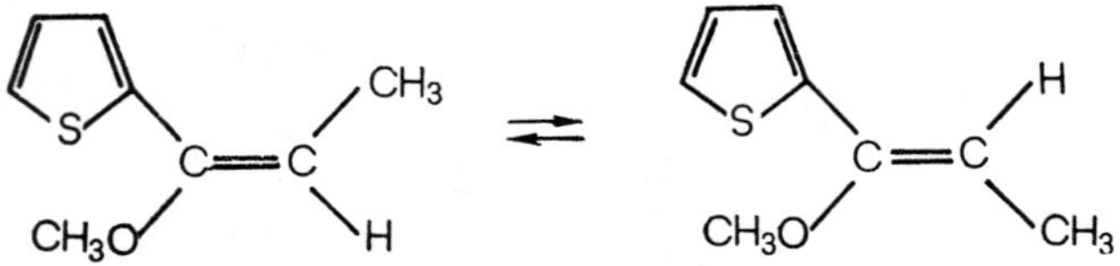 (cis-) ⇌ (trans-) methoxymethylene-thiophene	298 (l)	2.21	7.73	SE, CT, NMR	173
$C_8H_{10}O_2$	2.1	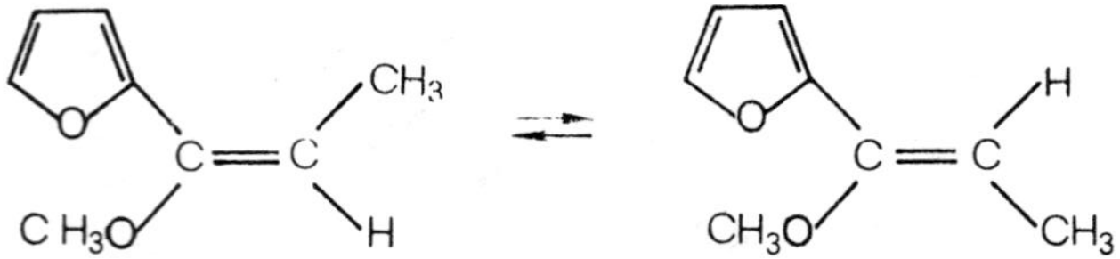 (cis-) ⇌ (trans-) methoxymethylene-furan	298 (l)	1.28	9.63	SE, CT, NMR	173

108

C₈H₁₁N	2.1	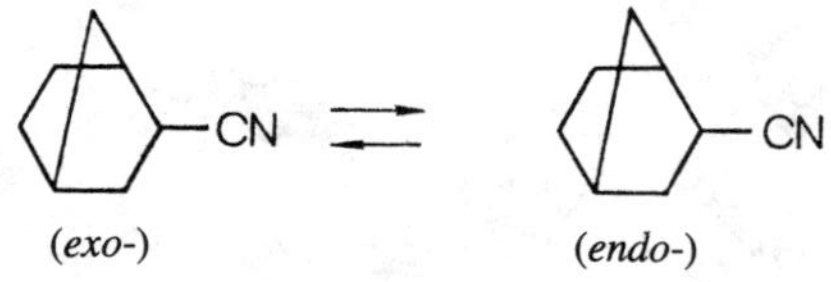 (exo-) ⇌ (endo-) CN structures	350 (l)	4.02 ±1.84	−5.94 ±1.67	CC, TP	174
C₈H₁₁N₅O	1.1 1.2	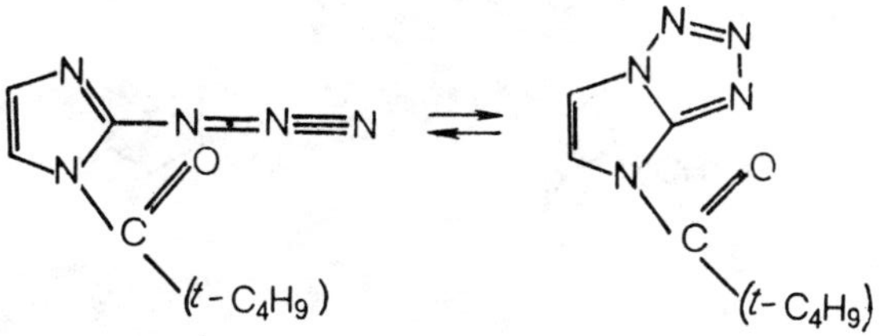 imidazole-N=N structures with C(=O)(t-C₄H₉)	293 (l)	−5.0 ±0.8	−20.9 ±2.9	SE, NMR; CH₃—C(=O)—CH₃	70
C₈H₁₂	1.3	CH₃ | CH=CH—CH=CH—CH=CH—CH₃ ⇌ CH₂=C—CH=CH—CH=CH—CH₃	550 (g)	4.60 ±0.08	0.79 ±0.17	SE, GLC	175
		CH₃ | CH₂=CH—C=CH—CH=CH—CH₃ ⇌ CH=CH—CH—CH=CH—CH=CH—CH₃	550 (g)	−3.47 ±0.08	−0.92 ±0.17	SE, GLC	175
		CH₃ | CH₂=CH—CH=CH—CH=CH—CH₂ ⇌ CH₂=CH—C=CH—CH=CH—CH₃ CH₃	550 (g)	−4.43 ±0.63	7.95 ±0.84	SE, GLC	175

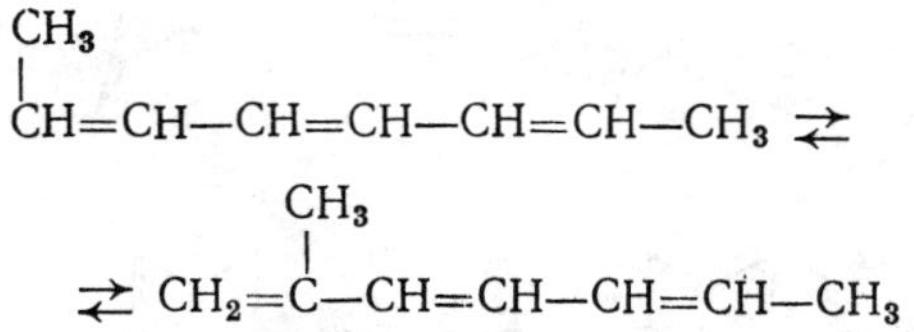

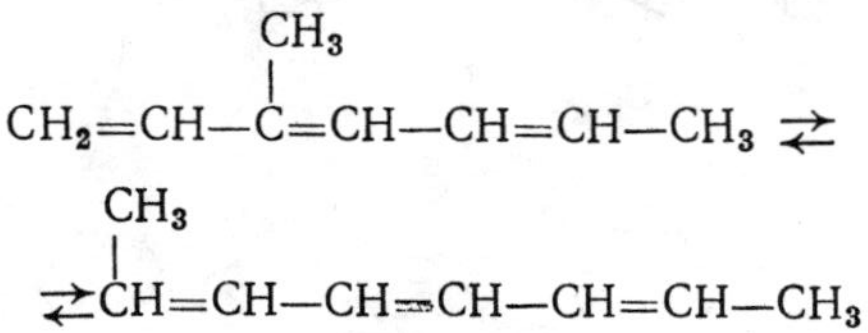

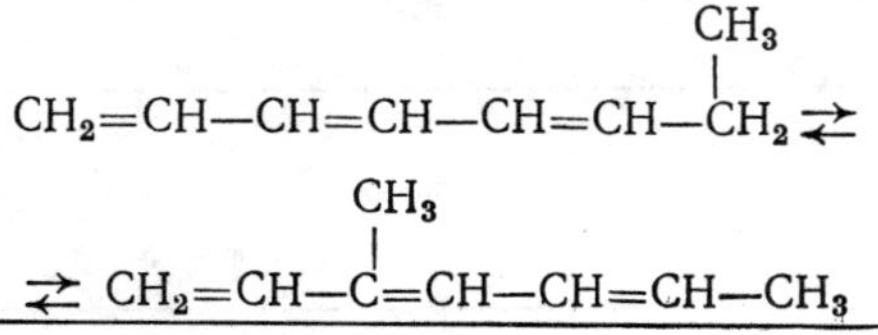

Table (Continued)

Formula	Type	Reaction	T, K (phase)	$\Delta_r H_T^\circ$ ($\Delta_r G_T^\circ$)	$\Delta_r S_T^\circ$	Technique	Reference
C_8H_{12}	1.3		298 (l)	(−4.56 ±0.13)	—	SE, CT, GLC	176
			298 (l)	−11.34 ±1.67	—	CC	177
	1.2		298 (l)	−8.53 ±2.22	—	CC	177
			298 (l)	−19.87 ±1.72	—	CC	146
$C_8H_{12}O_2$	1.3	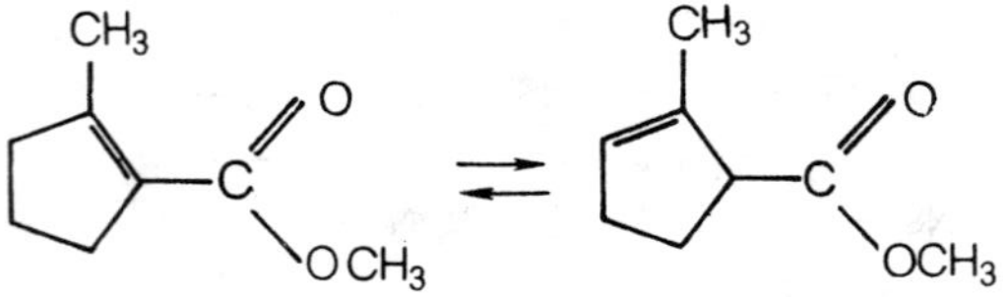	373 (l)	(8.78 ±0.84)	—	SE, CT, GLC CH_3OH	56

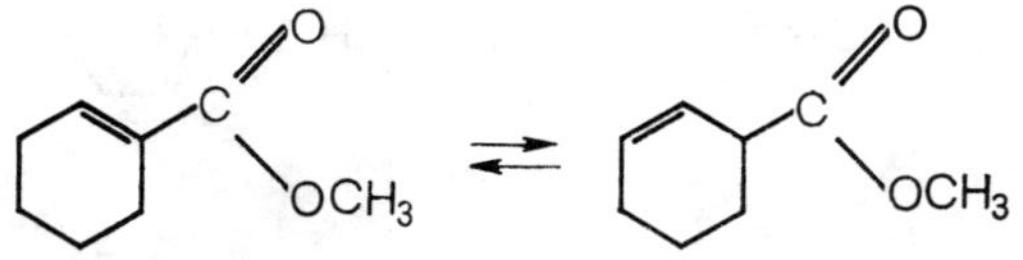

Formula		Equilibrium	T, K (phase)	ΔH (±)		Method, solvent	Ref.
	2.1	cyclohex-2-enyl-COOCH₃ ⇌ cyclohex-1-enyl-COOCH₃	373 (l)	(9.20 ±0.84)	—	SE, CT, GLC CH₃OH	56
C₈H₁₂O₃	1.2 1.3	(cis-cis) 2,6-dimethyl-4-ethynyl-1,3-dioxane ⇌ (trans-trans)	298 (l)	(−0.88 ±0.08)	—	SE, CT, GLC CCl₄	175
	2.1	1-OCH₃-2-COOCH₃-cyclopentene ⇌ 2-OCH₃-1-COOCH₃-cyclopentene	423 (l)	(2.68 ±0.17)	—	SE, GLC	56
C₈H₁₃NO₂	1.3	(cis-) 2-(i-C₃H₇)-5-NC-1,3-dioxane ⇌ (trans-)	303 (l)	(−0.88)	/	CT, GLC (C₂H₅)₂O	153
C₈H₁₄		CH≡C—CH₂—CH₂—C₄H₉ ⇌ CH₃—C≡C—CH₂—C₄H₉	298 (l)	16.94 ±3.43	—	RC, CT C₆H₁₄	112
		CH₃—C≡C—CH₂—C₄H₉ ⇌ CH₃—CH₂—C≡C—C₄H₉	298 (l)	−1.25 ±2.23	—	RC, CT C₆H₁₄	112

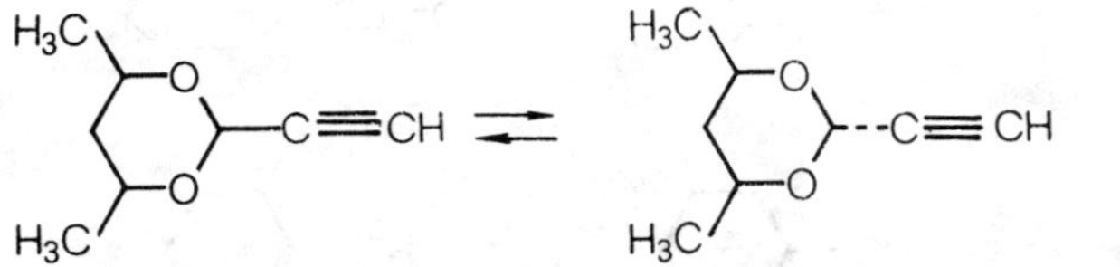

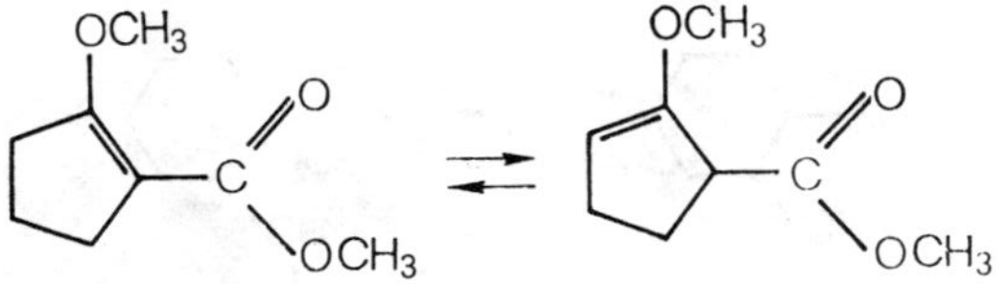

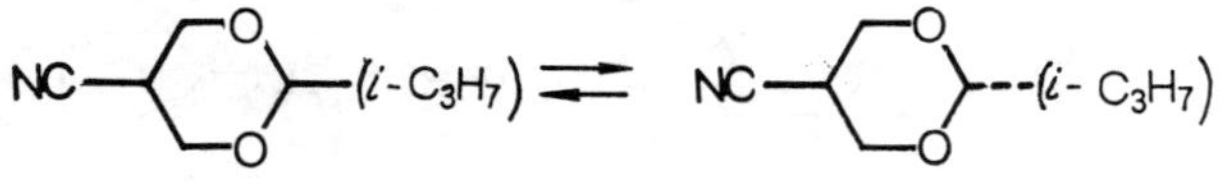

Table (Continued)

Formula	Type	Reaction	T, K (phase)	$\Delta_r H°_T$ ($\Delta_r G°_T$)	$\Delta_r S°_T$	Technique	Reference
C_8H_{14}	1.3	$CH_3-CH_2-C≡C-CH_2-C_3H_7 \rightleftarrows CH_3-CH_2-CH_2-C≡C-C_3H_7$	298 (l)	-2.43 ± 2.65	—	RC, CT C_6H_{14}	112
	1.2	[cyclopentyl-CH$_2$-CH=CH$_2$ ⇌ cyclopentyl-CH=CH$_2$]	298 (l)	-22.80 ± 1.00	—	CC	114
		[phenyl-CH=CH$_2$ type equilibrium]	298 (g)	-23.42 ± 1.34	—	CC	115
		[1,2-dimethylcyclohexene ⇌ 1-methyl-2-methylenecyclohexane]	583 (g)	16.31	2.93	FE, CT, GLC	149
		[2,3-dimethylcyclohexene ⇌ 1-methyl-2-methylenecyclohexane]	583 (g)	7.95	-7.95	FE, CT, GLC	149
		[phenyl-CH=CH-CH$_3$ ⇌ phenyl-CH=CH$_2$]	298 (l)	14.81 ± 1.09	—	CC	114
		[ethylcyclohexene ⇌ ethylidenecyclohexane]	298 (g)	12.55 ± 1.14	—	CC	115
		[ethylcyclohexene ⇌ ethylidenecyclohexane]	298 (l)	3.22 ± 1.20	—	CC	114

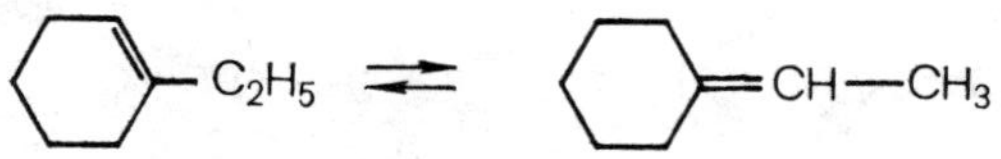	523 (g)	(8.49 ±0.17)	—	FE, CT, GLC	73
1.3	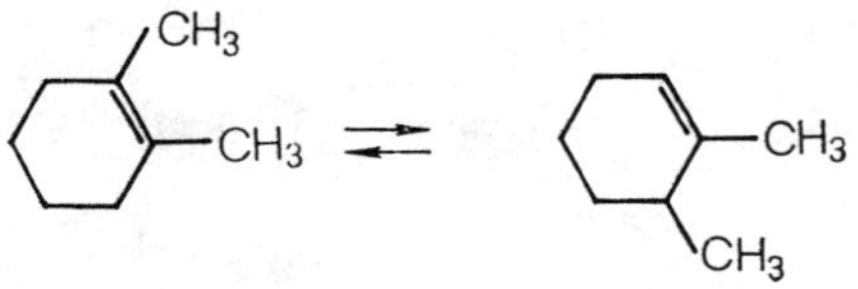 583 (g)	8.37	10.88	FE, CT, GLC	149
	352 (l)	−3.35 ±0.04	−3.35 ±0.04	SE, CT, GLC CH_3-S-CH_3 $\overset{O}{\parallel}$ $t-C_4H_9OH$	178
	523 (g)	(8.78 ±0.17)	—	FE, CT, GLC	73
	523 (g)	(8.74 ±0.17)	—	FE, CT, GLC	73
	411 (g, l)	−2.72	—	SE, FE, CT, GLC	116

113

Table (Continued)

Formula	Type	Reaction	T, K (phase)	$\Delta_r H_T^\circ$ ($\Delta_r G_T^\circ$)	$\Delta_r S_T^\circ$	Technique	Reference
C_8H_{14}	1.3	 (exo-) ⇌ (exo-) [methyl bicyclic]	298 (l)	8.37 ±1.17	—	CC	174
		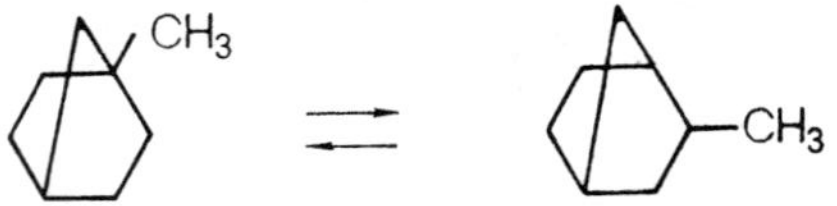 (endo-) ⇌ (endo-)	298 (l)	8.70 ±1.30	—	CC	174
	2.1	 (exo-) ⇌ (endo-)	298 (l)	0.33 ±1.30	−8.12 ±1.67	CC,TP	174
$C_8H_{14}O$	1.3	$CH_2=C\genfrac{}{}{0pt}{}{CH_2-(i-C_3H_7)}{O-CH-CH_2}$ ⇌ $i-C_3H_7\genfrac{}{}{0pt}{}{}{H}C=C\genfrac{}{}{0pt}{}{O-CH=CH_2}{CH_3}$ (cis-)	298 (g)	−0.3 ±0.4	0.1 ±0.7	SE, CT, GLC	74
		$CH_2=C\genfrac{}{}{0pt}{}{CH_2-(i-C_3H_7)}{O-CH-CH_2}$ ⇌ $i-C_3H_7\genfrac{}{}{0pt}{}{}{H}C=C\genfrac{}{}{0pt}{}{CH_3}{O-CH=CH_2}$ (trans-)	298 (g)	−0.1 ±0.4	−0.9 ±0.8	SE, CT, GLC	74

2.1	i—C$_3$H$_7$\C=C/O—CH=CH$_2$ ⇌ i—C$_3$H$_7$\C=C/CH$_3$ H/ \CH$_3$ (cis-) H/ \O—CH=CH$_2$ (trans-)	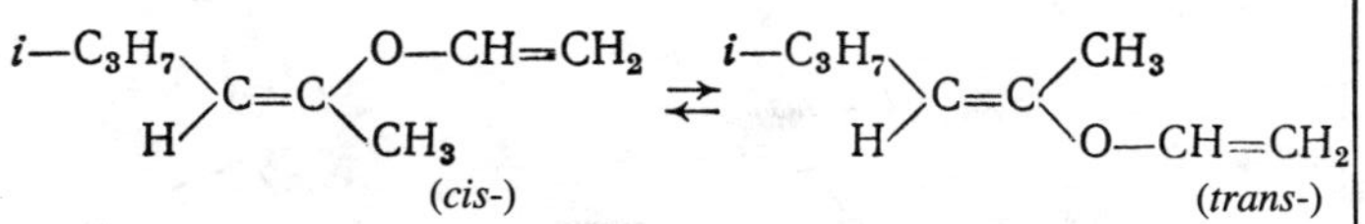 298 (g)	0.4 ±0.4	−1.0 ±0.9	SE, CT, GLC  ⌬	74
	i—C$_3$H$_7$\C=C/O\C=C/CH$_3$ ⇌ i—C$_3$H$_7$\C=C/CH$_3$\C=C/H H/ HH\ \H H/ HH\O \H (cis-,cis-) (trans-,trans-)	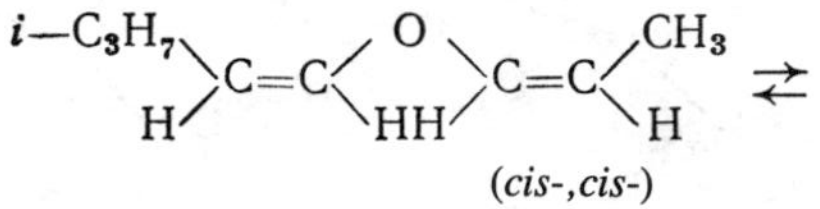 298 (g)	4.8 ±0.4	5.4 ±0.9	SE, CT, GLC	74
	i—C$_3$H$_7$\C=C/O\C=C/H ⇌ i—C$_3$H$_7$\C=C/CH$_3$\C=C/H H/ HH\ \CH$_3$ H/ HH\O \CH$_3$ (cis-,trans-) (trans-,trans-)	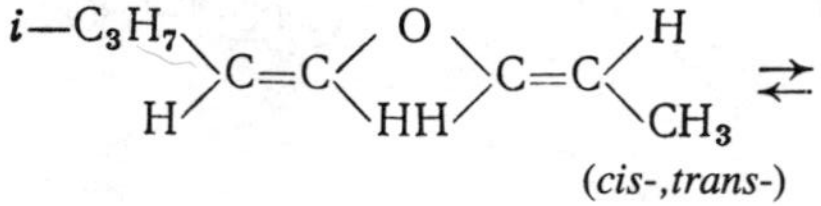 298 (g)	1.7 ±0.3	3.4 ±0.6	SE, CT, GLC	74
	i—C$_3$H$_7$\C=C/CH$_3$\C=C/H ⇌ i—C$_3$H$_7$\C=C/CH$_3$\C=C/H H/ HH\O \CH$_3$ H/ HH\O \CH$_3$ (trans-,cis-) (trans-,trans-)	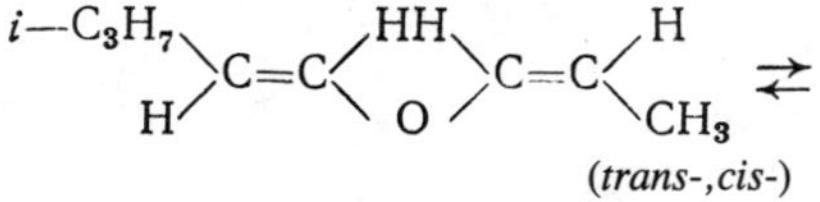 298 (g)	3.1 ±0.3	2.0 ±0.6	SE, CT, GLC	74

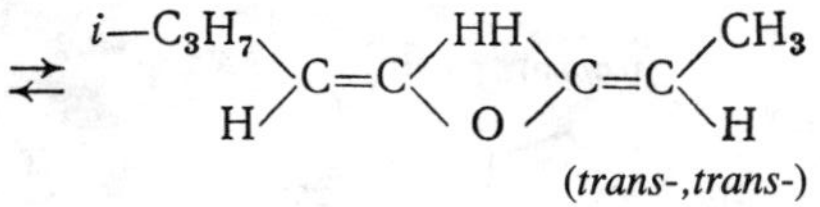

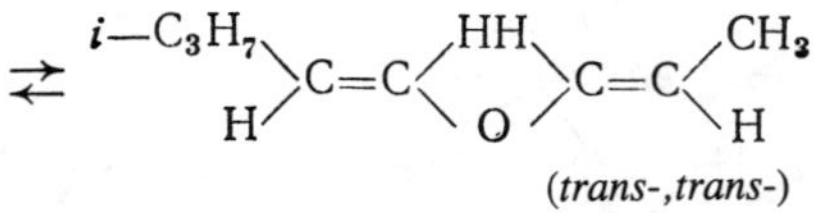

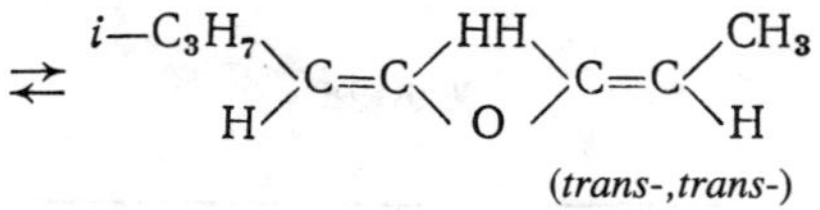

Table (Continued)

Formula	Type	Reaction	T, K (phase)	$\Delta_r H_T^\circ$ ($\Delta_r G_T^\circ$)	$\Delta_r S_T^\circ$	Technique	Reference
$C_8H_{14}O$	1.2	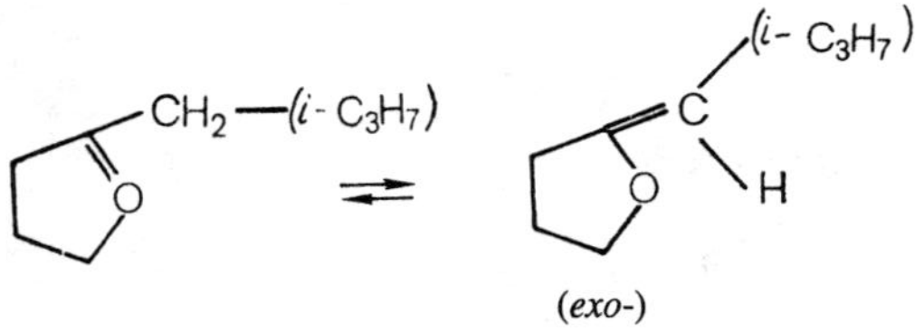 [CH$_2$–(i-C$_3$H$_7$) dihydrofuran] ⇌ [(i-C$_3$H$_7$)/H alkylidene tetrahydrofuran] (exo-)	298 (g)	5.9 ±0.6	−8.9 ±1.6	SE, CT, GLC	119
		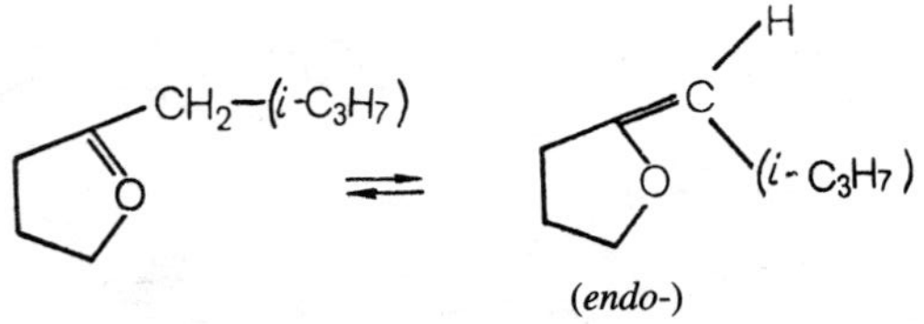 [CH$_2$–(i-C$_3$H$_7$) dihydrofuran] ⇌ [H/(i-C$_3$H$_7$) alkylidene tetrahydrofuran] (endo-)	298 (g)	2.0 ±0.4	1.9 ±0.9	SE, CT, GLC	119
	2.1	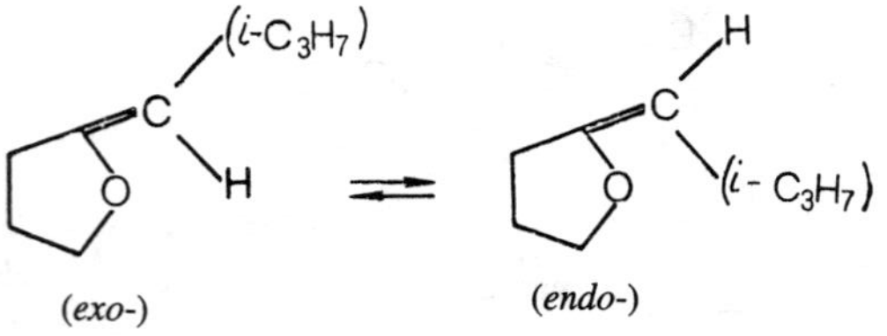 [(i-C$_3$H$_7$)/H alkylidene tetrahydrofuran] (exo-) ⇌ [H/(i-C$_3$H$_7$) alkylidene tetrahydrofuran] (endo-)	298 (g)	−3.9 ±0.4	10.8 ±1.0	SE, CT, GLC	119

1.3	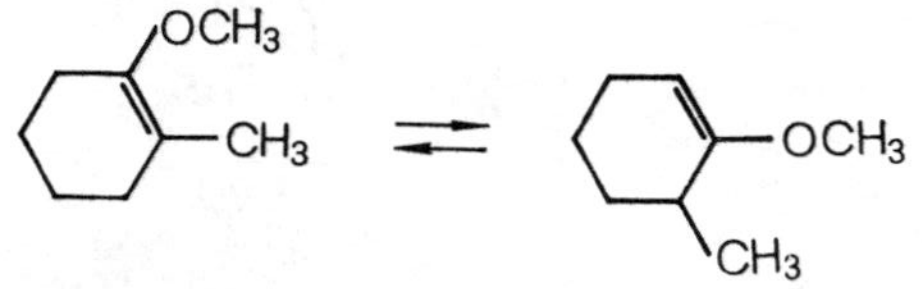	298 (l)	(−1.09 ±0.08)	—	SE, CT, GLC	56
	$C_8H_{14}O_2$ 1.3	298 (g)	−7.72 ±0.41	−14.1 ±1.0	SE, CT, GLC	160
		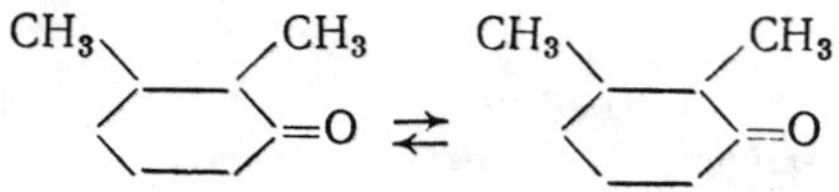 413 (l)	−5.15 ±0.42	−3.26 ±0.63	SE, CT, GLC	179
		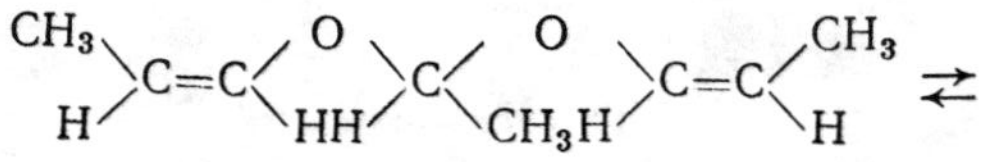 298 (l)	2.0 ±0.3	4.2 ±0.8	SE, CT, GLC	92
		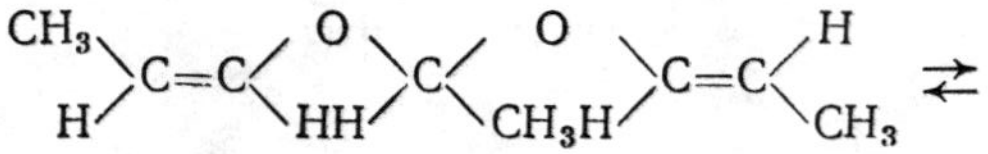 298 (l)	1.9 ±0.2	−7.3 ±0.7	SE, CT, GLC	92

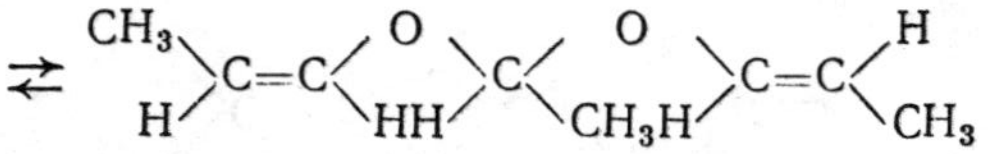

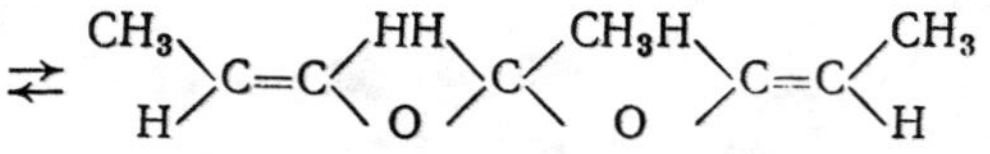

117

Table (Continued)

Formula	Type	Reaction	T, K (phase)	$\Delta_r H°_T$ ($\Delta_r G°_T$)	$\Delta_r S°_T$	Technique	Reference
$C_8H_{14}O_2$	1.3	$CH_3\!\!>\!\!C\!=\!\!C\!\!<\!\!{O \atop H}\!\!>\!\!C\!\!<\!\!{HH \atop CH_3H}\!\!>\!\!C\!=\!C\!\!<\!\!{CH_3 \atop H}$ (cis-,cis-) $\rightleftarrows$ $CH_3\!\!>\!\!C\!=\!C\!\!<\!\!{H \atop H}\!\!>\!\!C\!\!<\!\!{HH \atop CH_3}\!\!>\!\!C\!=\!C\!\!<\!\!{CH_3 \atop H}$ (trans-,trans-)	298 (l)	3.9 ±0.2	−3.1 ±0.7	SE, CT, GLC	92
		$CH_3\!\!>\!\!C\!=\!CH\!-\!CH_2\!-\!C\!\!<\!\!{O \atop OC_2H_5}$ $\rightleftarrows$ $CH_3\!\!>\!\!CH\!-\!CH\!=\!CH\!-\!C\!\!<\!\!{O \atop OC_2H_5}$ (cis- + trans-)	293 (l)	(3.92)	—	SE, CT, GLC $t-C_4H_9OH$	180
	2.1	[lactone structure, cis- ⇌ trans-]	298 (l)	(0.71)	—	SE, CT, GLC $t-C_4H_9OH$	121
	1.3	$CH_2\!=\!C\!-\!(CH_2)_2\!-\!C\!=\!CH_2$ with CH_3O and OCH_3 $\rightleftarrows$	298 (l)	−17.5 ±1.6	−6.0 ±4.4	SE, CT, GLC	152

118

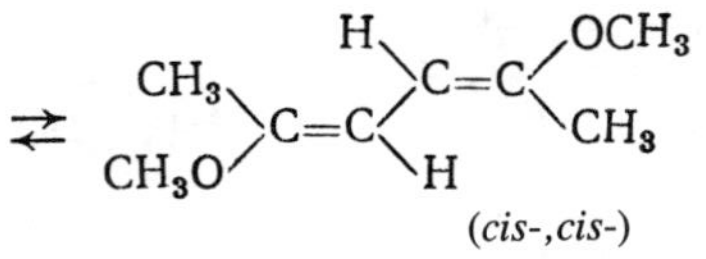

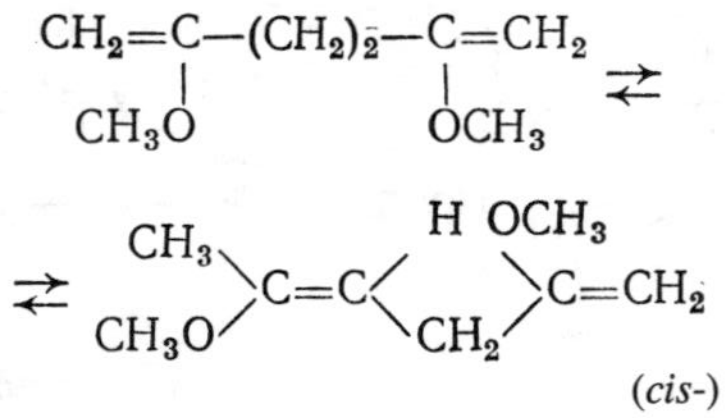

![structure](cis,cis CH3O/CH3 dimethoxy diene)	298 (l)	9.5 ±0.3	10.7 ±0.8	SE, CT, GLC	152
CH$_2$=C−(CH$_2$)$_2$−C=CH$_2$ with CH$_3$O groups (cis-)	298 (l)	−0.1 ±1.7	2.7 ±4.3	SE, CT, GLC	152
CH$_2$=C−(CH$_2$)$_2$−C=CH$_2$ with CH$_3$O groups (trans-)	298 (l)	−11.4 ±0.2	−12.1 ±0.7	SE, CT, GLC	81
1,2-dimethoxycyclohexene equilibrium				SE, CT, GLC	
Cl−CH$_2$−C(OC$_2$H$_5$)(CHO)=CH$_2$ ⇌ ... (cis-) C$_8$H$_{15}$ClO 1.3	298 (l)	5.1 ±0.6	2.7 ±1.7	SE, CT, GLC	35

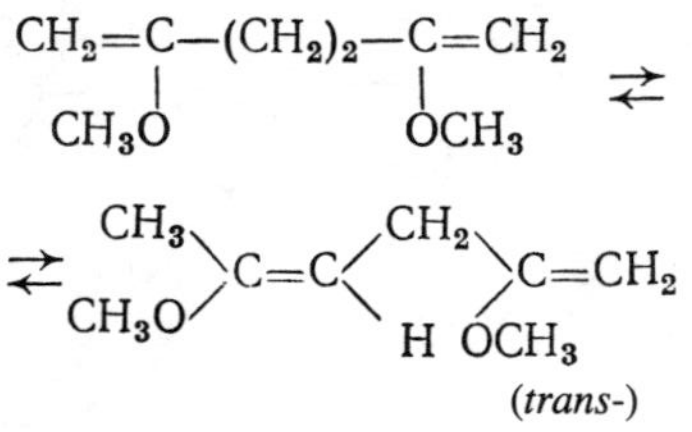

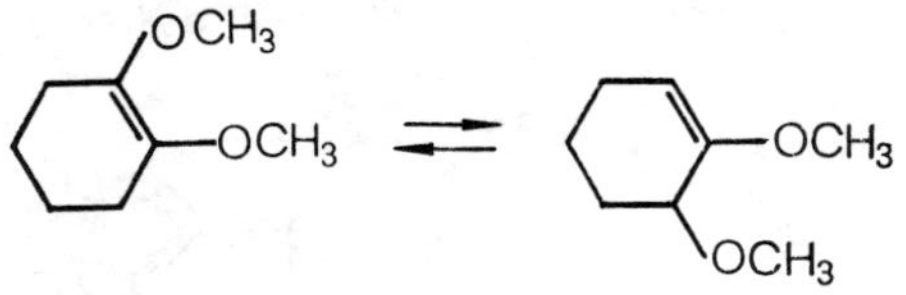

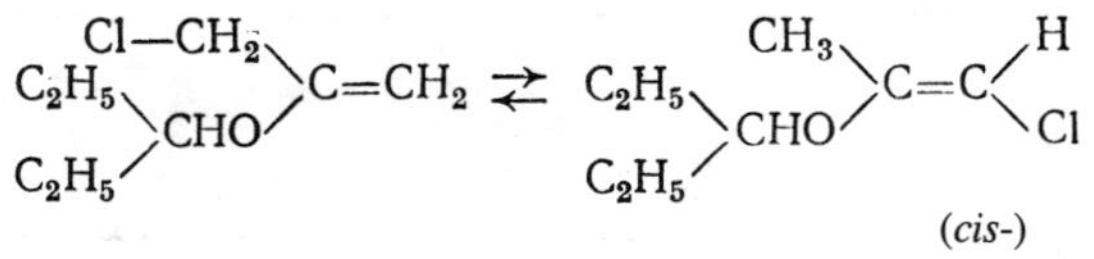

Table (Continued)

Formula	Type	Reaction	T, K (phase)	$\Delta_r H_T^\circ$ ($\Delta_r G_T^\circ$)	$\Delta_r S_T^\circ$	Technique	Reference
$C_8H_{15}ClO$	1,3	$\underset{C_2H_5}{\overset{Cl-CH_2}{\diagdown}}C=CH_2 \rightleftarrows \underset{C_2H_5}{\overset{C_2H_5}{\diagdown}}C=C\underset{H}{\overset{Cl}{\diagdown}}$ (cis-)	298 (l)	-0.7 ± 0.2	-3.0 ± 0.5	SE, CT, GLC ⌬	35
	2,1	$\underset{C_2H_5}{\overset{CH_3}{\diagdown}}C=C\underset{Cl}{\overset{H}{\diagdown}} \rightleftarrows \underset{C_2H_5}{\overset{C_2H_5}{\diagdown}}C=C\underset{H}{\overset{Cl}{\diagdown}}$ (cis-) (trans-)	298 (l)	-5.8 ± 0.7	-5.7 ± 2.0	SE, CT, GLC ⌬	35
$C_8H_{15}NO_4$	2,1	(cis-, trans-) ⇌ (trans-, cis-)	298 (l)	(-1.80)	—	SE, CT, NMR CCl_4	35
C_8H_{16}	1,3	$CH_2=CH-CH_2-C_5H_{11}i \rightleftarrows CH_2=C\underset{CH-C_2H_5}{\overset{CH_3\ C_2H_5}{\diagup}}$	298 (g)	-19.16 ± 0.75	—	CC	156
		$CH_2=C\underset{CH-C_2H_5}{\overset{CH_3\ C_2H_5}{\diagup}} \rightleftarrows CH_2=C\underset{C-CH_3}{\overset{CH_3}{\diagup}}\overset{CH_3}{\underset{CH_3}{|}}$	298 (g)	-9.54 ± 0.71	—	CC	156
		$CH_2=C\underset{CH-C_2H_5}{\overset{CH_3\ C_2H_5}{\diagup}} \rightleftarrows CH_3-C=C\underset{C_2H_5}{\overset{CH_3\ C_2H_5}{\diagup}}$	583 (g)	-5.86	-7.11	FE, CT, GLC	149

Structure	T (K)	ΔH	—	Method	Ref
$CH_2=C(CH_3)-CH_2-C(CH_3)_2-CH_3 \rightleftharpoons CH_3-C(CH_3)=CH-C(CH_3)_2-CH_3$	298 (g)	4.94 ±1.23	—	CC	156
$CH_2=C(CH_3)-CH_2-C(CH_3)_2-CH_3 \rightleftharpoons CH_3-C(CH_3)=CH-C(CH_3)_2-CH_3$	583 (g)	3.97	0.90	FE, CT, GLC	149
$CH_2=CH-CH_2-C_5H_{11} \rightleftharpoons \underset{H}{\overset{C_2H_5}{>}}C=C\underset{H}{\overset{(t-C_4H_9)}{<}}$ (cis-)	298 (g)	−8.16 ±2.38	—	CC	156
$\underset{CH_3}{\overset{C_2H_5}{>}}C=C\underset{C_2H_5}{\overset{(i-C_3H_7)}{<}} \rightleftharpoons \underset{CH_3}{\overset{C_2H_5}{>}}C=C\underset{H}{\overset{(i-C_3H_7)}{<}}$ (cis-)	583 (g)	3.55	5.44	FE, CT, GLC	149
$\underset{CH_3}{\overset{C_2H_5}{>}}C=C\underset{C_2H_5}{\overset{(i-C_3H_7)}{<}} \rightleftharpoons \underset{CH_3}{\overset{C_2H_5}{>}}C=C\underset{H}{\overset{(i-C_3H_7)}{<}}$ (trans-)	583 (g)	5.23	1.67	FE, CT, GLC	149
$\underset{H}{\overset{CH_3}{>}}C=C\underset{C_2H_5}{\overset{(i-C_3H_7)}{<}} \rightleftharpoons \underset{H}{\overset{CH_3}{>}}C=C\underset{H}{\overset{(i-C_3H_7)}{<}}$ (cis- ⇌ trans-)	583 (g)	1.68	−3.77	FE, CT, GLC	149
$\underset{H}{\overset{C_2H_5}{>}}C=C\underset{H}{\overset{(t-C_4H_9)}{<}} \rightleftharpoons \underset{H}{\overset{C_2H_5}{>}}C=C\underset{H}{\overset{(t-C_4H_9)}{<}}$ (cis- ⇌ trans-)	298 (g)	−18.36 ±1.80	—	CC	156

2.1

Table (*Continued*)

Formula	Type	Reaction	T, K (phase)	$\Delta_r H_T^\circ$ ($\Delta_r G_T^\circ$)	$\Delta_r S_T^\circ$	Technique	Reference
C_8H_{16}	1.3	cyclopentyl-CH₂-CH₂-CH₃ ⇌ 1-methyl-1-ethylcyclopentane	298 (l)	−5.02 ±1.28	—	CC	181
	2.1	1-methyl-1-ethylcyclopentane ⇌ 1-methyl-2-ethylcyclopentane (*cis*-)	298 (l)	2.93 ±1.34	—	CC	181
		1-methyl-2-ethylcyclopentane (*cis*-) ⇌ 1-methyl-3-ethylcyclopentane (*cis*-)	298 (l)	−3.56 ±1.34	—	CC	181
		1-methyl-2-ethylcyclopentane (*cis*-) ⇌ 1-methyl-2-ethylcyclopentane (*trans*-)	298 (l)	−4.27 ±1.34	—	CC	181
		1-methyl-3-ethylcyclopentane (*cis*-) ⇌ 1-methyl-3-ethylcyclopentane (*trans*-)	298 (l)	−1.59 ±1.28	—	CC	181

$C_8H_{16}O$	1.3	$t-C_4H_9-CH_2\diagdown_{CH_3O}C=CH_2 \rightleftarrows \diagup^{CH_3}_{CH_3O}C=C\diagdown^{(t-C_4H_9)}_{H}$ (cis-)	298 (g)	14.4 ±0.4	2.4 ±0.8	SE, CT, GLC	183
		$t-C_4H_9-CH_2\diagdown_{CH_3O}C=CH_2 \rightleftarrows \diagup^{CH_3}_{CH_3O}C=C\diagdown^{H}_{(t-C_4H_9)}$ (trans-)	298 (g)	16.4 ±0.5	9.5 ±1.2	SE, CT, GLC	183
		$C_3H_7\diagdown_{CH_3O}C=C\diagup^{CH_3}_{CH_3} \rightleftarrows \diagup^{i-C_3H_7}_{CH_3O}C=C\diagdown^{C_2H_5}_{H}$ (cis-)	298 (g)	−7.1 ±0.5	−19.6 ±1.3	SE, CT, GLC	183
		$C_3H_7\diagdown_{CH_3O}C=C\diagup^{CH_3}_{CH_3} \rightleftarrows \diagup^{i-C_3H_7}_{CH_3O}C=C\diagdown^{H}_{C_2H_5}$ (trans-)	298 (g)	4.2 ±1,3	−2.2 ±3.4	SE, CT, GLC	183

Additional entries (cyclohexane isomerizations):

Structure	T (K)	ΔH	ΔS	Methods	Ref
cis-1,2-dimethylcyclohexane ⇌ trans-1,2-dimethylcyclohexane	500 (g)	(−7.53)	—	FE, CT, GLC	159
cis-1,2-dimethylcyclohexane ⇌ trans-1,2-dimethylcyclohexane	536 (g)	−7.11	−2.50	FE, CT, GLC	182

Table (*Continued*)

Formula	Type	Reaction	T, K (phase)	$\Delta_r H_T^\circ$ ($\Delta_r G_T^\circ$)	$\Delta_r S_T^\circ$	Technique	Reference
$C_8H_{16}O$	2.1	$i\text{-}C_3H_7\text{-}(CH_3O)C{=}C(C_2H_5)H$ (*cis-*) ⇌ $i\text{-}C_3H_7\text{-}(CH_3O)C{=}C(H)C_2H_5$ (*trans-*)	298 (g)	11.3 ±1.3	17.4 ±3.6	SE, CT, GLC	183
		$CH_3\text{-}(CH_3O)C{=}C(t\text{-}C_4H_9)H$ (*cis-*) ⇌ $CH_3\text{-}(CH_3O)C{=}C(H)(t\text{-}C_4H_9)$ (*trans-*)	298 (g)	1.9 ±0.6	6.9 ±1.0	SE, CT, GLC	183
		$t\text{-}C_4H_9\text{-}(CH_3O)C{=}C(CH_3)H$ (*cis-*) ⇌ $t\text{-}C_4H_9\text{-}(CH_3O)C{=}C(H)CH_3$ (*trans-*)	298 (g)	−7.1 ±0.4	5.5 ±0.9	SE, CT, GLC	183
		cyclobutanol with $t\text{-}C_4H_9$ (*cis-*) ⇌ (*trans-*)	400 (l)	6.69 ±0.84	4.60 ±2.09	SE, CT, NMR $i\text{-}C_3H_7OH$	184
		cyclohexanol with C_2H_5 (*cis-*) ⇌ (*trans-*)	353 (l)	(−3.55)	—	SE, CT, GLC $i\text{-}C_3H_7OH$	162

Formula		Compound (equilibrium)	T, K (phase)	ΔH (kJ/mol)	ΔS (J/mol·K)	Method, solvent	Ref.
		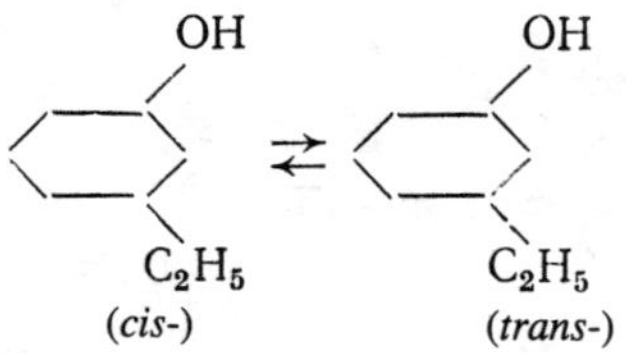 cis-3-ethylphenol ⇌ trans-3-ethylphenol (OH, C$_2$H$_5$)	353 (l)	(3.72)	—	SE, CT, GLC i—C$_3$H$_7$OH	162
		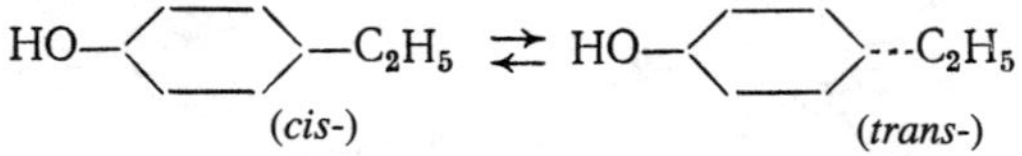 cis-4-ethylphenol ⇌ trans-4-ethylphenol	353 (l)	(−3.14)	—	SE, CT, GLC i—C$_3$H$_7$OH	162
		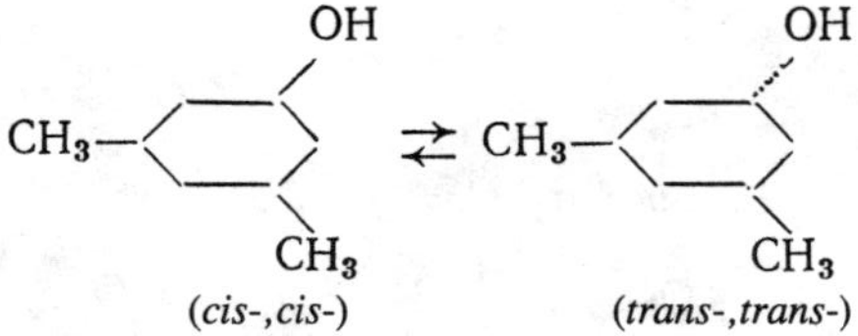 cis,cis-3,5-dimethylcyclohexanol ⇌ trans,trans-	353 (l)	(3.72)	—	SE, CT, GLC i—C$_3$H$_7$OH	185
C$_8$H$_{16}$O$_2$	1.3	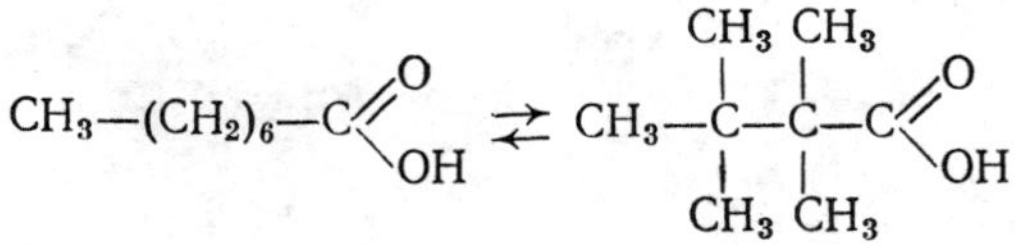 CH$_3$—(CH$_2$)$_6$—COOH ⇌ (CH$_3$)$_3$C—C(CH$_3$)$_2$—COOH	298 (l)	−11.55	—	CC	58
	2.1	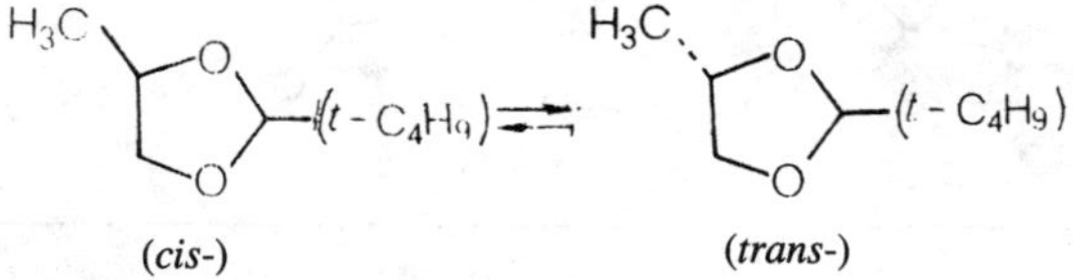 cis-2-(t-C$_4$H$_9$)-4-methyl-1,3-dioxolane ⇌ trans-	341 (l)	1.06	−0.42	SE, CT, NMR CCl$_4$	84
			298 (l)	(1.15 ±0.05)	—	SE, CT, GLC (C$_2$H$_5$)$_2$O	93

Table (Continued)

Formula	Type	Reaction	T, K (phase)	$\Delta_r H_T^\circ$ ($\Delta_r G_T^\circ$)	$\Delta_r S_T^\circ$	Technique	Reference
$C_8H_{16}O_2$	2.1	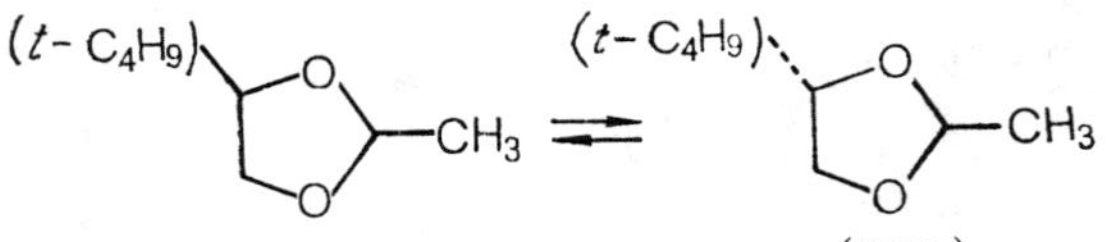 (cis-, 2-t-C$_4$H$_9$, 4-CH$_3$-1,3-dioxolane) ⇌ (trans-)	298 (l)	(2.05 ±0.07)	—	SE, CT, GLC $(C_2H_5)_2O$	93
		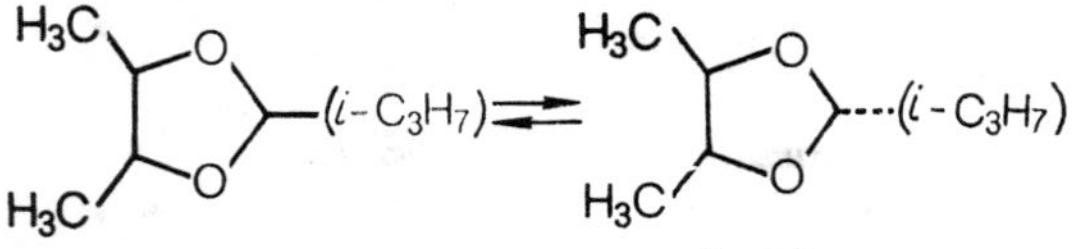 (syn-, 2-i-C$_3$H$_7$, 4,5-(CH$_3$)$_2$-1,3-dioxolane) ⇌ (anti-)	298 (l)	(2.49 ±0.11)	—	SE, CT, GLC $(C_2H_5)_2O$	93
		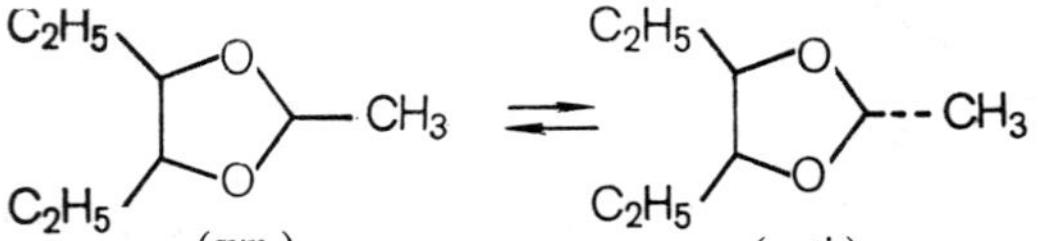 (syn-, 2-CH$_3$, 4,5-(C$_2$H$_5$)$_2$-1,3-dioxolane) ⇌ (anti-)	298 (l)	(2.96 ±0.13)	—	SE, CT, GLC $(C_2H_5)_2O$	93
		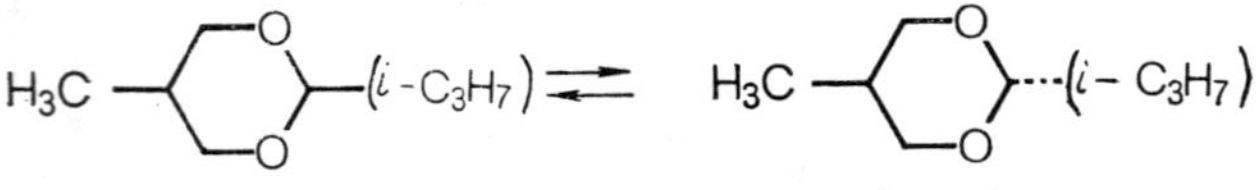 (cis-, 2-i-C$_3$H$_7$, 5-CH$_3$-1,3-dioxane) ⇌ (trans-)	298 (l)	(−3.35)	—	SE, CT, GLC $(C_2H_5)_2O$	186

		T (K)		—	Method	Ref.
	![cis i-C3H7/CH3 dioxane] (cis-) ⇌ (trans-)	333 (l)	(−3.94 ±0.04)	—	SE, CT, GLC	133
	 ![cis C2H5 dioxane] (cis-) ⇌ (trans-)	333 (l)	(−4.60)	—	SE, CT, GLC	165
	![2,4,6-trimethyl dioxane] (cis-,cis-) ⇌ (trans-,trans-)	333 (l)	(−3.51 ±0.04)	—	SE, CT, GLC	165
	H₃C—S—(i-C₃H₇) (cis-) ⇌	 298 (l)	(16.73)	—	SE, CT, NMR (C₂H₅)₂O	166
$C_8H_{16}O_2S$ 2.1	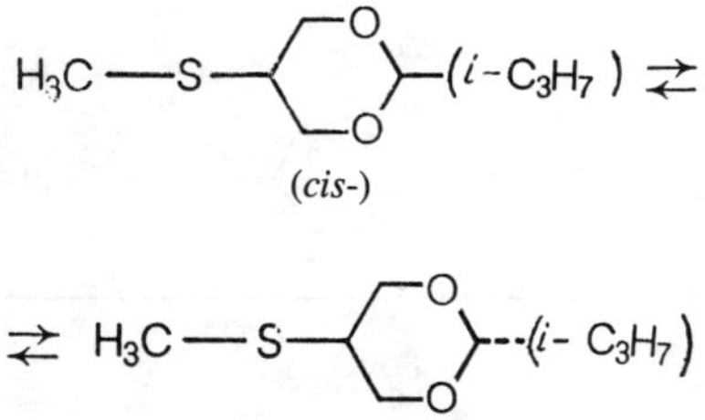 ⇌ H₃C—S—(i-C₃H₇) (trans-)	300 (l)	(−7.28 ±0.08)	—	SE, CT, GLC CCl₄	187

Table (Continued)

Formula	Type	Reaction	T, K (phase)	$\Delta_r H_T^\circ$ ($\Delta_r G_T^\circ$)	$\Delta_r S_T^\circ$	Technique	Reference
$C_8H_{14}O_2S$	2.1	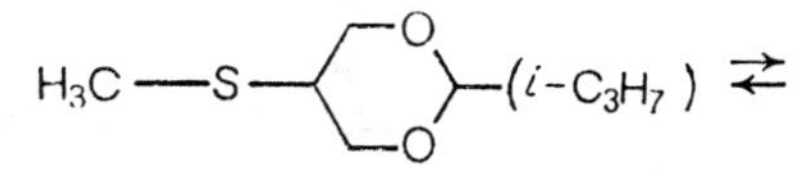 (cis-) 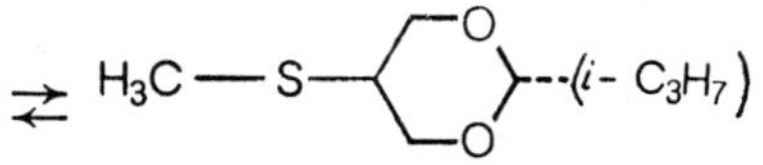 ⇌ (trans-)	298 (l)	(−4.73)	—	SE, CT, GLC CH_3CN	188
$C_8H_{16}O_3$	2.1	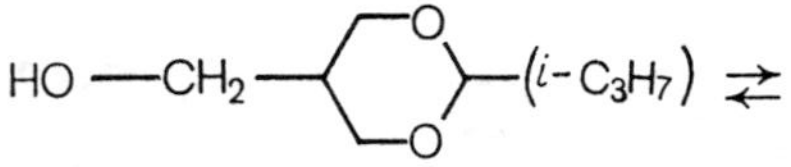 (cis-) 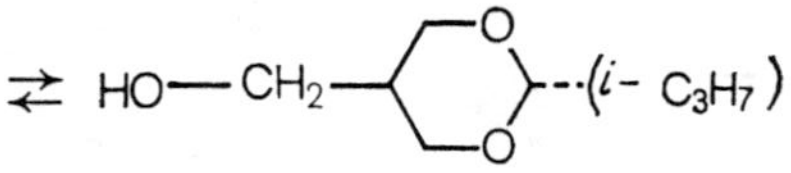 ⇌ (trans-)	300 (l)	(1.13)	—	SE, CT, GLC CCl_4	153
			298 (l)	(−0.13 ±0.17)	—	SE, CT, GLC $(C_2H_5)_2O$	167
		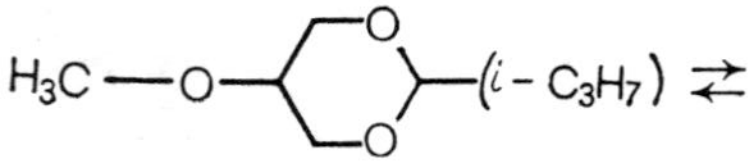 (cis-)	298 (l)	(−3.76)	—	SE, CT, GLC CCl_4	189

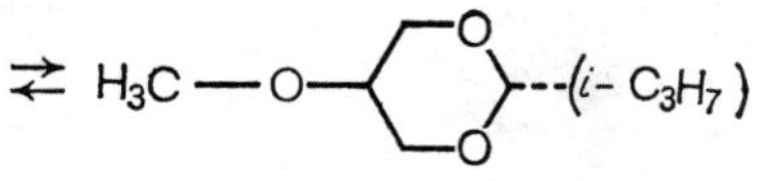

⇌ H₃C—O—[trans-2-isopropyl-5-methoxy-1,3-dioxane] (trans-)	298 (l)	(0.04)	—	SE, CT, GLC CH₃CN	188
	323 (l)	(−3.47)	—	SE, CT, GLC (C₂H₅)₂O	153
 [trans,cis- 2-isopropyl-5-hydroxy-5-methyl-1,3-dioxane] ⇌ [cis,trans- isomer]	298 (l)	(−1.72)		SE, CT, GLC (C₂H₅)₂O	190
 [trans-4,6-dimethyl-2-methoxy-1,3-dioxane] ⇌ [cis- isomer]	298 (l)	(0.21)	—	SE, CT, NMR (C₂H₅)₂O	166

Table (Continued)

Formula	Type	Reaction	T, K (phase)	$\Delta_r H^\circ_T$ ($\Delta_r G^\circ_T$)	$\Delta_r S^\circ_T$	Technique	Reference
$C_8H_{16}O_3S$	2.1	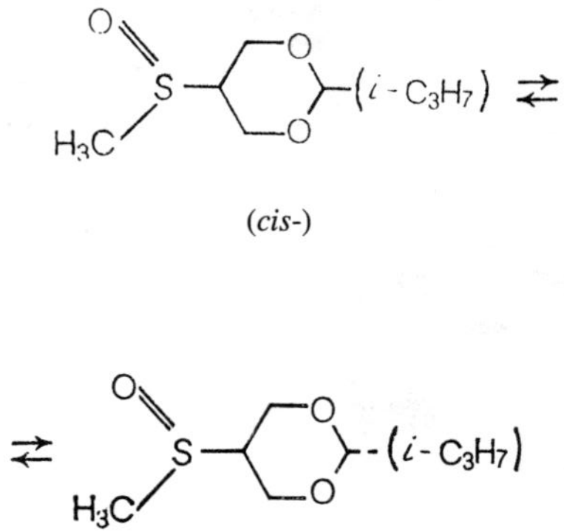 (cis-) ⇌ (trans-) methyl 2-isopropyl-1,3-dioxane-5-sulfoxide	327 (l)	(2.51)	—	SE, CT, NMR CCl_4	187
$C_8H_{16}O_4S$	2.1	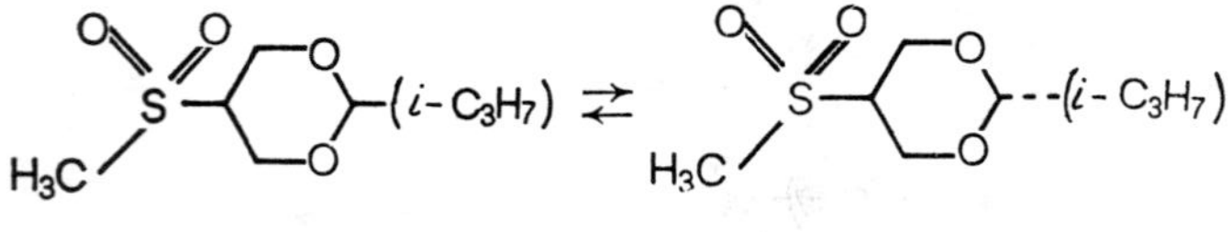 (cis-) ⇌ (trans-) methyl 2-isopropyl-1,3-dioxane-5-sulfone	323 (l)	(4.85 ±0.42)	—	SE, CT, NMR  C_6H_6	187

$C_8H_{16}S_2$	2.1	 (cis-) H₃C–[1,3-dithiane]–(i-C₃H₇) ⇌ (trans-) H₃C–[1,3-dithiane]–(i-C₃H₇)	298 (l)	(6.07 ±0.08)	—	SE, CT, GLC CHCl₃	137
			368 (l)	7.16 ±0.11	4.27 ±0.29	SE, CT, GLC CCl₄	136
		 (cis-,cis-) H₃C–[1,3-dithiane]–C₂H₅ / H₃C ⇌ (trans-,trans-) H₃C–[1,3-dithiane]–C₂H₅ / H₃C	342 (l)	(6.44 ±0.04)	—	SE, CT, GLC CHCl₃	137
$C_8H_{17}Cl$	1.3	Cl–CH₃–CH–CH₂–CH₂–C₅H₁₁ ⇌ CH₃–CH₂–CH–CH₂–CH₂–C₅H₁₁ Cl	390 (g, l)	−0.13 ±0.42	−1.46 ±1.25	SE, FE, CT, GLC	96
		Cl–CH₃–CH–CH₂–CH₂–C₄H₉ ⇌ CH₃–CH₂–CH–CH₂–CH₂–C₄H₉ Cl	390 (g, l)	0.27 ±0.42	−2.93 ±1.25	SE, FE, CT, GLC	96
C_8H_{18}	1.3	CH₃–CH₃–CH–CH₂–CH₂–C₃H₇ ⇌ CH₃–CH₂–CH₂–CH–C₃H₇ / CH₃	368 (g, l)	1.34 ±0.38	−5.31 ±1.25	SE, FE, CT, GLC	139

Table (Continued)

Formula	Type	Reaction	T, K (phase)	$\Delta_r H_T^\circ$ ($\Delta_r G_T^\circ$)	$\Delta_r S_T^\circ$	Technique	Reference
C_8H_{18}	1.3	$CH_3-\overset{\overset{CH_3}{\|}}{CH}-CH_2-C_4H_9 \rightleftarrows CH_3-CH_2-CH_2-\overset{\overset{CH_3}{\|}}{CH}-C_4H_9$	368 (g, l)	1.86 ±0.42	4.14 ±1.25	SE, FE, CT, GLC	139
		$CH_3-CH_2-CH_2-C_5H_{11} \rightleftarrows CH_3-\overset{\overset{CH_3}{\|}}{\underset{\underset{CH_3}{\|}}{C}}-\overset{\overset{CH_3}{\|}}{\underset{\underset{CH_3}{\|}}{C}}-CH_3$	298 (g)	−16.94 ±1.98	—	CC	191
$C_9H_7ClO_3$	1.1 1.2	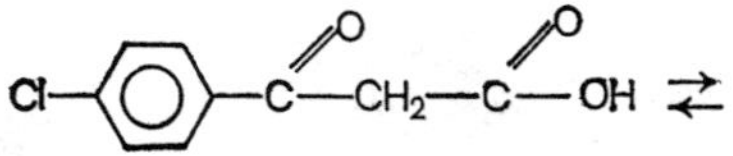 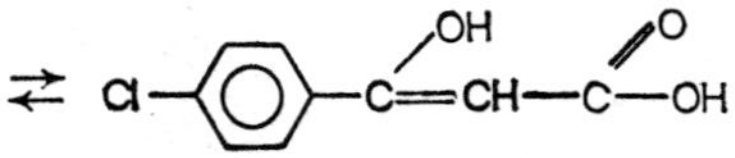 4-Cl-C$_6$H$_4$-CO-CH$_2$-COOH $\rightleftarrows$ 4-Cl-C$_6$H$_4$-C(OH)=CH-COOH	288 (l)	2.9	16.7	SE, NMR	192
$C_9H_7NO_5$	1.1 1.2	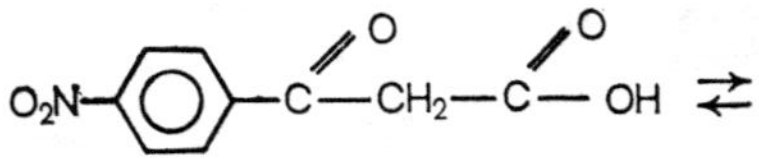 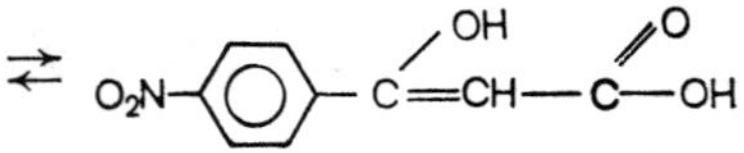 4-O$_2$N-C$_6$H$_4$-CO-CH$_2$-COOH $\rightleftarrows$ 4-O$_2$N-C$_6$H$_4$-C(OH)=CH-COOH	288 (l)	7.5	20.9	SE, NMR	192

Formula		Structure	T (K)	ΔH	ΔS	Method	Ref
$C_9H_8O_3$	1.1 1.2	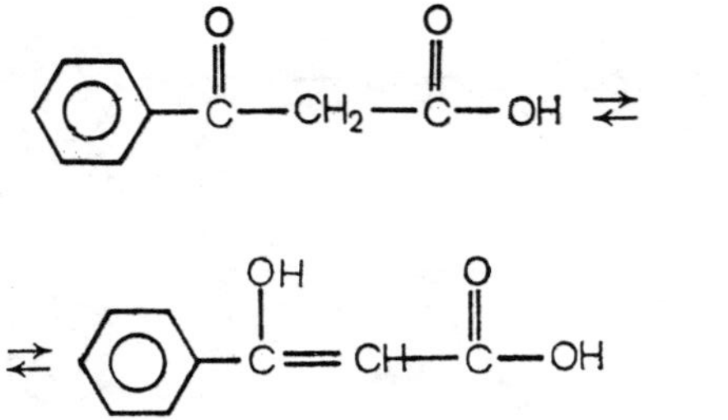 PhCO-CH$_2$-COOH ⇌ (enol form)	278 (l)	2.9	16.7	SE, NMR 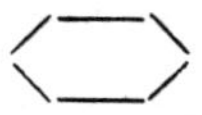	192
$C_9H_{10}O$	2.1	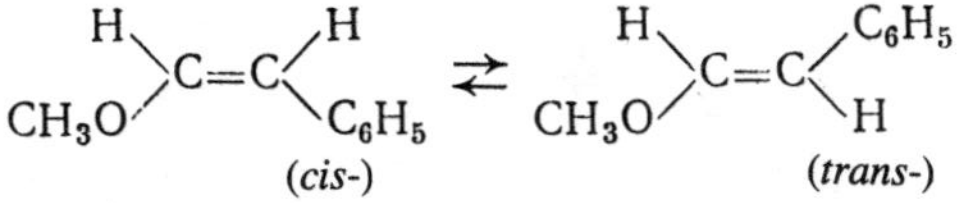 trans/cis methoxystyrene	298 (g)	−1.2 ±0.4	±2.1 ±0.7	SE, CT, GLC	193
$C_9H_{10}O_2$	1.3	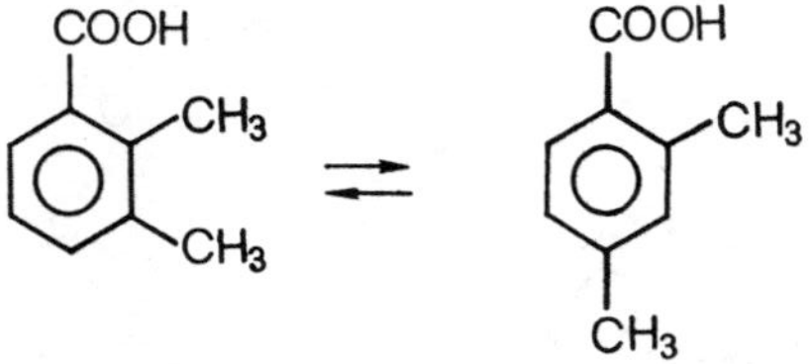 2,4-dimethylbenzoic acid ⇌ 2,3-dimethylbenzoic acid	298 (g)	−9.2 ±2.2	—	CC	259
		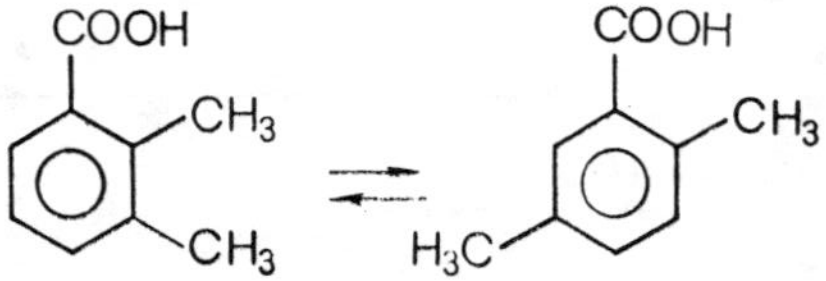 2,5-dimethylbenzoic acid ⇌ 2,3-dimethylbenzoic acid	298 (g)	−5.3 ±2.2	—	CC	259

Table (Continued)

Formula	Type	Reaction	T, K (phase)	$\Delta_r H_T^\circ$ ($\Delta_r G_T^\circ$)	$\Delta_r S_T^\circ$	Technique	Reference
$C_9H_{10}O_2$	1.3	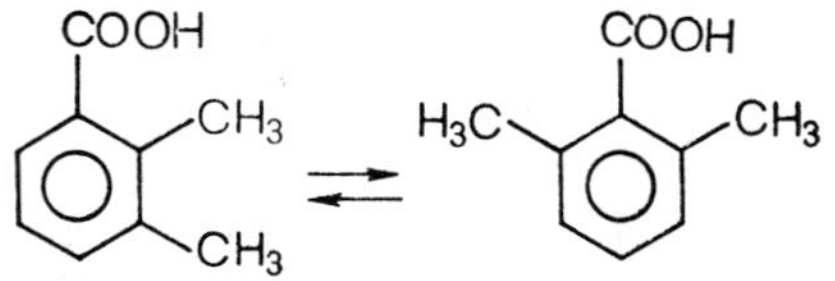 2,3-dimethylbenzoic acid ⇌ 2,6-dimethylbenzoic acid	298 (g)	4.2 ±2.2	—	CC	259
		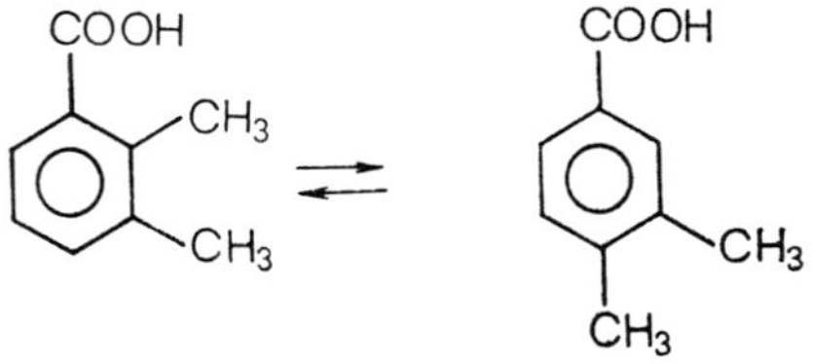 2,3-dimethylbenzoic acid ⇌ 2,4-dimethylbenzoic acid	298 (g)	−16.6 ±2.4	—	CC	259
		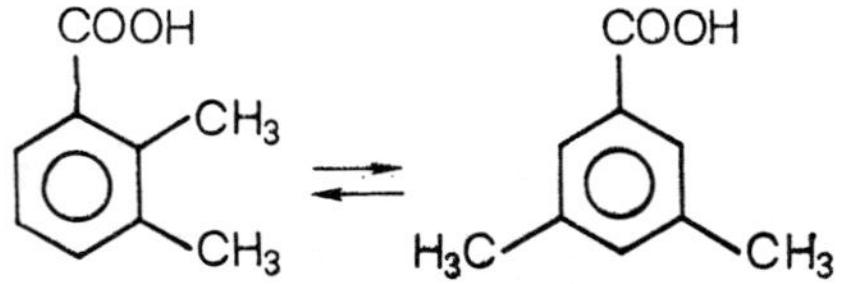 2,3-dimethylbenzoic acid ⇌ 3,5-dimethylbenzoic acid	298 (g)	−18.7 ±2.2	—	CC	259
$C_9H_{10}O_4$	1.3	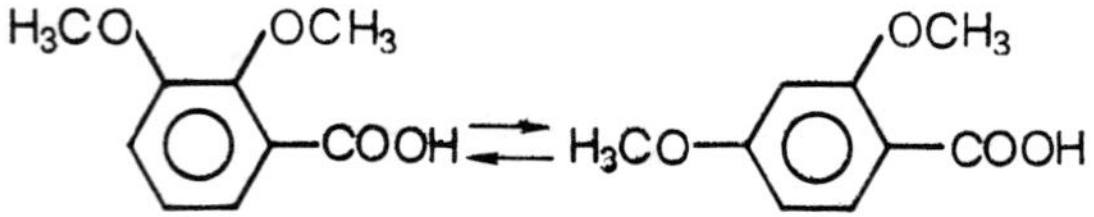 2,3-dimethoxybenzoic acid ⇌ 2,4-dimethoxybenzoic acid	298 (g)	−18.4 ±2.0	—	CC	260

134

Formula	No.	Reaction	T(K) (state)	ΔH (±)		Method	Ref.
		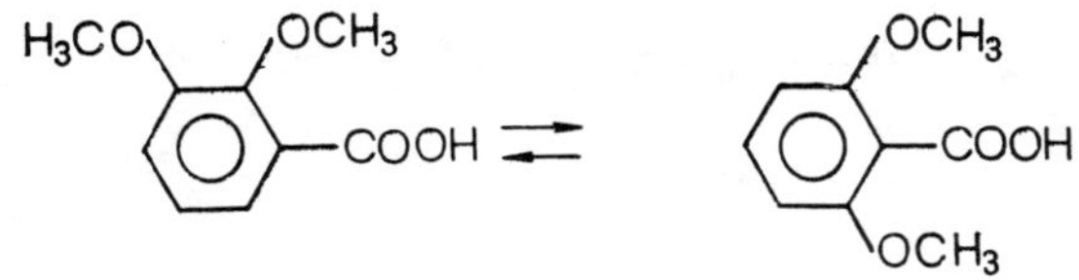 2,3-(OCH₃)₂-C₆H₃-COOH ⇌ 2,6-(OCH₃)₂-C₆H₃-COOH	298 (g)	−1.7 ±2.0	—	CC	260
		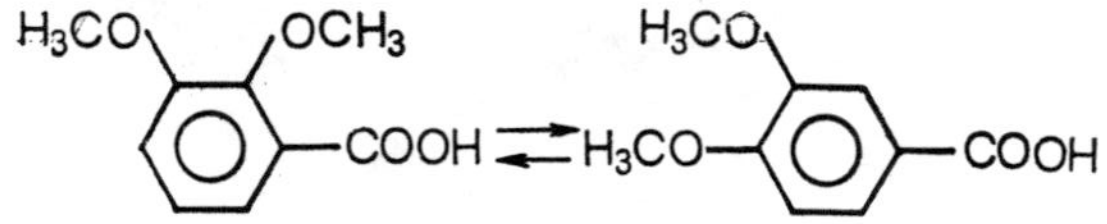 2,3-(OCH₃)₂-C₆H₃-COOH ⇌ 3,4-(OCH₃)₂-C₆H₃-COOH	298 (g)	−13.8 ±2.0	—	CC	260
		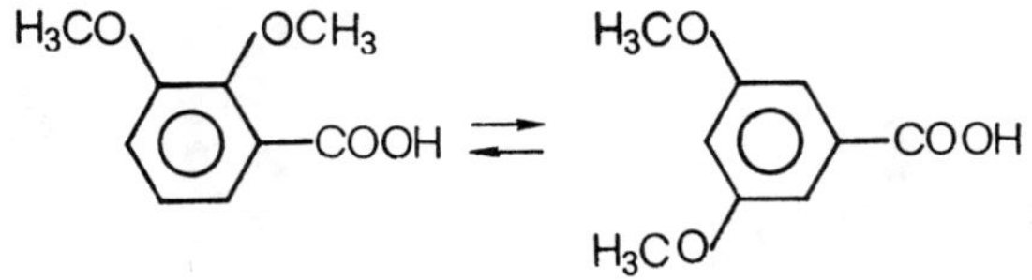 2,3-(OCH₃)₂-C₆H₃-COOH ⇌ 3,5-(OCH₃)₂-C₆H₃-COOH	298 (g)	−26.5 ±2.0	—	CC	260
C₉H₁₁Br	1.3	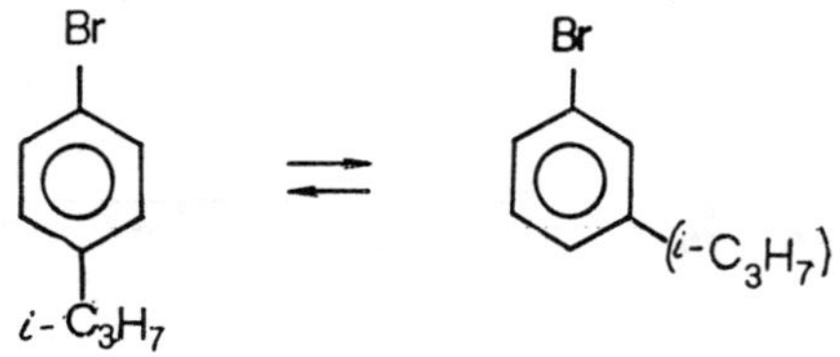 4-Br-C₆H₄-i-C₃H₇ ⇌ 3-Br-C₆H₄-(i-C₃H₇)	325 (l)	−1.63 ±0.30	4.00 ±0.90	SE, CT, GLC H₂O	194

Table (*Continued*)

Formula	Type	Reaction	T, K (phase)	$\Delta_r H_T^\circ$ ($\Delta_r G_T^\circ$)	$\Delta_r S_T^\circ$	Technique	Reference
$C_9H_{11}F$	1.3	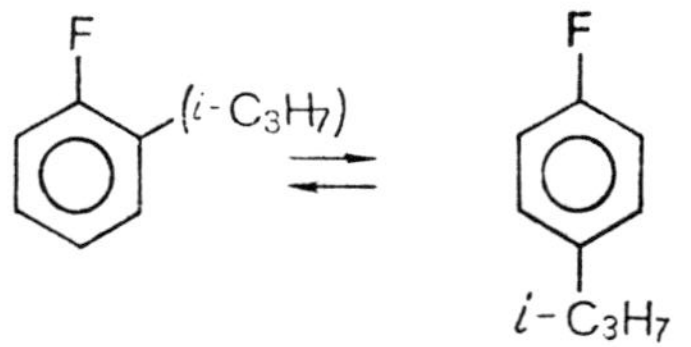 F–C₆H₄–(i-C₃H₇) ⇌ F–C₆H₄–(i-C₃H₇) (para)	330 (l)	0.04 ±0.18	−2.86 ±0.56	SE, CT, GLC	195
	1.3	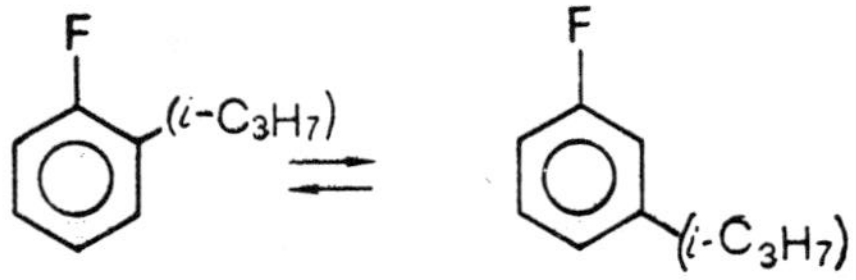 F–C₆H₄–(i-C₃H₇) ⇌ F–C₆H₄–(i-C₃H₇) (meta)	330 (l)	−1.79 ±0.23	2.76 ±0.72	SE, CT, GLC	195
$C_9H_{11}NO$	1.1 1.2	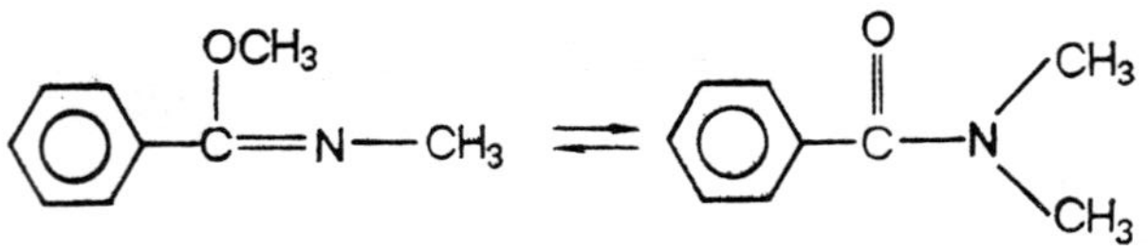 C₆H₅–C(OCH₃)=N–CH₃ ⇌ C₆H₅–C(=O)–N(CH₃)₂	298 (g)	−69.44 ±13.39	—	RC	63
		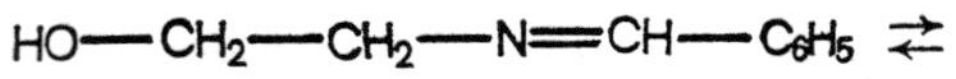 HO–CH₂–CH₂–N=CH–C₆H₅ ⇌	380 (g)	7.1		SE, MS	196

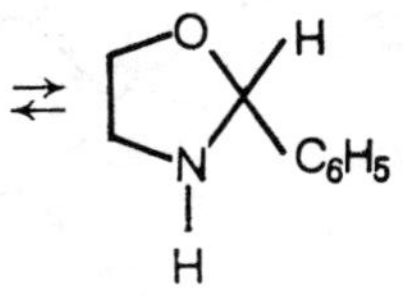

Formula		Structure	T (K)	ΔH	ΔS	Method	Ref
$C_9H_{11}NS$	1.1 1.2	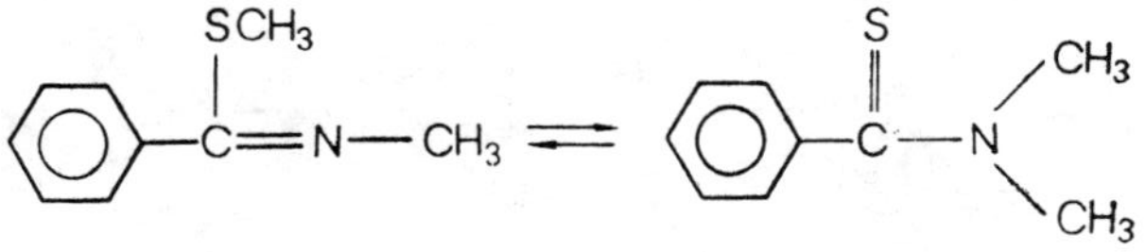 (oxazolidine ⇌ thioamide structures)	298 (g)	−11.29 ±10.04	—	RC	63
C_9H_{12}	1.3	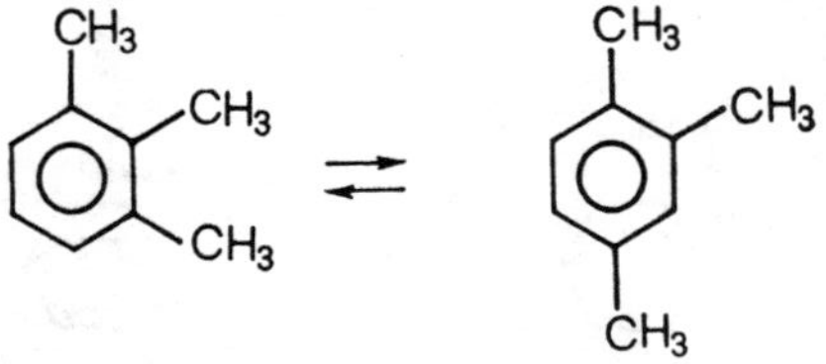 (1,2,4-trimethylbenzene ⇌ 1,2,3-trimethylbenzene)	298 (g)	−4.39 ±1.52	10.15 ±1.20	CC, ST	258
		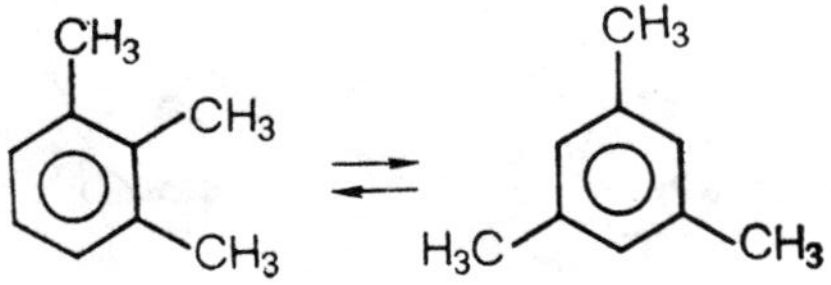 (1,3,5-trimethylbenzene ⇌ 1,2,3-trimethylbenzene)	298 (g)	−6.48 ±1.70	−0.49 ±1.00	CC, ST	258
C_9H_{14}	1.2	 (bicyclic diene isomers)	298 (g)	52.7 ±1.8	—	RC	197

Table (*Continued*)

Formula	Type	Reaction	T, K (phase)	$\Delta_r H_T^\circ$ ($\Delta_r G_T^\circ$)	$\Delta_r S_T^\circ$	Technique	Reference
C_9H_{14}	1.2	(bicyclic diene ⇌ bicyclic diene)	298 (g)	45.6 ±2.5	—	RC	197
	1.2, 1.3	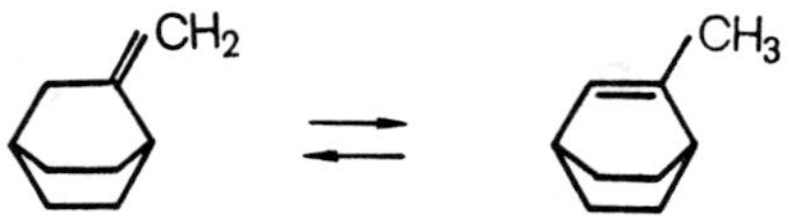 (methylbicyclic ⇌ methylenebicyclic)	298 (l)	−7.53 ±2.93	—	RC	198
$C_9H_{14}BrNO_4S$	1.1	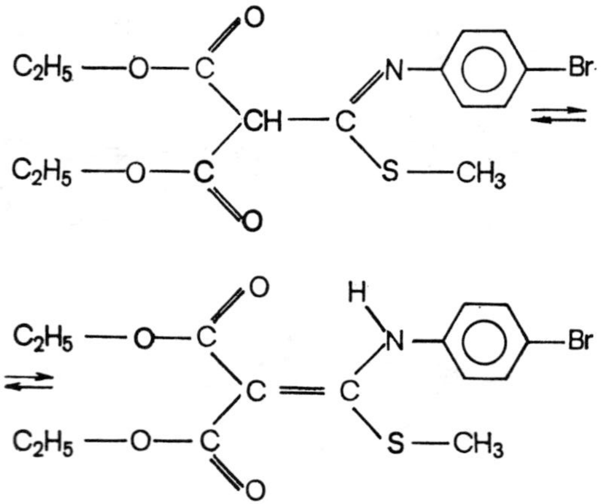 (tautomer A ⇌ tautomer B)	326 (l)	−15.02	−39.11	SE, NMR $CDCl_3$	199
			326 (l)	−10.88	−28.65	SE, GLC $C_6D_5NO_2$	199

$C_9H_{14}O_2$	1.3	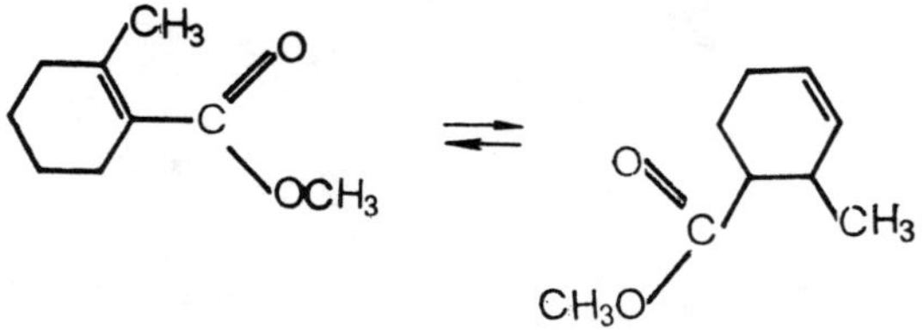 (structures shown below)	373 (l)	3.55 ±0.17	—	SE, CT, GLC CH$_3$OH	56
$C_9H_{14}O_3$	1.3	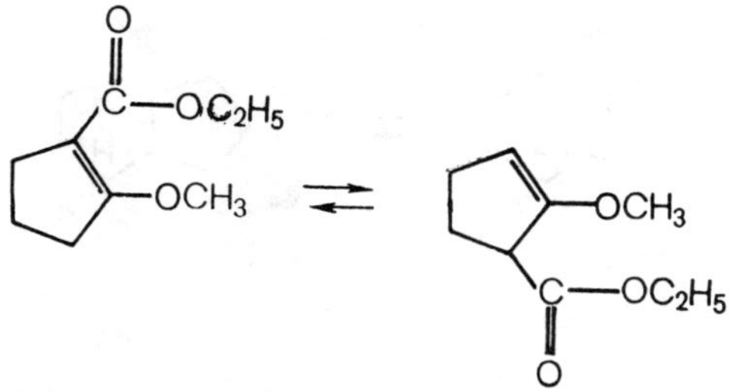 (structures shown below)	298 (g)	3.5 ±0.2	4.1 ±0.5	SE, CT, NMC ⌬	108
		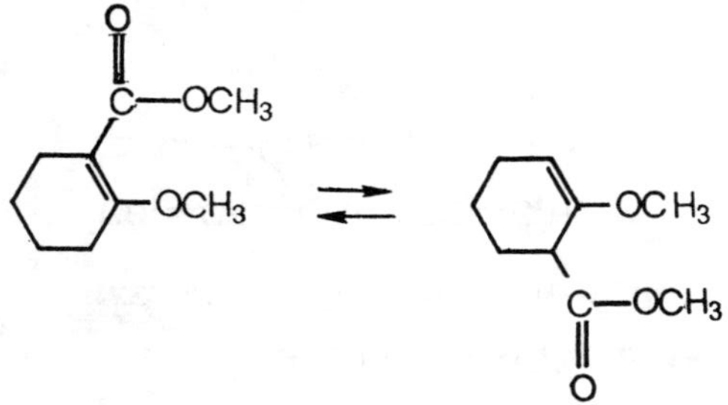 (structures shown below)	423 (l)	(−5.65 ±0.21)	—	SE, GLC	56
C_9H_{16}	1.3	$CH\equiv C-CH_2-CH_2-CH_2-C_4H_9 \rightleftarrows$ $\rightleftarrows CH_3-C\equiv C-CH_2-CH_2-C_4H_9$	298 (l)	−18.66 ±3.99	—	RC, CT ⌬	112

Table (Continued)

Formula	Type	Reaction	T, K (phase)	$\Delta_r H_T^\circ$ ($\Delta_r G_T^\circ$)	$\Delta_r S_T^\circ$	Technique	Reference
C_9H_{16}	1.3	$CH_3-C\equiv C-CH_2-CH_2-C_4H_9 \rightleftarrows$ $\rightleftarrows CH_3-CH_2-C\equiv C-CH_2-C_4H_9$	298 (l)	−1.63 ±3.65	—	RC, CT	112
		$CH_3-CH_2-C\equiv C-CH_2-C_4H_9 \rightleftarrows$ $\rightleftarrows CH_3-CH_2-CH_2-C\equiv C-C_4H_9$	298 (l)	0.00 ±3.54	—	RC, CT	112
	2.1	[cyclohexene-(i-C_3H_7)] $\rightleftarrows$ [cyclohexene-(i-C_3H_7)]	355 (l)	2.93 ±0.42	7.95 ±1.25	SE, CT, GLC $CH_3-\overset{O}{\underset{\parallel}{S}}-CH_3$ $t-C_4H_9OH$	178
		(cis-) bicyclic $\rightleftarrows$ (trans-) bicyclic	298 (l)	−3.01 ±2.18	—	CC	200
			298 (g)	−4.35 ±2.22	—	CC	200
			552 (l)	−4.48 ±0.38	−9.62 ±0.42	SE, CT, GLC	201
			286 (l)	−2.43 ±0.21	−4.18 ±0.25	SE, CT, GLC	202

Formula		Compound	T, K (state)	ΔH	ΔG	Method	Ref.
	2.1	(bicyclic structure equilibrium)	379 (g)	−5.94	−2.63	SE, CT, GLC	203
			298 (l)	—	−6.61 ±1.09	TP	204
			298 (g)	—	−8.53 ±2.51	TP	204
	1.3	(dimethylbicyclic equilibrium)	298 (l)	−17.23 ±2.09	—	CC	177
$C_9H_{16}O$	1.3	$CH_2=C\genfrac{}{}{0pt}{}{CH_2-(i-C_3H_7)}{O-C\genfrac{}{}{0pt}{}{CH_2}{CH_3}} \rightleftarrows i-C_3H_7\genfrac{}{}{0pt}{}{CH_3}{H}\!C=C\genfrac{}{}{0pt}{}{O-C\genfrac{}{}{0pt}{}{CH_2}{CH_3}}{CH_3}$ (cis-)	298 (g)	−4.3 ±0.4	−6.7 ±1.0	SE, CT, GLC	74
		$CH_2=C\genfrac{}{}{0pt}{}{CH_2-(i-C_3H_7)}{O-C\genfrac{}{}{0pt}{}{CH_2}{CH_3}} \rightleftarrows i-C_3H_7\genfrac{}{}{0pt}{}{H}{CH_3}\!C=C\genfrac{}{}{0pt}{}{CH_3}{O-C\genfrac{}{}{0pt}{}{CH_2}{CH_3}}$ (trans-)	298 (g)	−10.9 ±0.6	−13.3 ±1.2	SE, CT, GLC	74
	2.1	$i-C_3H_7\genfrac{}{}{0pt}{}{CH_3}{H}\!C=C\genfrac{}{}{0pt}{}{O-C\genfrac{}{}{0pt}{}{CH_2}{CH_3}}{CH_3}$ (cis-) $\rightleftarrows i-C_3H_7\genfrac{}{}{0pt}{}{H}{CH_3}\!C=C\genfrac{}{}{0pt}{}{CH_3}{O-C\genfrac{}{}{0pt}{}{CH_2}{CH_3}}$ (trans-)	298 (g)	−6.6 ±0.4	−6.6 ±0.8	SE, CT, GLC	74

Table (Continued)

Formula	Type	Reaction	T, K (phase)	$\Delta_r H_T^\circ$ ($\Delta_r G_T^\circ$)	$\Delta_r S_T^\circ$	Technique	Reference
$C_9H_{16}O$	1.3	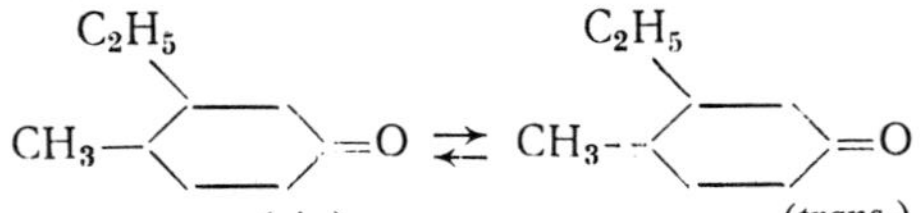 CH₃–⟨C₂H₅⟩–=O ⇌ CH₃–⟨C₂H₅⟩–=O (cis-) (trans-)	413 (l)	−3.76 ±0.42	−0.38 ±0.63	SE, CT, GLC	179
		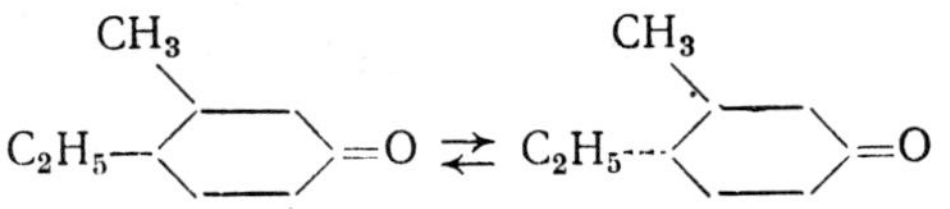 CH₃–⟨CH₃⟩–=O ⇌ C₂H₅–⟨CH₃⟩–=O (cis-) (trans-)	413 (l)	−2.93 ±0.42	−0.04 ±0.63	SE, CT, GLC	179
$C_9H_{16}O_2$	2.1	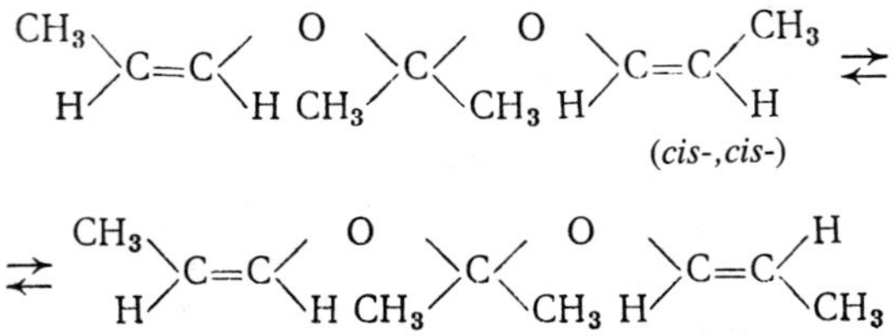 (cis-,cis-) ⇌ (cis-,trans-)	298 (l)	(0.9 ±0.1)	—	SE, CT, GLC	92
		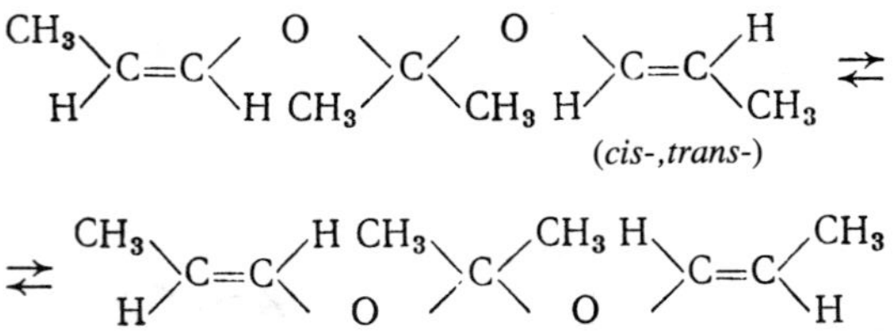 (cis-,trans-) ⇌ (trans-,trans-)	298 (l)	(4.7 ±0.2)	—	SE, CT, GLC	92

$C_9H_{16}O_3$	1.3	$\begin{array}{c}CH_3\\ \diagdown\\ H\diagup C=C\diagdownO\diagup CH_3\\ HCH_3\diagup C\diagdown O\diagup C=C\diagdown H\\ CH_3\diagupH\\ (cis\text{-},cis\text{-})\end{array}$ ⇌ $\begin{array}{c}CH_3\\ \diagdown\\ H\diagup C=C\diagdownH CH_3\diagup CH_3\\ C\diagdown O\diagup C=C\diagdown H\\ (trans\text{-},trans\text{-})\end{array}$	298 (l)	(5.6 ±0.2)	—	SE, CT, GLC	92
		$\begin{array}{c}CH_3OO\\ \diagdown\|\|\\ CH_3-C-C-CH_2-C-CH_2-CH_3\\ \diagup\\ CH_3\end{array}$ ⇌ $\begin{array}{c}CH_3OCH_3\\ \|\|\\ CH_3-CH-C-CH_2-C-CH-CH_3\end{array}$	298 (l)	0.3 ±2.8	—	CC	205
	2.1	$\begin{array}{c}O\\ \|\!\!\!\!\diagup\diagdown\\ C_4H_9\diagdownCH_3\\ (trans\text{-})\end{array}$ ⇌ $\begin{array}{c}O\\ \|\!\!\!\!\diagup\diagdown\\ \diagdownCH_3\\ C_4H_9\\ (cis\text{-})\end{array}$	298 (l)	(0.79)	—	SE, CT, GLC $t\text{-}C_4H_9OH$	121
	2.1	$\begin{array}{c}CH_3\\ \diagdown\\ H\diagup C=C\diagdownOCH_3\\ HCH_3\diagup C\diagdown O\diagup C=C\diagdown H\\ (cis\text{-},cis\text{-})\end{array}$ ⇌ $\begin{array}{c}CH_3\\ \diagdown\\ H\diagup C=C\diagdownOCH_3 H\\ CH_3\diagup C\diagdown O\diagup C=C\diagdown CH_3\\ (trans\text{-},cis\text{-})\end{array}$	298 (l)	2.2 ±0.2	3.8 ±0.5	SE, CT, GLC	117

Table (*Continued*)

Formula	Type	Reaction	T, K (phase)	$\Delta_r H_T^\circ$ ($\Delta_r G_T^\circ$)	$\Delta_r S_T^\circ$	Technique	Reference
$C_9H_{16}O_3$	2.1	(*trans-,cis-*) ⇌ (*trans-,trans-*)	298 (l)	2.4 ±0.2	−6.7 ±0.5	SE, CT, GLC	117
		(*trans-,cis-*) ⇌ (*cis-,cis-*)	298 (l)	4.6 ±0.4	−2.9 ±1.2	SE, CT, GLC	117
		(*trans-,cis-*) ⇌ (*trans-,trans-*)					
		(*cis*-) ⇌ (*trans*-)	298 (l)	(−1.72 ±0.25)	—	SE, CT, GLC	206

$C_9H_{16}O_4$	2.1	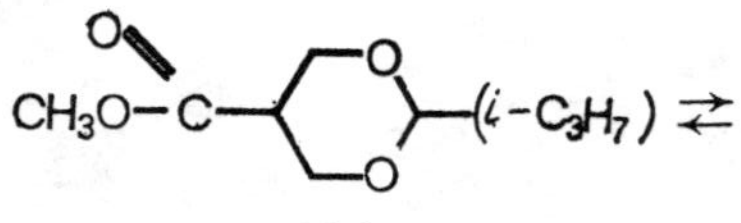 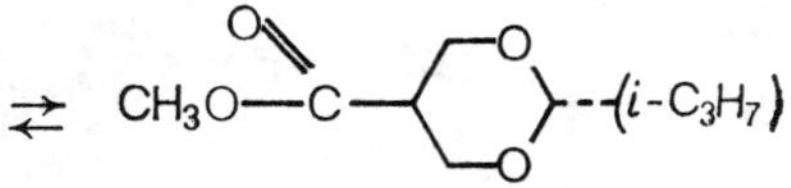 (structures)	303 (l)	(−3.43)	—	SE, CT, GLC ($C_2H_5)_2O$	153
		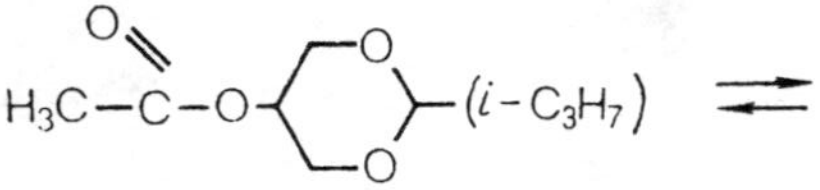 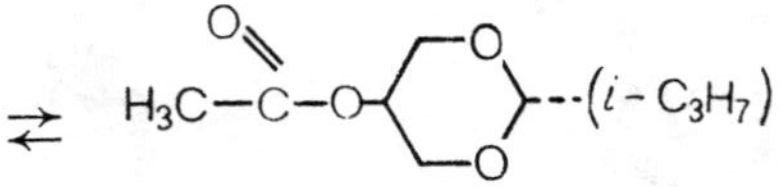	298 (l)	(0.00)	—	SE, CT, GLC ($C_2H_5)_2O$	153
		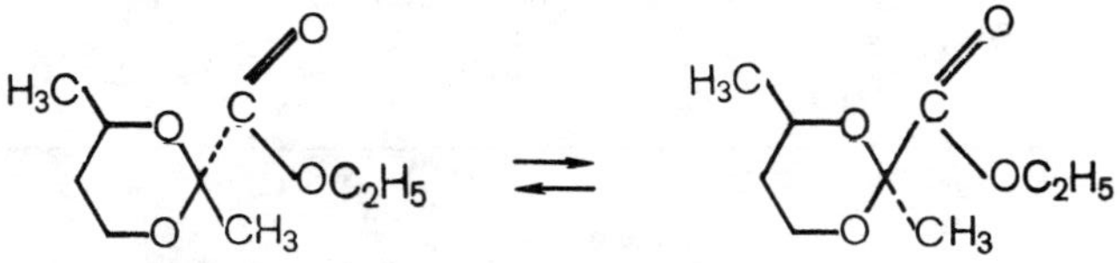	298 (l)	(13.22 ±0.17)	—	SE, CT, GLC	154

Table (Continued)

Formula	Type	Reaction	T, K (phase)	$\Delta_r H_T^\circ$ ($\Delta_r G_T^\circ$)	$\Delta_r S_T^\circ$	Technique	Reference
$C_9H_{17}ClO$	1.3	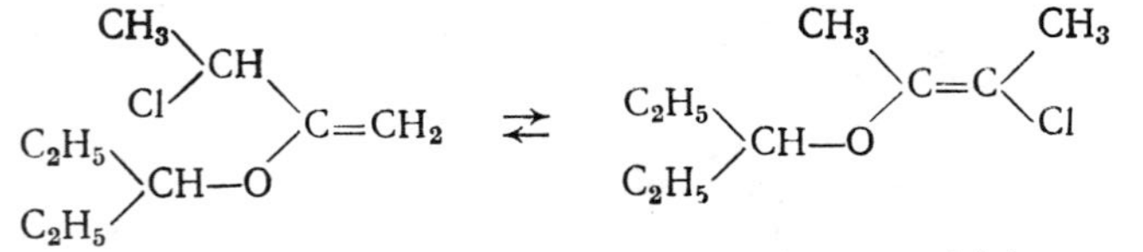 (cis-isomerization reaction structure)	298 (l)	6.3 ±0.5	3.9 ±1.4	SE, CT, GLC	35
		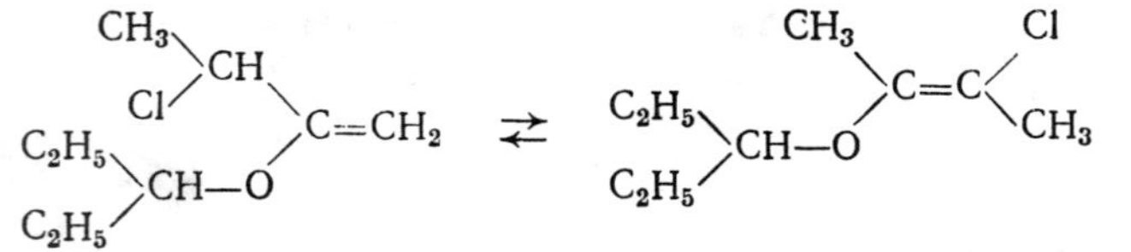 (cis-)	298 (l)	0.4 ±0.6	−1.7 ±1.6	SE, CT, GLC	35
	2.1	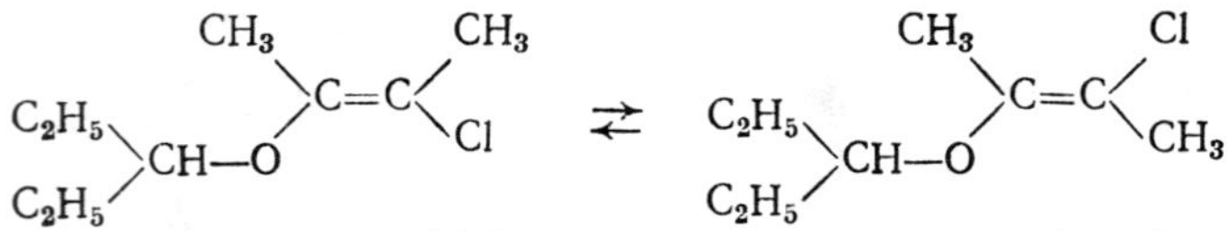 (cis- ⇌ trans-)	298 (l)	−6.1 ±1.3	−5.8 ±3.6	SE, CT, GLC	35
C_9H_{18}	1.3	 (C₆H₅–C₃H₇ ⇌ o-C₂H₅,CH₃ substituted benzene)	298 (l)	−2.72 ±1.39	—	CC	207
		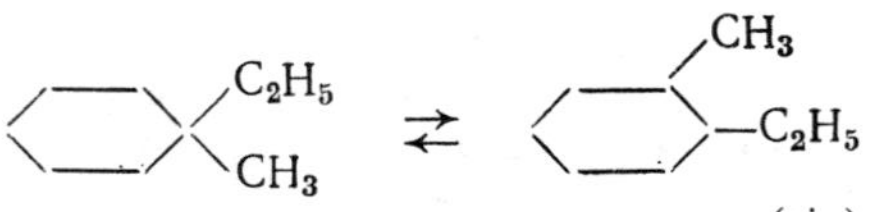 (cis-)	298 (l)	3.97 ±1.39	—	CC	207

Formula		Reaction	T, K (state)	ΔH	ΔS	Method	Ref.
	2.1	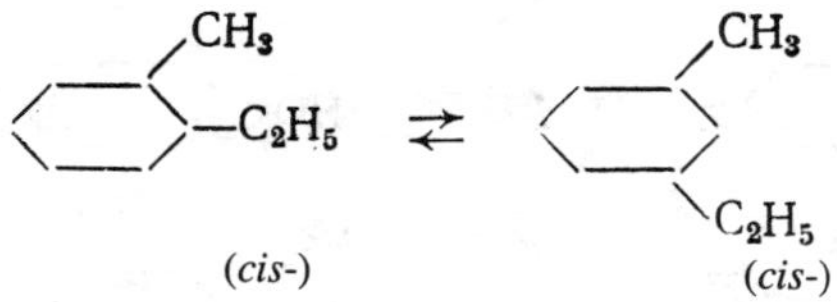 cis-1-methyl-2-ethylcyclohexane ⇌ cis-1-methyl-3-ethylcyclohexane	298 (l)	−10.50 ±1.56	—	CC	207
		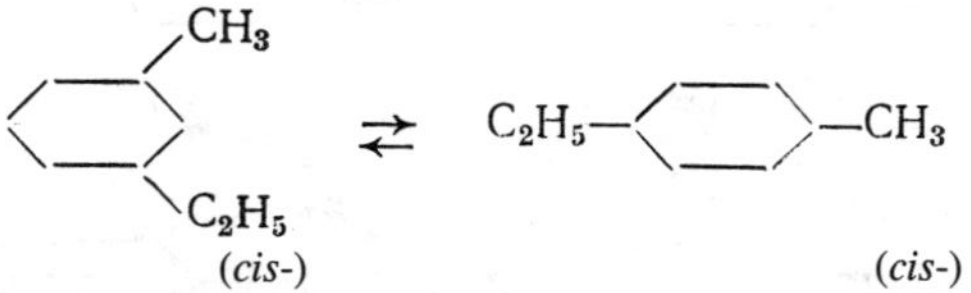 1,4-diethylcyclohexane ⇌ 1,3-diethylcyclohexane (cis-)	298 (l)	7.74 ±1.62	—	CC	207
		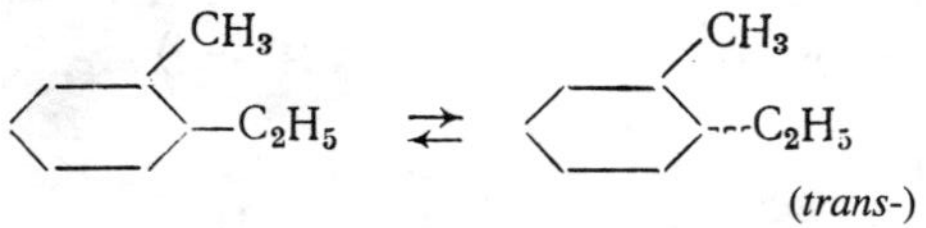 cis-1-methyl-2-ethylcyclohexane ⇌ trans-1-methyl-2-ethylcyclohexane	298 (l)	−4.06 ±1.39	—	CC	207
		 cis-1-ethyl-4-methylcyclohexane ⇌ trans-1-ethyl-4-methylcyclohexane	298 (l)	−7.45 ±1.37	—	CC	207
		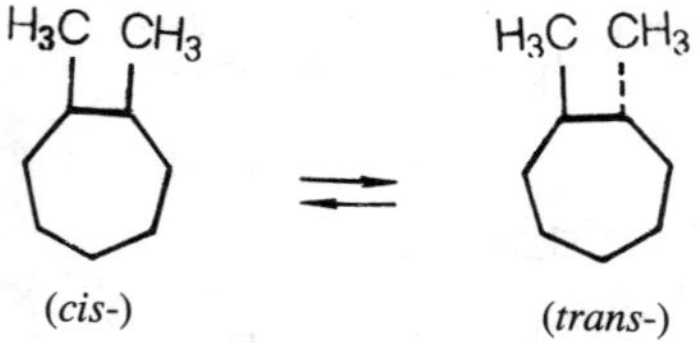 cis-1,2-dimethylcycloheptane ⇌ trans-1,2-dimethylcycloheptane	448 (g)	−2.93	−1.67	FE, CT, GLC	182
$C_9H_{18}O$	1.3	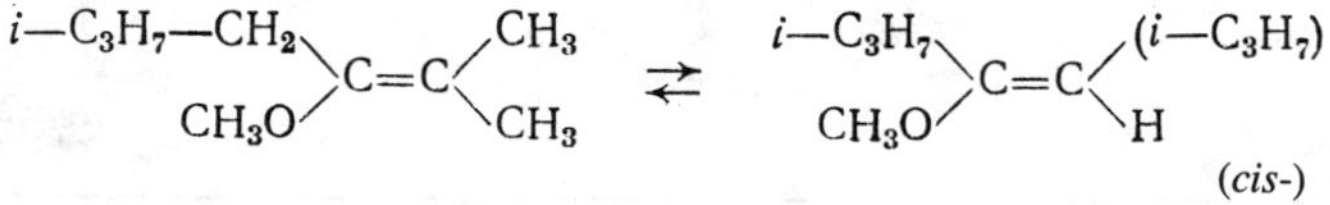 $i\text{-}C_3H_7\text{-}CH_2\text{-}C(OCH_3)=C(CH_3)_2$ ⇌ $i\text{-}C_3H_7\text{-}C(OCH_3)=CH\text{-}CH(i\text{-}C_3H_7)$ (cis-)	298 (g)	−7.2 ±0.4	−23.2 ±1.0	SE, CT, GLC	183

Table (Continued)

Formula	Type	Reaction	T, K (phase)	$\Delta_r H_T^\circ$ ($\Delta_r G_T^\circ$)	$\Delta_r S_T^\circ$	Technique	Reference
$C_9H_{18}O$	1.3	$i\text{-}C_3H_7\text{-}CH_2\underset{CH_3O}{\overset{CH_3}{>}}C=C\underset{CH_3}{<}$ (cis-) $\rightleftarrows$ $i\text{-}C_3H_7\text{-}CH_2\underset{CH_3O}{\overset{H}{>}}C=C\underset{(i\text{-}C_3H_7)}{<}$ (trans-)	298 (g)	5.9 ±0.4	−2.7 ±1.1	SE, CT, GLC	183
	2.1	$i\text{-}C_3H_7\underset{CH_3O}{\overset{(i\text{-}C_3H_7)}{>}}C=C\underset{H}{<}$ (cis-) $\rightleftarrows$ $i\text{-}C_3H_7\underset{CH_3O}{\overset{H}{>}}C=C\underset{(i\text{-}C_3H_7)}{<}$ (trans-)	298 (g)	13.0 ±0.6	20.4 ±1.5	SE, CT, GLC	183
		$C_5H_{11}\underset{H}{\overset{OC_2H_5}{>}}C=C\underset{H}{<}$ (cis-) $\rightleftarrows$ $C_5H_{11}\underset{H}{\overset{H}{>}}C=C\underset{OC_2H_5}{<}$ (trans-)	311 (l)	−3.85 ±0.25	−11.84 ±0.75	SE, CT, GLC	40
		$\underset{i\text{-}C_3H_7}{\text{(OH, }i\text{-}C_3H_7\text{ on benzene ring, cis-)}}$ $\rightleftarrows$ $\underset{i\text{-}C_3H_7}{\text{(OH, }i\text{-}C_3H_7\text{ on benzene ring, trans-)}}$	353 (l)	(−3.05)	—	SE, CT, GLC $i\text{-}C_3H_7OH$	162
		(ortho isopropylphenol cis-/trans-)	353 (l)	(4.27)	—	SE, CT, GLC $i\text{-}C_3H_7OH$	162
		(para isopropylphenol cis-/trans-)	353 (l)	(−3.55)	—	SE, CT, GLC $i\text{-}C_3H_7OH$	162

148

		373 (l)	(3.14 ±0.21)	—	SE, CT, IR ⌬	161
		375 (l)	6.48 ±0.08	1.09 ±0.17	SE, CT, GLC ⌬	208
		382 (l)	7.95 ±0.04	3.22 ±0.04	SE, CT, GLC CH_3OCH_2— —CH_2OCH_3	208
		382 (l)	10.33 ±0.29	4.85 ±0.75	SE, CT, GLC $i-C_3H_7OH$	208
$C_9H_{18}OS_2$	2.1	298 (l)	(−5.10 ±0.75)	—	SE, CT, GLC CH_3CN	188
$C_9H_{18}O_2$	1.3	298 (l)	(0.24 ±0.10)	—	SE, CT, GLC ⌬	81
		298 (l)	(−0.32 ±0.10)	—	SE, CT, GLC ⌬	81

Table (Continued)

Formula	Type	Reaction	T, K (phase)	$\Delta_r H°_T$ ($\Delta_r G°_T$)	$\Delta_r S°_T$	Technique	Reference
$C_9H_{18}O_2$	2.1	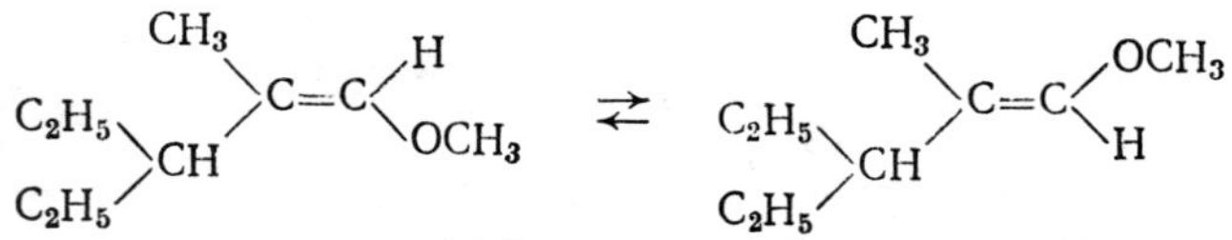 cis- ⇌ trans- (methoxy alkene)	298 (l)	(−0.55 ±0.10)	—	SE, CT, GLC (C$_6$H$_6$)	81
		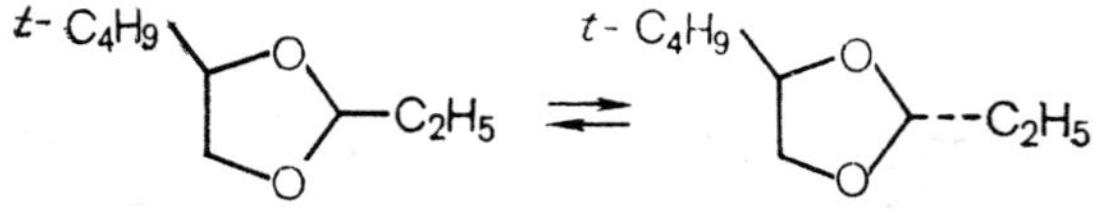 cis-dioxolane ⇌ trans-dioxolane (t-C$_4$H$_9$, C$_2$H$_5$)	298 (l)	(2.09)	—	SE, CT, NMR (C$_2$H$_5$)$_2$O	93
		syn- ⇌ anti- (dioxolane, t-C$_4$H$_9$, H$_3$C)	323 (l)	1.86	2.5	SE, CT, NMR CCl$_4$	84
		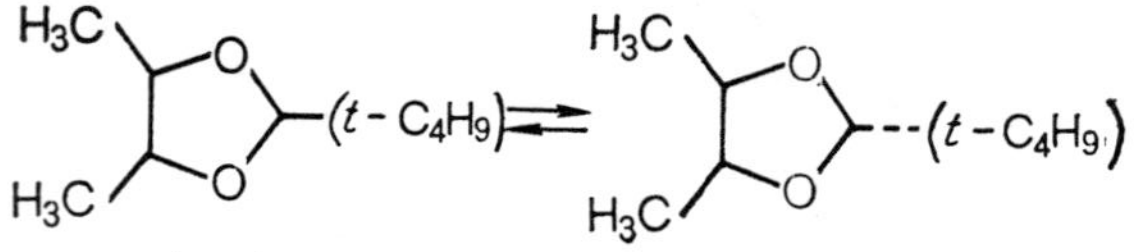 cis- ⇌ trans- (dioxolane, t-C$_4$H$_9$)	298 (l)	(2.46 ±0.16)	—	SE, CT, GLC (C$_2$H$_5$)$_2$O	93
		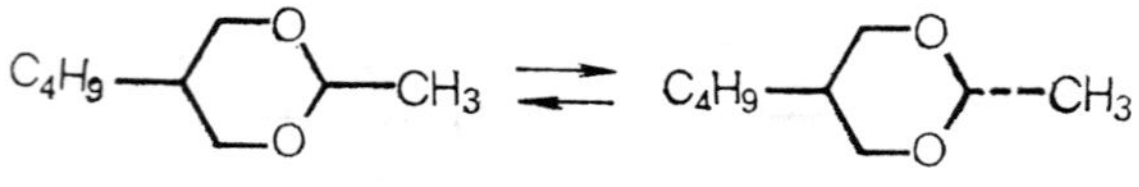 cis- ⇌ trans- (dioxane, C$_4$H$_9$, CH$_3$)	333 (l)	(−3.81 ±0.13)	—	SE, CT, GLC	165

Compound	Equilibrium	T (K) (phase)	ΔG (kJ/mol)		Method (solvent)	Ref.
	(cis-) ⇌ (trans-) with t-C₄H₉ and H₃C substituents	 298 (l)	(−6.39 ±0.13)	—	SE, CT, GLC CCl₄	209
	 (cis-,trans-) ⇌ (trans-,cis-) with i-C₃H₇, CH₃, H₃C	333 (l)	(−1.25 ±0.17)	—	SE, CT, GLC	165
	(cis-,cis-) ⇌ (trans-,trans-) with i-C₃H₇ and H₃C	363 (l)	(−1.27 ±0.04)	—	SE, CT, GLC	133
C₉H₁₈O₃ 2.1 (cis-) ⇌ (trans-) with OCH₃ and t-C₄H₉		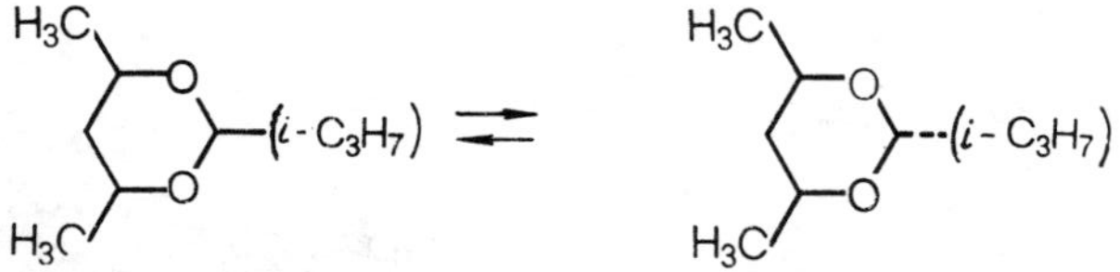 298 (l)	(17.36)	—	SE, CT, NMR (C₂H₅)₂O	166
	(cis-) ⇌ (trans-) with OCH₃ and t-C₄H₉	 298 (l)	(2.09)	—	SE, CT, NMR (C₂H₅)₂O	166
	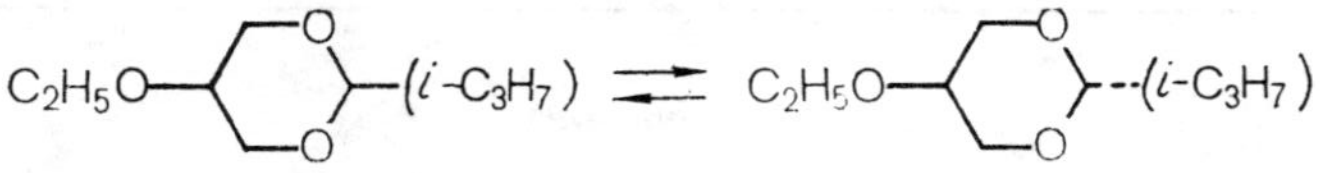 (cis-) ⇌ (trans-) with i-C₃H₇ and C₂H₅O	298 (l)	(−4.56)	—	SE, CT, GLC CCl₄	189

Table (Continued)

Formula	Type	Reaction	T, K (phase)	$\Delta_r H°_T$ ($\Delta_r G°_T$)	$\Delta_r S°_T$	Technique	Reference
$C_9H_{18}O_3$	2.1	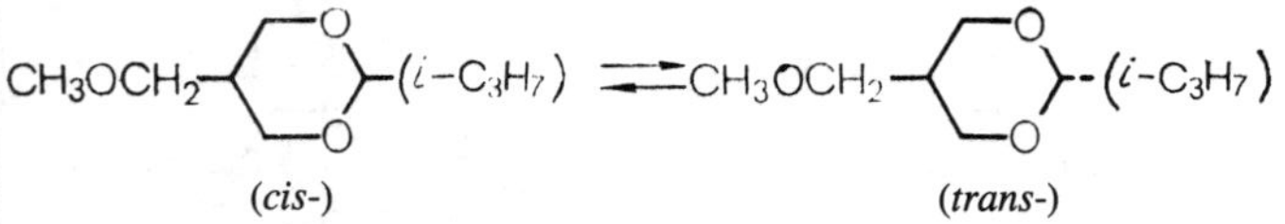 CH_3OCH_2-(1,3-dioxane)-(i-C_3H_7) (cis-) ⇌ CH_3OCH_2-(1,3-dioxane)-(i-C_3H_7) (trans-)	303 (l)	(−0.21)	—	SE, CT, GLC ($C_2H_5)_2O$	153
		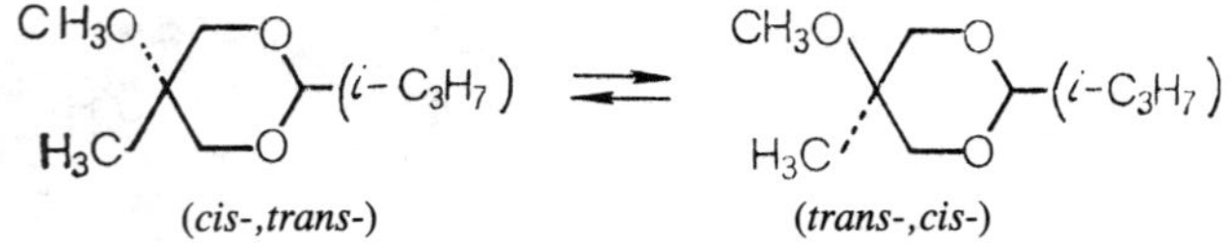 (cis-, trans-) 1,3-dioxane with CH_3O, H_3C, (i-C_3H_7) ⇌ (trans-, cis-)	298 (l)	(1.42)	—	SE, CT, GLC ($C_2H_5)_2O$	190
			298 (l)	(1.88 ±0.13)	—	SE, CT, GLC CCl_4	167
		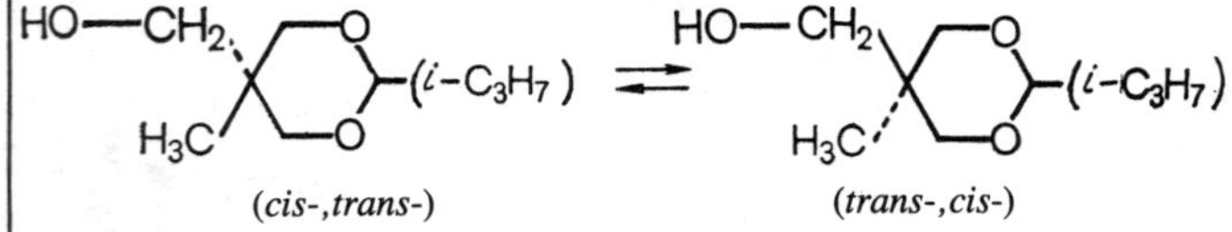 (cis-, trans-) 1,3-dioxane with $HO-CH_2$, H_3C, (i-C_3H_7) ⇌ (trans-, cis-)	298 (l)	(−2.84)	—	SE, CT, GLC ($C_2H_5)_2O$	190
			298 (l)	(−4.31 ±0.17)	—	SE, CT, GLC CCl_4	167

$C_9H_{18}S$	1.3	$\begin{array}{c}H_3C\\H_3C\end{array}\!\!>\!\!C=CH-CH_2-S-C_4H_9 \rightleftarrows \begin{array}{c}H_3C\\H_3C\end{array}\!\!>\!\!CH-CH=CH-S-C_4H_9$ (cis- + trans-)	293 (l)	(6.51)	—	SE, CT, GLC $t-C_4H_9OH$, $CH_3-\overset{\underset{\|}{O}}{S}-CH_3$	180
$C_9H_{18}S_2$	2.1	[cis- 4-methyl-2-(t-C₄H₉)-1,3-dithiane] ⇌ [trans- 4-methyl-2-(t-C₄H₉)-1,3-dithiane]	361 (l)	6. ±0.02	−0.55 −0.05	SE, CT, GLC CCl_4	136
		[cis- 5-methyl-2-(t-C₄H₉)-1,3-dithiane] ⇌ [trans- 5-methyl-2-(t-C₄H₉)-1,3-dithiane]	342 (l)	(7.07 ±0.04)	—	SE, CT, GLC $CHCl_3$	137
		[cis- 5-methyl-2-(t-C₄H₉)-1,3-dithiane (other)] ⇌ [trans- (other)]	342 (l)	(−4.35 ±0.04)	—	SE, CT, GLC $CHCl_3$	137
		[cis-,cis- 4,6-dimethyl-2-(i-C₃H₇)-1,3-dithiane] ⇌ [trans-,trans- 4,6-dimethyl-2-(i-C₃H₇)-1,3-dithiane]	342 (l)	(8.16 ±0.08)	—	SE, CT, GLC $CHCl_3$	137

Table (Continued)

Formula	Type	Reaction	T, K (phase)	$\Delta_r H_T°$ ($\Delta_r G_T°$)	$\Delta_r S_T°$	Technique	Reference
$C_9H_{18}S_3$	2.1	(cis-) ⇌ (trans-) [dithiane with t-C_4H_9 and H_3C–S– substituents]	298 (l)	(−4.57 ±0.71)	—	SE, CT, GLC CH_3CN	188
$C_9H_{19}Cl$	1.3	CH_3–CH(Cl)–CH_2–C_6H_{13} ⇌ CH_3–CH_2–CH(Cl)–C_6H_{13}	355 (l)	−0.40 ±0.42	−3.35 ±1.25	SE, CT, GLC	96
	1.3	CH_3–CH(Cl)–CH_2–C_6H_{13} ⇌ { CH_3–CH_2–CH_2–CH(Cl)–C_5H_{11} ; CH_3–CH_2–CH_2–CH_2–CH(Cl)–C_4H_9 }	355 (l)	−0.65 ±0.42	—	SE, CT, GLC	96
C_9H_{20}	1.3	CH_3–CH_2–C(C_2H_5)$_2$–CH_2–CH_3 ⇌ CH_3–C(CH_3)(CH_2CH_3)–C(CH_3)$_3$	298 (g)	−8.62 ±2.15	—	CC	115
$C_{10}H_{10}O$	1.2	[cage ketone] ⇌ [tricyclic dienone]	298 (g)	−68.64	—	CC	210

Formula		Structure	T (K) (S/l)	ΔH	ΔS	Method	Ref
$C_{10}H_{10}O_2$	2.1	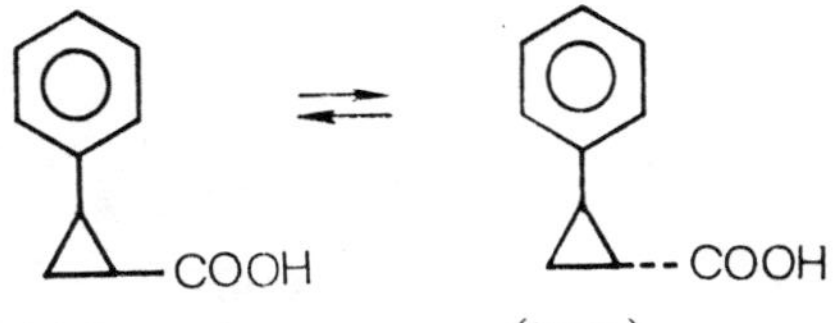 (cis-) ⇌ (trans-) phenylcyclopropane-COOH	298 (S)	−11.46 ±2.38	—	CC	211
$C_{10}H_{10}O_3$	1.1 1.2	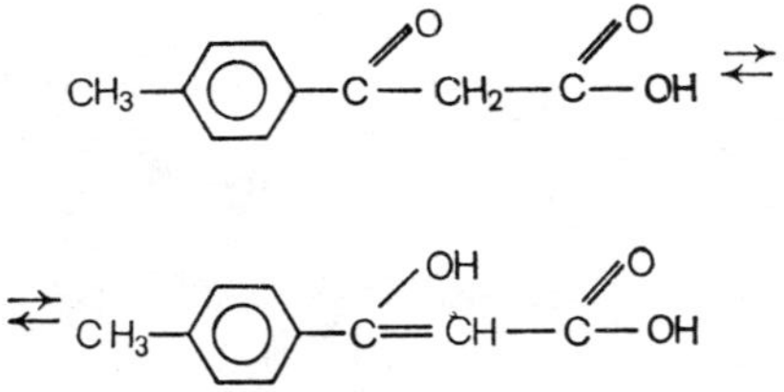 CH_3-C$_6$H$_4$-C(=O)-CH$_2$-C(=O)-OH ⇌ CH_3-C$_6$H$_4$-C(OH)=CH-C(=O)-OH	278 (l)	2.1	16.7	SE, CT, NMR	192
$C_{10}H_{10}O_4$	1.1 1.2	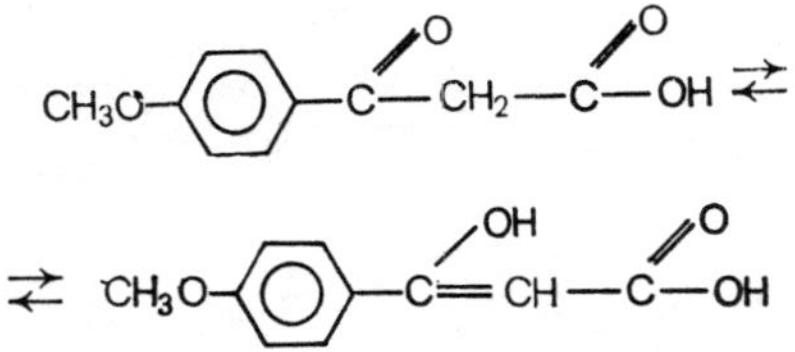 CH_3O-C$_6$H$_4$-C(=O)-CH$_2$-C(=O)-OH ⇌ CH_3O-C$_6$H$_4$-C(OH)=CH-C(=O)-OH	278 (l)	1.7	20.9	SE, CT, NMR	192
$C_{10}H_{11}ClO$	2.1	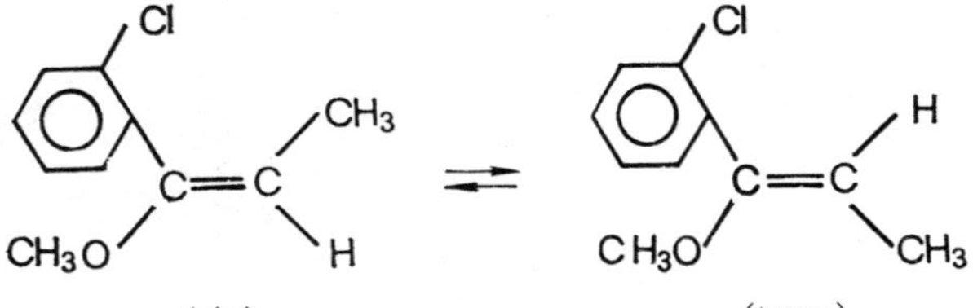 (cis-) ⇌ (trans-) 2-Cl-C$_6$H$_4$-C(OCH$_3$)=CH-CH$_3$	298 (l)	5.46	8.12	SE, CT, NMR	173

Table (Continued)

Formula	Type	Reaction	T, K (phase)	$\Delta_r H_T^\circ$ ($\Delta_r G_T^\circ$)	$\Delta_r S_T^\circ$	Technique	Reference
$C_{10}H_{11}FO$	2.1	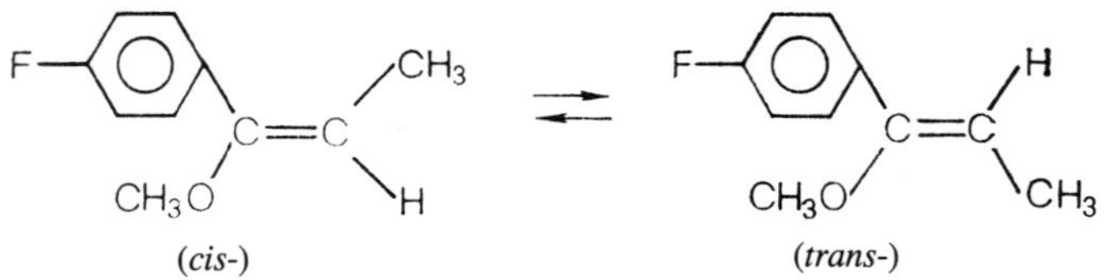 (cis-) ⇌ (trans-) [4-F-C6H4-C(OCH3)=CHCH3]	298 (l)	2.65	12.52	SE, CT, NMR 	173
$C_{10}H_{12}O$	1.3	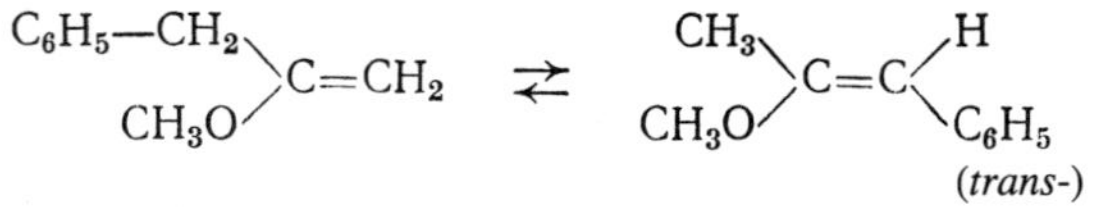 C_6H_5–CH_2\C(OCH3)=CH2 ⇌ (trans-) CH3\C(OCH3)=CH(C6H5)	298 (g)	0.0 ±0.6	0.5 ±1.5	SE, CT, GLC	193
		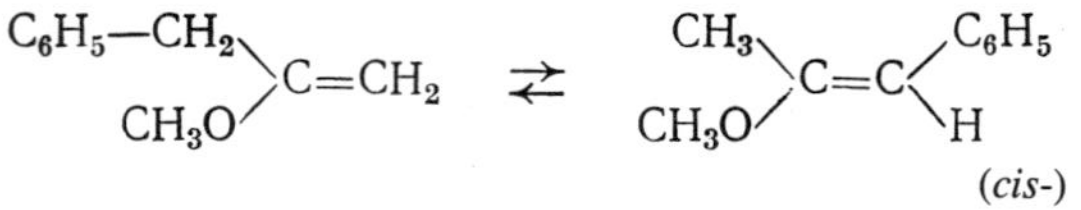 C_6H_5–CH_2\C(OCH3)=CH2 ⇌ (cis-) CH3\C(OCH3)=CH(C6H5)	298 (g)	−7.3 ±0.6	−6.8 ±1.3	SE, CT, GLC	193
	2.1	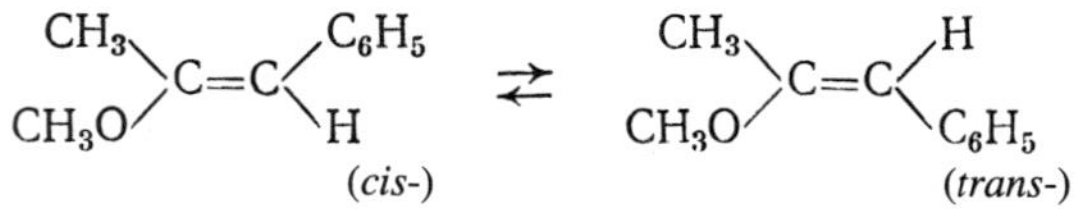 (cis-) CH3\C(C6H5)=CH(OCH3) ⇌ (trans-) CH3\C(OCH3)=CH(C6H5)	298 (g)	7.3 ±0.5	7.3 ±1.1	SE, CT, GLC	193
		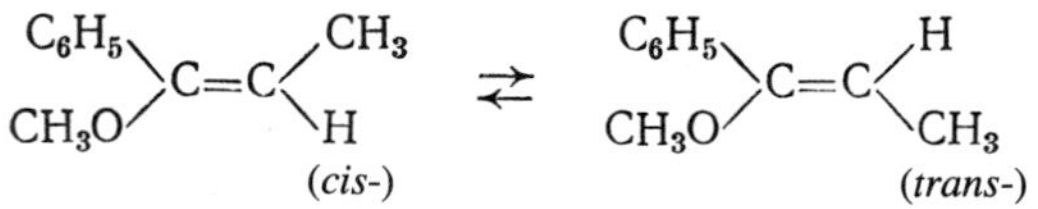 (cis-) C6H5\C(CH3)=CH(OCH3) ⇌ (trans-) C6H5\C(OCH3)=CH(CH3)	298 (l)	1.88	11.53	SE, CT, NMR 	193
		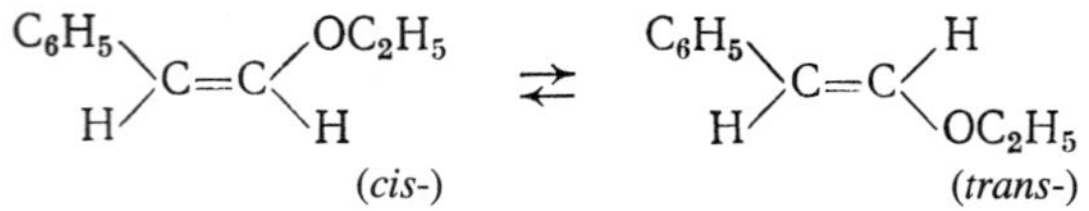 (cis-) C6H5\C(OC2H5)=CH(H) ⇌ (trans-) C6H5(H)\C=C(H)(OC2H5)	311 (l)	−1.59 ±0.33	−2.72 ±1.30	SE, CT, GLC	40

$C_{10}H_{13}Cl$	1.3	![structure: o-chloro-tert-butylbenzene ⇌ p-chloro-tert-butylbenzene]	400 (l)	−21.4 ±1.7	−21.8 ±4.2	SE, CT, GLC H_2O	212
		![structure: m-chloro-tert-butylbenzene ⇌ p-chloro-tert-butylbenzene]	313 (l)	0.84 ±0.42	−5.3 ±1.5	SE, CT, GLC H_2O	212
$C_{10}H_{13}N$	1.1 1.2	![structure: 2-chloro-5-isopropyl-toluene ⇌ 4-chloro-3-isopropyl-toluene]	353 (l)	−3.12 ±1.08	4.04 ±3.18	SE, CT, GLC	261
		$\begin{array}{c}C_6H_5\\CH_3\end{array}\!\!\!>\!\!CH-CH=N-CH_3 \rightleftarrows \begin{array}{c}C_6H_5\\CH_3\end{array}\!\!\!>\!\!C=CH-NH-CH_3$	333 (l)	−0.8	−10.0	SE, NMR ![benzene ring]	213
$C_{10}H_{14}$	1.3	![structure: m-diethylbenzene ⇌ p-diethylbenzene]	298 (l)	0.67 ±1.37	—	CC	214

Table (Continued)

Formula	Type	Reaction	T, K (phase)	$\Delta_r H_T^\circ$ ($\Delta_r G_T^\circ$)	$\Delta_r S_T^\circ$	Technique	Reference
$C_{10}H_{14}$	1.3	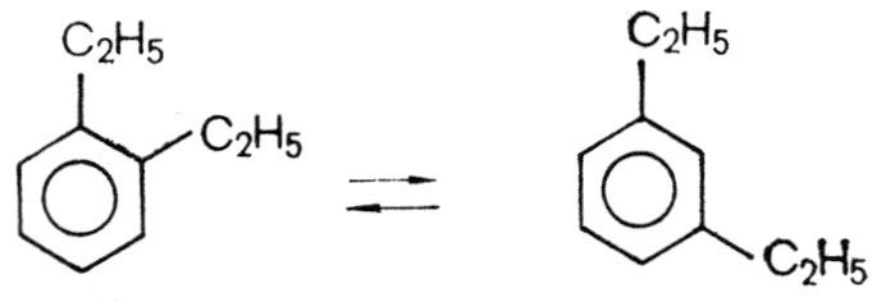 1,2-diethylbenzene ⇌ 1,3-diethylbenzene	298 (l)	−5.02 ±1.42	—	CC	214
		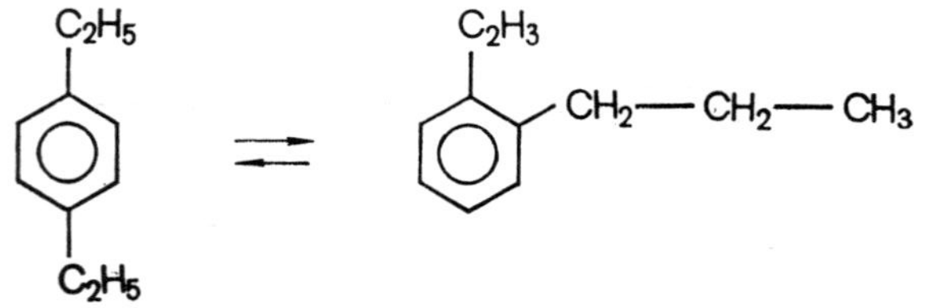 1,2-diethylbenzene ⇌ n-butylbenzene / 1,4-diethylbenzene	298 (l)	0.38 ±1.31	—	CC	214
		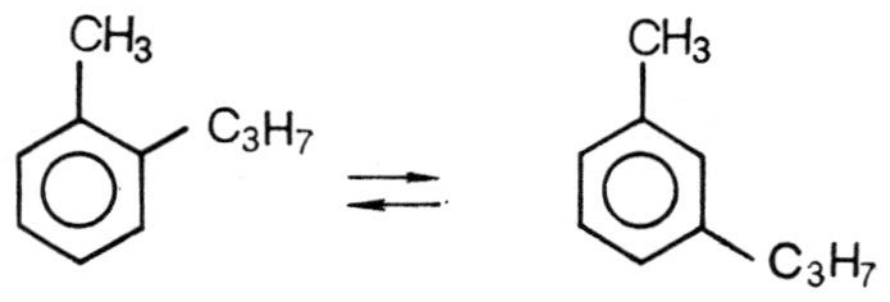 1,2-methyl-propylbenzene ⇌ 1,3-methyl-propylbenzene	298 (l)	−3.76 ±1.48	—	CC	214
		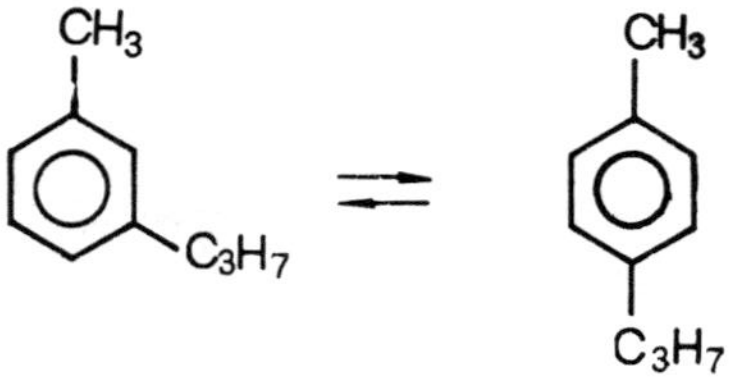 1,3-methyl-(n-propyl)benzene ⇌ 1,4-methyl-(n-propyl)benzene	298 (l)	1.17 ±1.51	—	CC	214

Reaction	T (K)	ΔH (kcal/mol)		Method	Ref
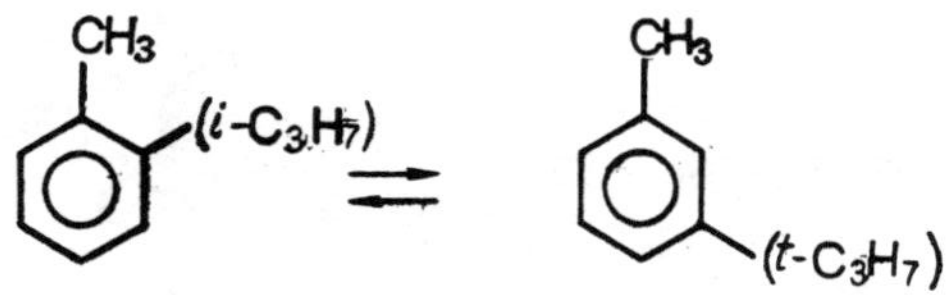 2-isopropyltoluene ⇌ 3-tert-butyltoluene	298 (l)	−5.35 ±1.42	—	CC	214
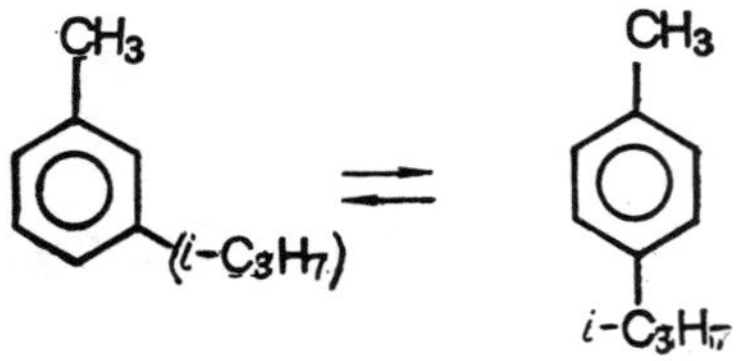 3-isopropyltoluene ⇌ 4-isopropyltoluene (with CH₃)	298 (l)	0.63 ±1.51	—	CC	214
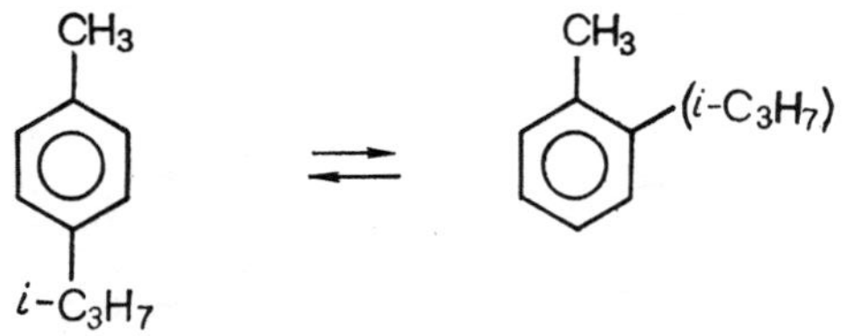 2-isopropyltoluene ⇌ 4-isopropyltoluene	298 (l)	1.76 ±1.37	—	CC	214
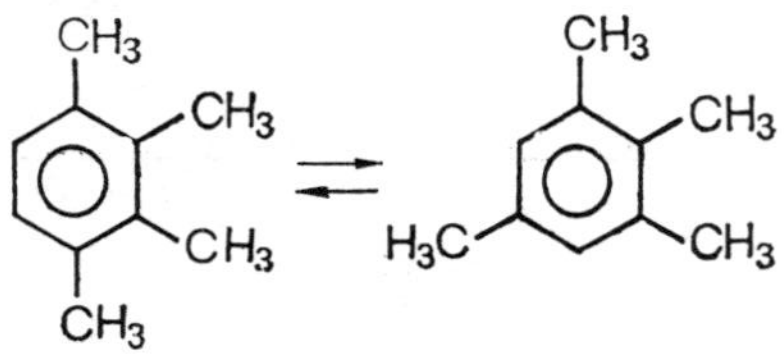 2,3,6-trimethyl-isopropylbenzene ⇌ 2,3,5-trimethyl	298 (l)	−6.15 ±1.62	—	CC	215

Table (Continued)

Formula	Type	Reaction	T, K (phase)	$\Delta_r H_T^\circ$ ($\Delta_r G_T^\circ$)	$\Delta_r S_T^\circ$	Technique	Reference
$C_{10}H_{14}$	1.3	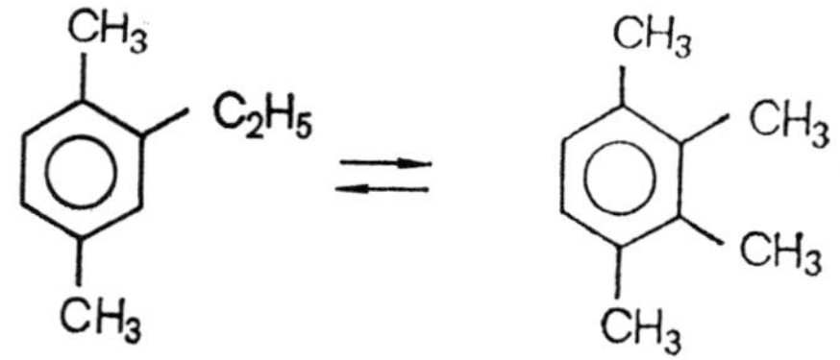 1,2,4-Me₃-3-Et-benzene ⇌ 1,2,3-Me₃-4-Et-benzene	298 (l)	−5.40 ±1.62	—	CC	215
		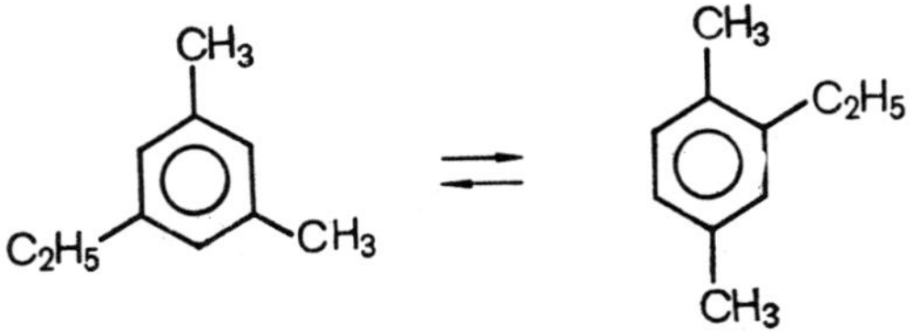 1,3,5-Me₃-benzene + Et ⇌ 2,4-Me₂-1-Et... (see structures)	298 (l)	2.97 ±1.59	—	CC	215
		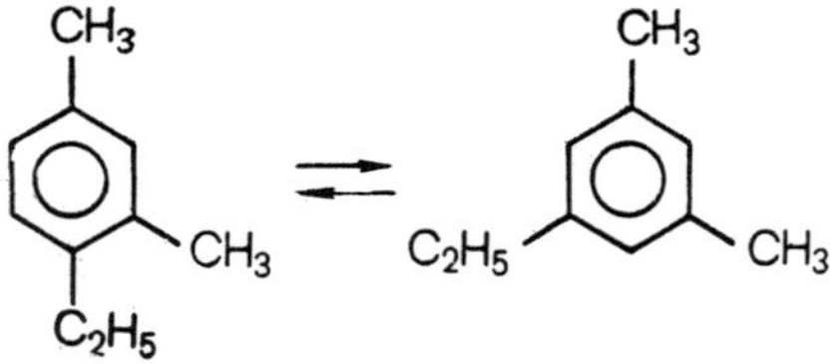 1,2,4-Me₃-3-Et / 1,3-Me₂-5-Et rearrangement	298 (l)	−3.68 ±1.62	—	CC	215

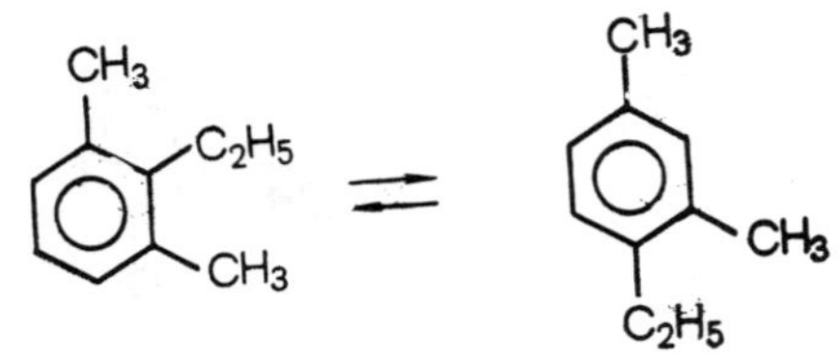 (structure: 1,2-C2H5, 2-CH3, on benzene ⇌ 1-C2H5, 2-CH3 isomer)	298 (l)	−3.97 ±1.67	—	cc	215
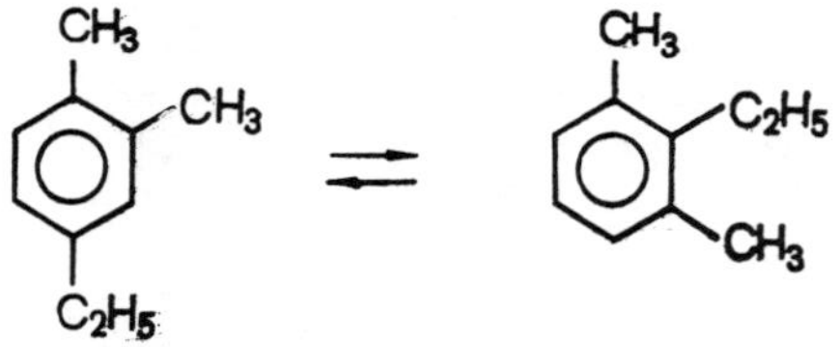 (structure: trimethyl/ethyl benzene isomer pair)	298 (l)	5.90 ±1.59	—	cc	215
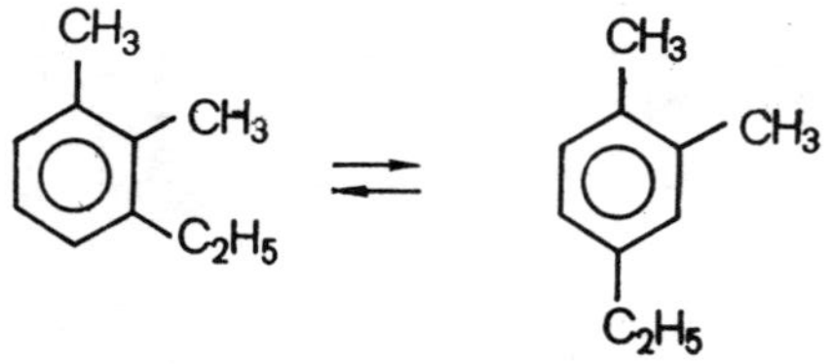 (structure: CH3/C2H5 isomer pair)	298 (l)	−5.62 ±1.62	—	cc	215
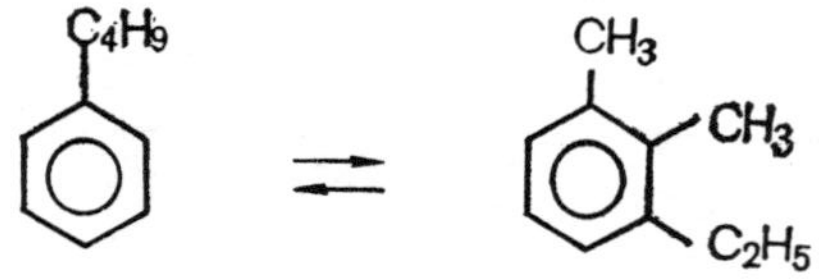 (structure: CH3/C2H5 benzene ⇌ C4H9-benzene)	298 (l)	−17.57 ±1.67	—	cc	215

Table (Continued)

Formula	Type	Reaction	T, K (phase)	$\Delta_r H_T^\circ$ ($\Delta_r G_T^\circ$)	$\Delta_r S_T^\circ$	Technique	Reference
$C_{10}H_{14}$	1.3	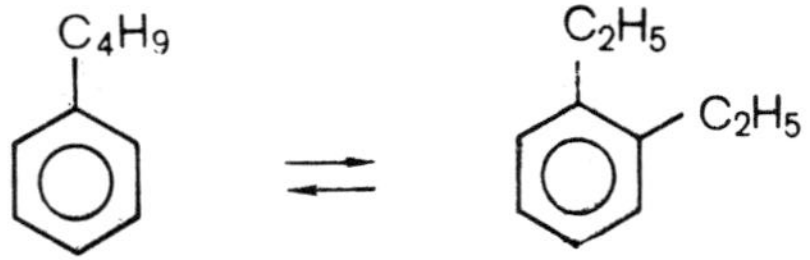 C_4H_9–C$_6$H$_5$ ⇌ C$_2$H$_5$–C$_6$H$_4$–C$_2$H$_5$	298 (l)	−5.06 ±1.39	—	CC	214
		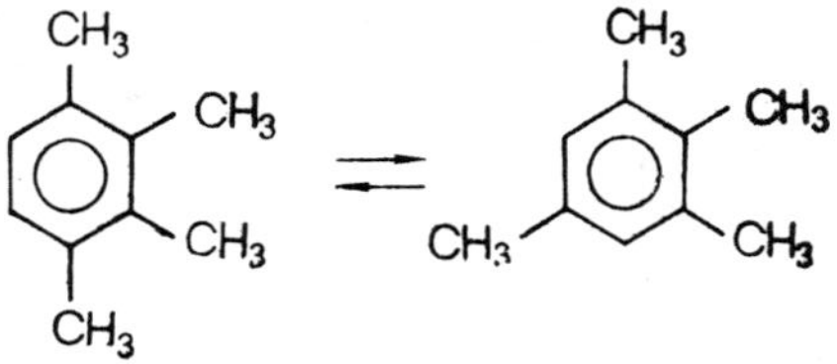 trimethylbenzene ⇌ trimethylbenzene	298 (g)	−7.17 ±1.76	1.2	CC, ST	258
		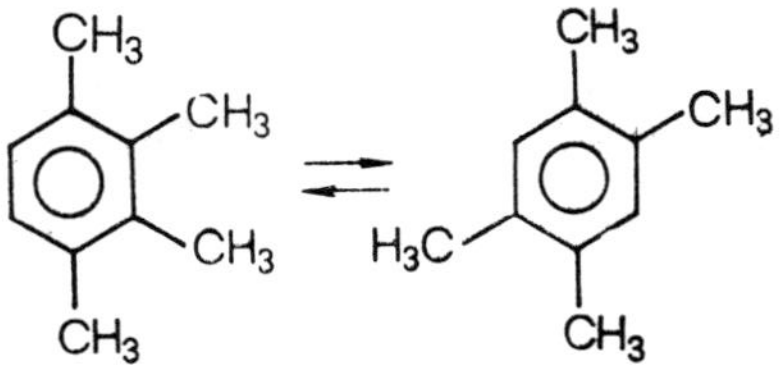 trimethylbenzene ⇌ trimethylbenzene	298 (g)	−10.89 ±2.13	−5.0	CC, ST	258
$C_{10}H_{14}O$	1.3	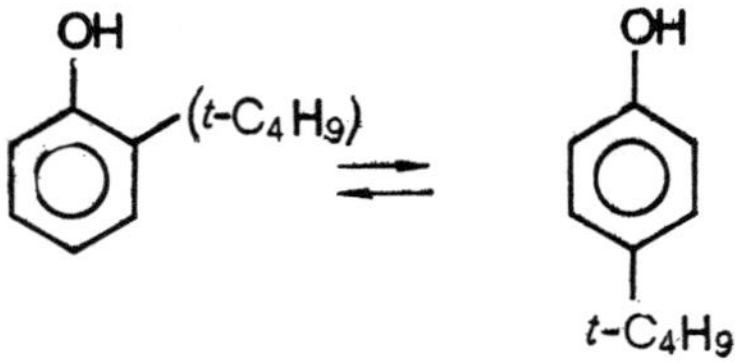 o-(t-C$_4$H$_9$)-phenol ⇌ p-(t-C$_4$H$_9$)-phenol	388 (l)	−18.57 ±0.79	−17.32 ±2.05	SE, CT, GLC (C$_2$H$_5$)$_2$O	216

Formula		Reaction				Method	Ref.
		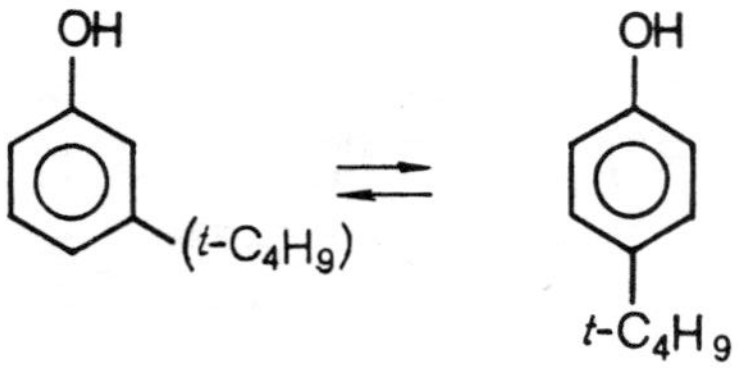 3-(t-C$_4$H$_9$)-C$_6$H$_4$OH ⇌ 4-(t-C$_4$H$_9$)-C$_6$H$_4$OH	413 (l)	0.96 ±1.34	−8.62 ±3.30	SE, CT, GLC (C$_2$H$_5$)$_2$O	216
C$_{10}$H$_{14}$O$_2$	1.3	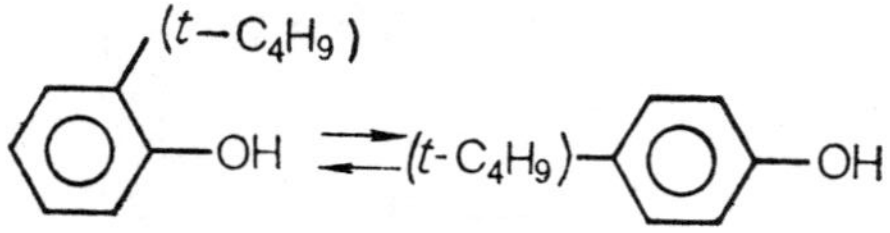 2-(t-C$_4$H$_9$)-C$_6$H$_4$OH ⇌ 4-(t-C$_4$H$_9$)-C$_6$H$_4$OH	408 (l)	−16.9 ±1.6	−12.5 ±3.9	SE, CT, GLC	262
		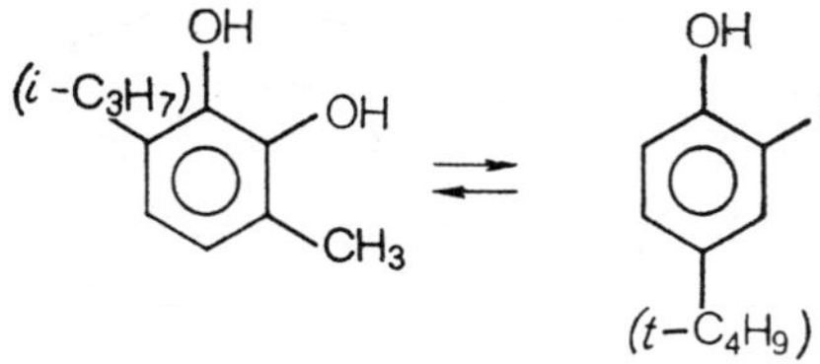 3-(i-C$_3$H$_7$)-2-CH$_3$-C$_6$H$_3$(OH)$_2$ ⇌ 4-(t-C$_4$H$_9$)-C$_6$H$_3$(OH)$_2$	298 (g)	4.4 ±2.6	—	CC	256
C$_{10}$H$_{16}$	1.2 1.3	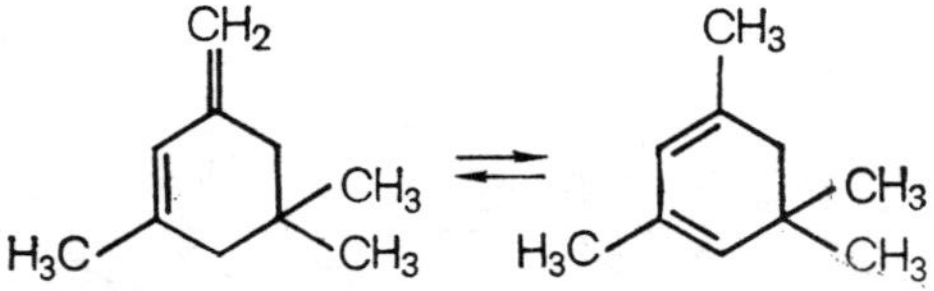 (1,3,3,5-tetramethyl-5-methylenecyclohexene) ⇌ (1,3,5,5-tetramethyl-1,3-cyclohexadiene)	629 (g)	1.67 ±1.25	5.86 ±3.35	SE, CT, GLC	217
	1.2	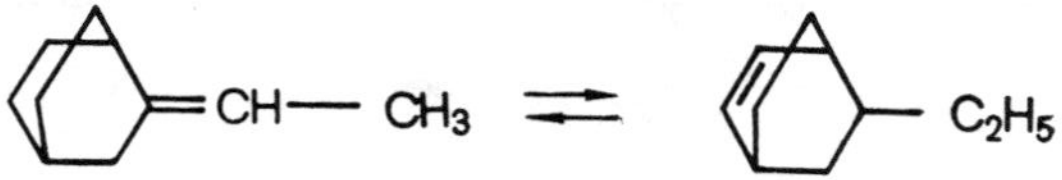 (ethylidene-norbornane) ⇌ (endo-2-ethyl-norbornene)	323 (l)	5.45 ±0.13	5.44 ±0.42	SE, CT, GLC	218

Table (Continued)

Formula	Type	Reaction	T, K (phase)	$\Delta_r H_T^\circ$ ($\Delta_r G_T^\circ$)	$\Delta_r S_T^\circ$	Technique	Reference
$C_{10}H_{16}O$	2.1	(cis-) ⇌ (trans-) cyclohexanone decalone	531 (l)	−9.41 ±1.00	−28.19 ±1.88	SE, CT, GLC	219
			498 (l)	−9.20 ±2.09	−7.5 ±4.2	SE, CT, GLC	220
			488 (l)	−10.49 ±1.30	−2.51 ±2.50	SE, CT, GLC	221
$C_{10}H_{18}$	1.2	bicyclopentyl ⇌ (trans-) decalin	298 (l)	−55.2	—	CC	222
	1.3	$CH{\equiv}C{-}CH_2{-}CH_2{-}CH_2{-}CH_2{-}C_4H_9 \rightleftarrows CH_3{-}C{\equiv}C{-}CH_2{-}CH_2{-}CH_2{-}C_4H_9$	298 (l)	−18.32 ±4.57	—	RC, CT, GLC	112
		$CH_3{-}C{\equiv}C{-}CH_2{-}CH_2{-}CH_2{-}C_4H_9 \rightleftarrows CH_3{-}CH_2{-}C{\equiv}C{-}CH_2{-}CH_2{-}C_4H_9$	298 (l)	−1.72 ∓4.47	—	RC, CT	112
		$CH_3{-}CH_2{-}C{\equiv}C{-}CH_2{-}CH_2{-}C_4H_9 \rightleftarrows CH_3{-}CH_2{-}CH_2{-}C{\equiv}C{-}CH_2{-}C_4H_9$	298 (l)	−1.97 ±4.18	—	RC, CT	112
		$CH_3{-}CH_2{-}CH_2{-}C{\equiv}C{-}CH_2{-}C_4H_9 \rightleftarrows CH_3{-}CH_2{-}CH_2{-}CH_2{-}C{\equiv}C{-}C_4H_9$	298 (l)	−1.21 ±4.18	—	RC, CT	112

$C_{10}H_{18}O$	2.1	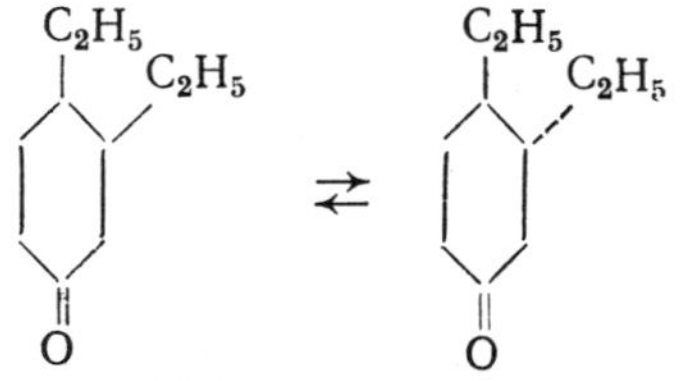 (structures: cis-/trans- diethyl cyclohexenone)	413 (l)	−1.88 ±0.42	2.84 ±0.63	SE, CT, GLC	179
$C_{10}H_{18}O_2$	2.1	 (structures: bicyclic diol isomers)	503 (g)	(9.4 ±1.3)	—	FE, CT, GLC; $C_{12}H_{26}$; 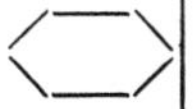 ⬡	142
		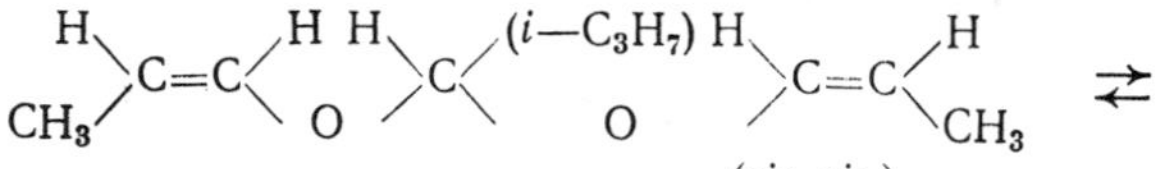 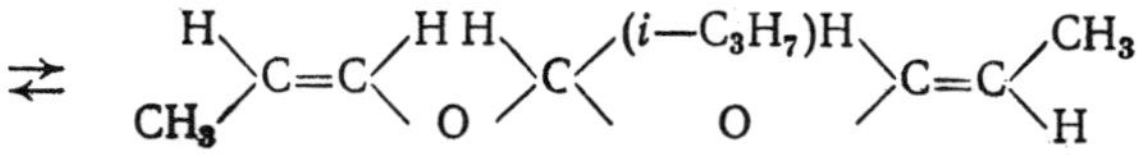 (cis-,cis- acetal)	298 (l)	2.3 ±0.4	4.0 ±1.4	SE, CT, GLC	92
		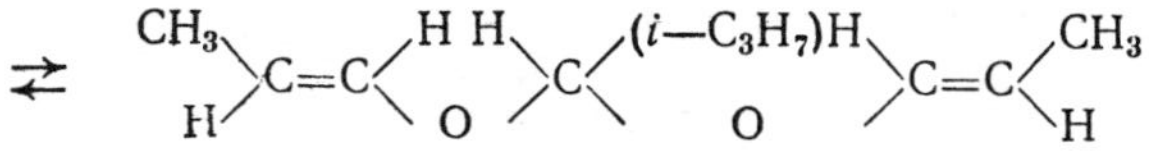 (cis-,trans- and trans-,trans- acetals)	298 (l)	2.9 ±0.2	−5.1 ±0.7	SE, CT, GLC	92

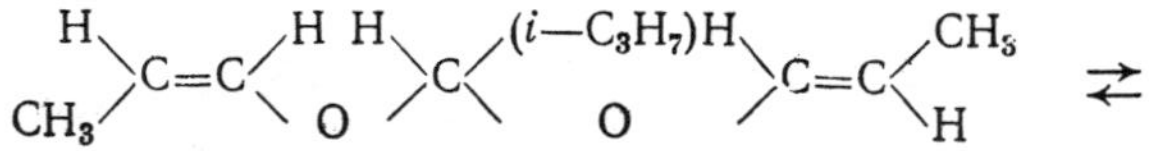

Table *(Continued)*

Formula	Type	Reaction	T, K (phase)	$\Delta_r H_T^\circ$ ($\Delta_r G_T^\circ$)	$\Delta_r S_T^\circ$	Technique	Reference
$C_{10}H_{18}O_2$	2.1	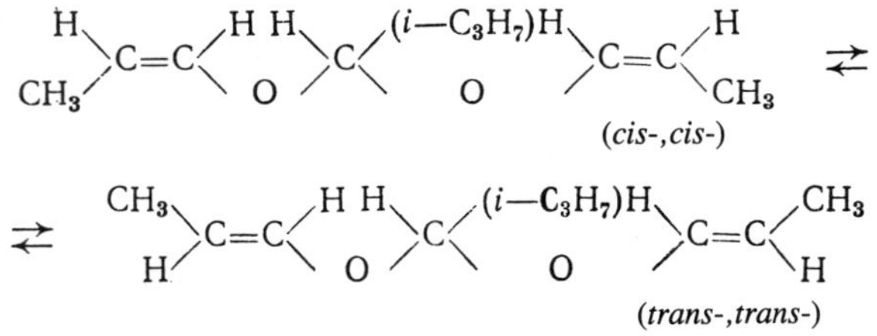 (cis-,cis-) ⇌ (trans-,trans-)	298 (l)	5.2 ±0.6	−3.1 ±0.7	SE, CT, GLC	92
$C_{10}H_{18}O_4$	2.1	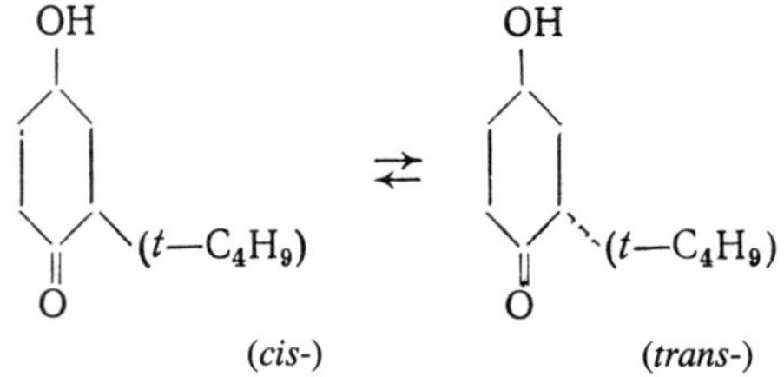 (cis-) ⇌ (trans-)	348 (l)	−0.54 ±0.13	−0.54 ±0.42	SE, CT, GLC  (⌬); HCl	223
		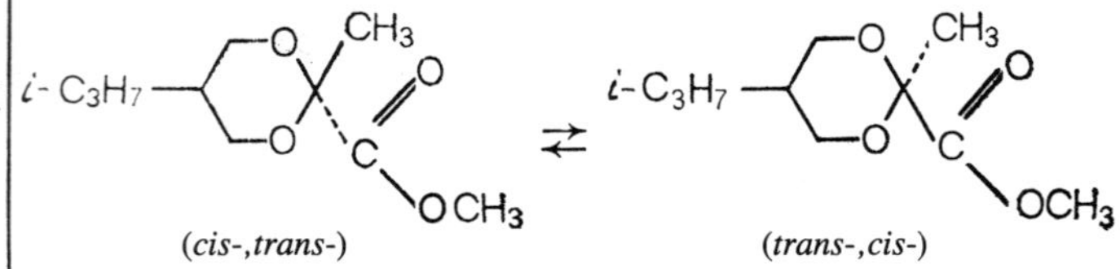 (cis-,trans-) ⇌ (trans-,cis-)	298 (l)	(−0.38 ±0.13)	—	SE, CT, GLC $(C_2H_5)_2O$	167
$C_{10}H_{20}$	1.3	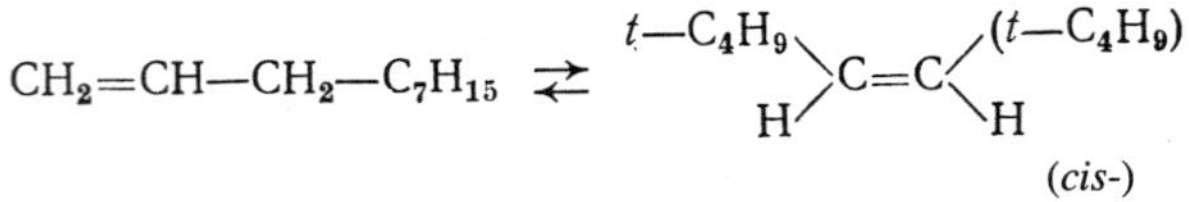 $CH_2=CH-CH_2-C_7H_{15}$ ⇌ (cis-)/(trans-)	298 (l)	10.25 ±1.72	—	CC	156

2.1	t-C$_4$H$_9$\C=C/(t-C$_4$H$_9$) / H H \ (cis-) ⇌ t-C$_4$H$_9$\C=C/H / H \(t-C$_4$H$_9$) (trans-)	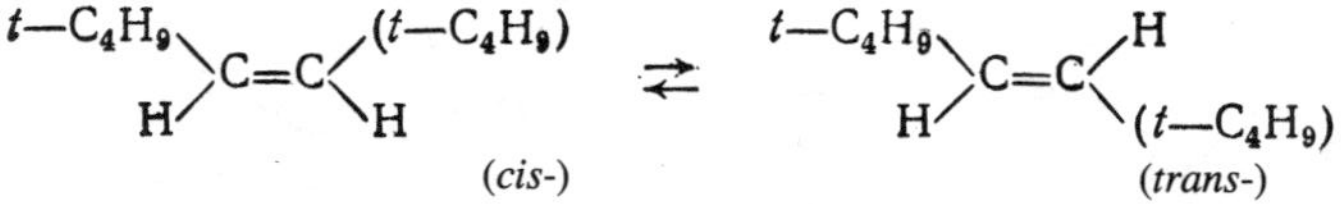 298 (l)	−43.96 ±2.09	—	CC	156
2.1	2-t-butylphenol (cis-) ⇌ (trans-)	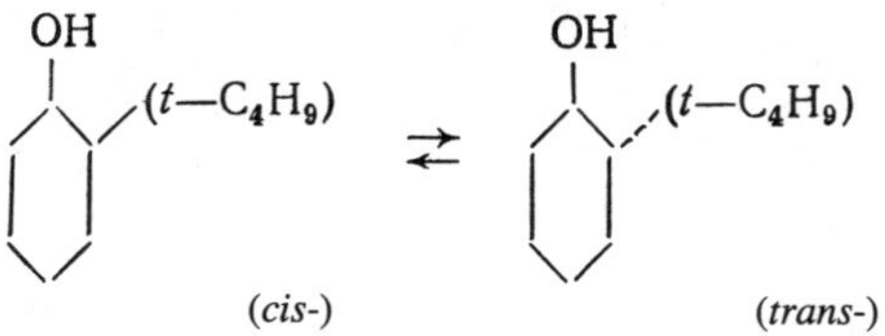 384 (l)	−3.35 ±0.42	−3.64 ±0.59	SE, CT, GLC ⬡	224
	(cis-) ⇌ (trans-)	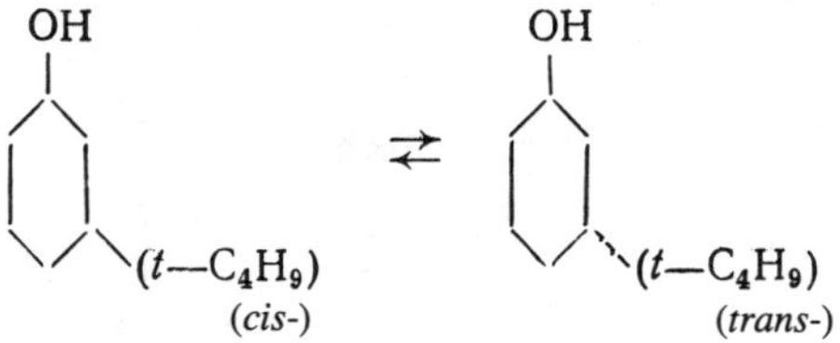 353 (l)	(3.39)	—	SE, CT, GLC  ⌬	162
	3-t-butylphenol (cis-) ⇌ (trans-)	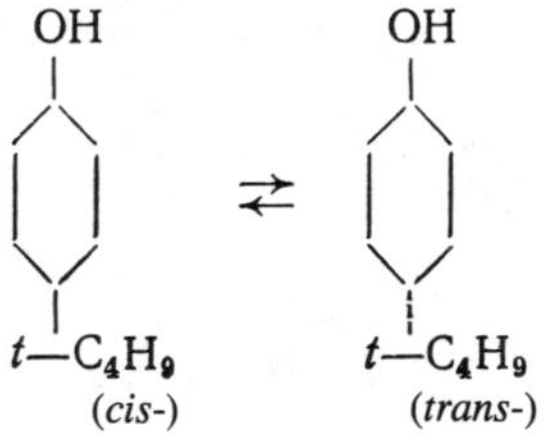 353 (l)	(−2.51 ±0.13)	—	SE, CT, GLC ⬡	185
	4-t-butylphenol (cis-) ⇌ (trans-)	371 (l)	−2.43 ±0.08	0.25 ±0.17	SE, CT, GLC ⬡	208

C$_{10}$H$_{20}$O

Table *(Continued)*

Formula	Type	Reaction	T, K (phase)	$\Delta_r H_T^\circ$ ($\Delta_r G_T^\circ$)	$\Delta_r S_T^\circ$	Technique	Reference
$C_{10}H_{20}O$	2.1	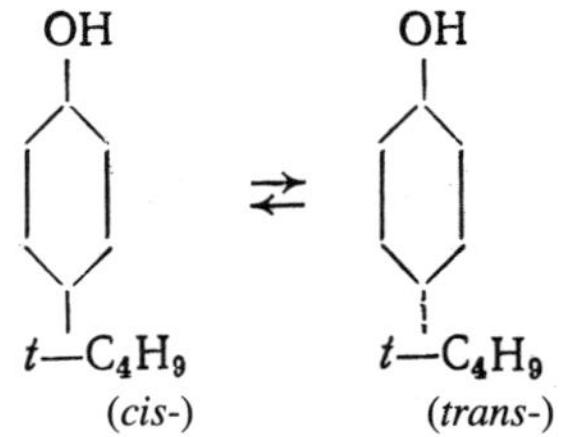 OH–C₆H₄–t-C₄H₉ (*cis*-) ⇌ OH–C₆H₄–t-C₄H₉ (*trans*-)	382 (l)	−3.47 ±0.08	−1.25 ±0.21	SE, CT, GLC CH₃OCH₂—CH₂OCH₃	208
		(cis,cis-) ⇌ (trans,trans-) [i-C₃H₇, CH₃ cyclohexanol]	382 (l)	−4.56 ±0.08	−1.92 ±0.17	SE, CT, GLC i–C₃H₇OH	208
			375 (l)	−4.94 ±0.08	−3.18	SE, CT, GLC t–C₄H₉OH	208
		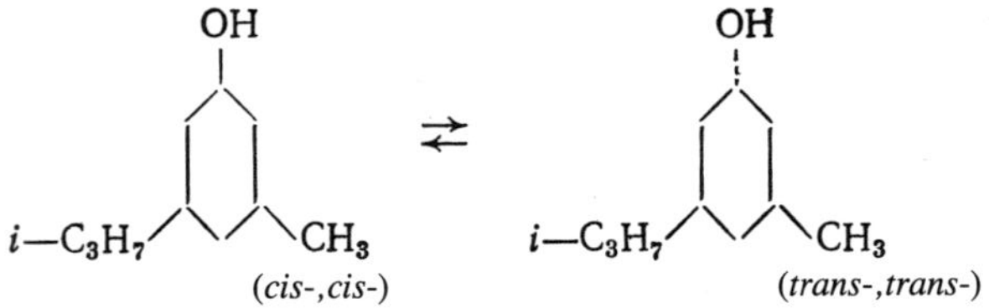	353 (l)	(4.68)	—	SE, CT, GLC i–C₃H₇OH	208
		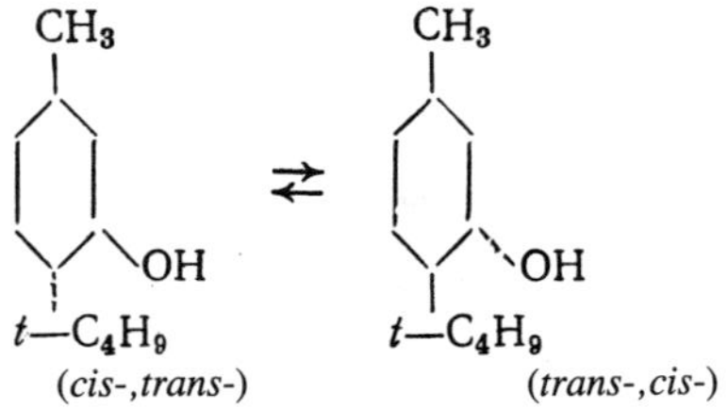 (cis,trans-) ⇌ (trans,cis-) [t-C₄H₉, CH₃, OH cyclohexanol]	423 (l)	(2.3 ±0.2)	—	SE, CT, GLC	225

$C_{10}H_{20}O_2$	2.1	(cis-) i-C₃H₇ ⇌ (trans-) i-C₃H₇	503 (g)	(7.4 ±0.8)	—	SE, CT, GLC ⬡	142
			333 (l)	(−3.76 ±0.04)	—	SE, CT, GLC	165
		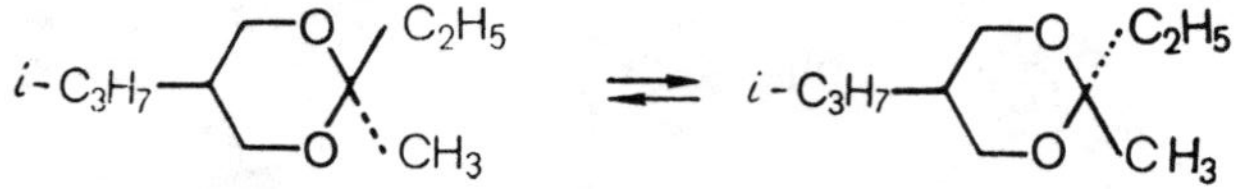 (cis-,trans-) ⇌ (trans-,cis-) with C₂H₅, CH₃, i-C₃H₇	298 (l)	(0.25)	—	SE, CT, GLC $(C_2H_5)_2O$	190
$C_{10}H_{20}O_3$	2.1	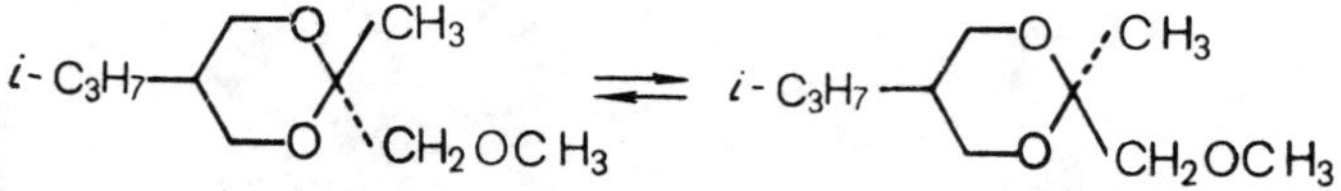 (cis-,trans-) ⇌ (trans-,cis-) with CH₃, CH₂OCH₃, i-C₃H₇	298 (l)	(−2.64)	—	SE, CT, GLC $(C_2H_5)_2O$	190
			298 (l)	(−3.05 ±0.13)	—	SE, CT, GLC CCl_4	167

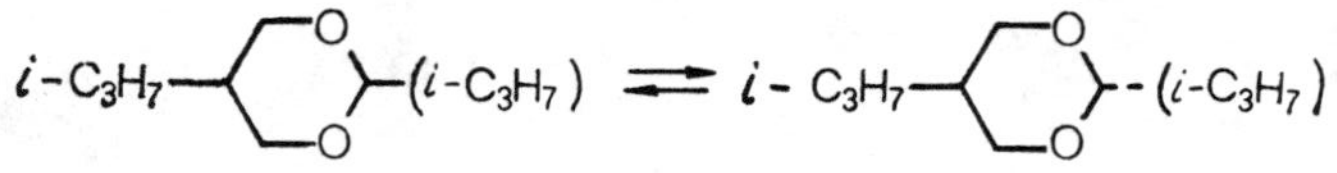

Table (Continued)

Formula	Type	Reaction	T, K (phase)	$\Delta_r H_T^\circ$ ($\Delta_r G_T^\circ$)	$\Delta_r S_T^\circ$	Technique	Reference
$C_{10}H_{20}S_2$	2.1	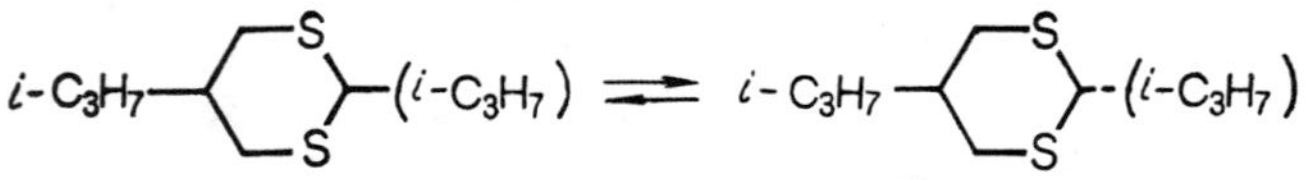 *i*-C_3H_7—[1,3-dithiane](cis-) ⇌ [1,3-dithiane]—(*i*-C_3H_7) (trans-)	342 (l)	(−3.26 ±0.04)	—	SE, CT, GLC $CHCl_3$	137
		 *t*-C_4H_9—[1,3-dithiane]—C_2H_5 (cis-) ⇌ [1,3-dithiane]—C_2H_5 (trans-)	342 (l)	(−3.22 ±0.08)	—	SE, CT, GLC $CHCl_3$	137
		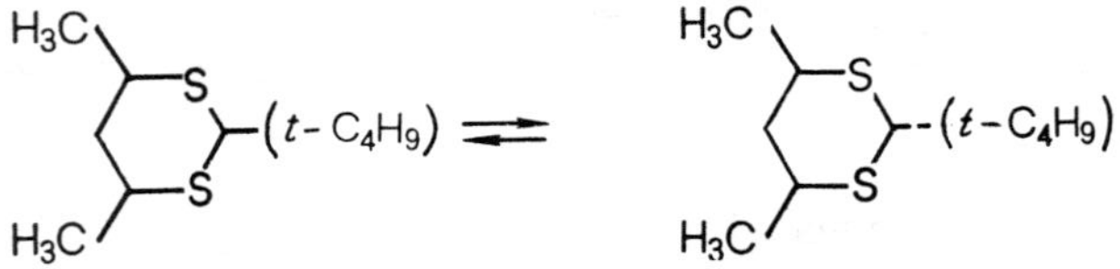 H_3C—[1,3-dithiane]—CH_3 (*t*-C_4H_9) (cis-,cis-) ⇌ H_3C—[1,3-dithiane]—CH_3 (*t*-C_4H_9) (trans-,trans-)	342 (l)	(11.38 ±0.08)	—	SE, CT, GLC $CHCl_3$	137
$C_{10}H_{21}Cl$	1.3	$CH_3-CH-CH_2-CH_2-CH_2-C_5H_{11}$ with Cl ⇌ $CH_3-CH_2-CH_2-CH-CH_2-CH_2-C_5H_{11}$ with Cl	343 (l)	0.02 ±0.42	−2.93 ±1.25	SE, CT, GLC	96

170

Formula	Structure	T (K)			Method	Ref.
	CH₃—CH(Cl)—CH₂—CH₂—CH₂—C₅H₁₁ ⇌ CH₃—CH₂—CH(Cl)—CH₂—CH₂—C₅H₁₁	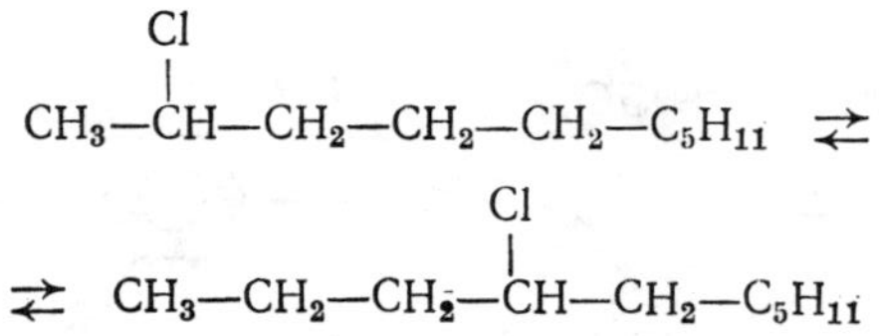 343 (l)	0.08 ±0.42	−5.4 ±1.3	SE, CT, GLC	96
	CH₃—CH₂—CH(Cl)—CH₂—CH₂—C₅H₁₁ ⇌ CH₃—CH₂—CH₂—CH₂—CH(Cl)—C₅H₁₁	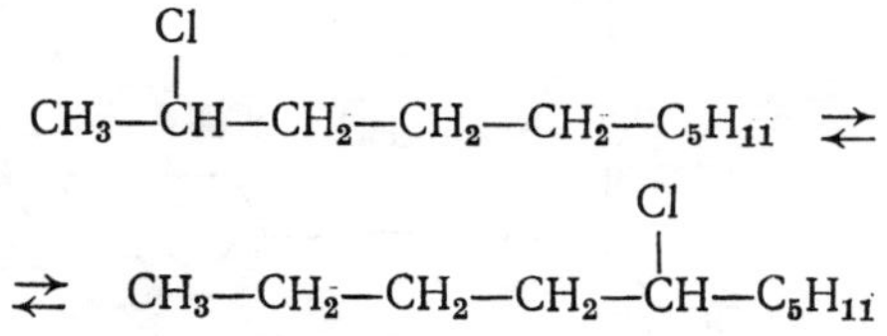 343 (l)	1.17 ±0.42	−4.20 ±1.24	SE, CT, GLC	96
$C_{11}H_{10}$ 1.3	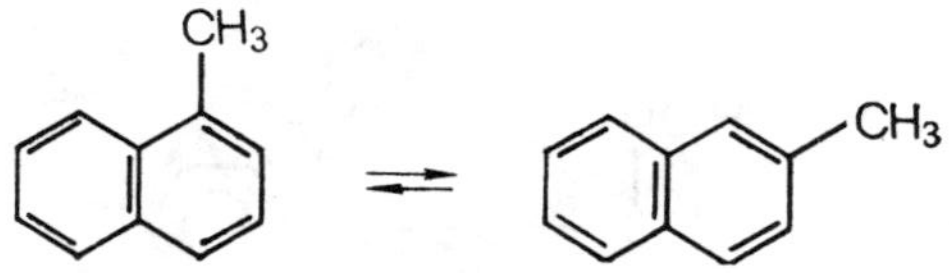 2-methylnaphthalene ⇌ 1-methylnaphthalene	298 (g)	−0.75 ±3.68	—	CC	226
$C_{11}H_{10}N_2O_2S$ 1.1 1.2	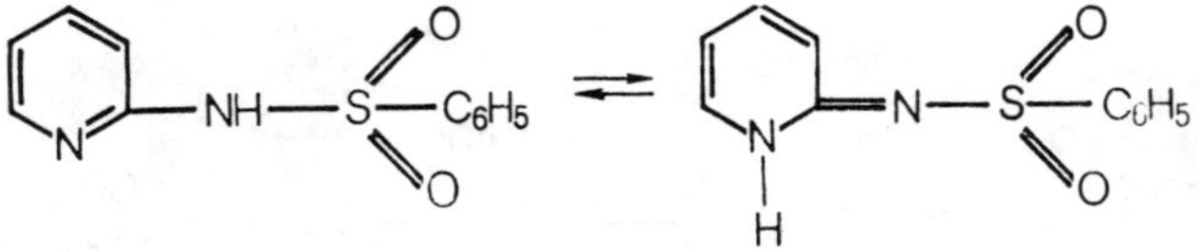 pyridine-N=S(O)₂C₆H₅ (H) ⇌ pyridine-NH—S(O)₂C₆H₆	321 (l)	−2.04	−7.86	SE, UV H₂O	69
$C_{11}H_{10}O_2$ 1.2	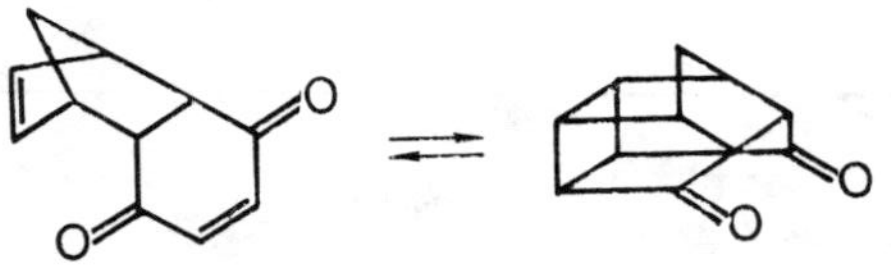 (diketone isomerization)	298 (g)	−47.35	—	CC	210

Table (Continued)

Formula	Type	Reaction	T, K (phase)	$\Delta_r H_T^\circ$ ($\Delta_r G_T^\circ$)	$\Delta_r S_T^\circ$	Technique	Reference
$C_{11}H_{11}BrO_3$	1.1 / 1.2	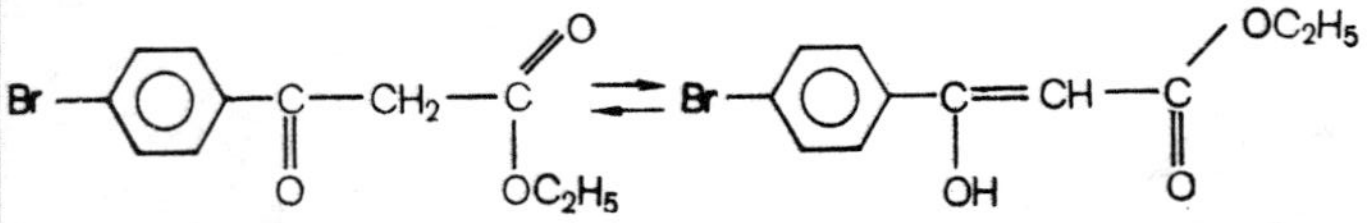 (4-Br-C₆H₄-CO-CH₂-COOC₂H₅ ⇌ 4-Br-C₆H₄-C(OH)=CH-COOC₂H₅)	293 (l)	−5.02 ±0.42	−16.7	SE, NMR CCl₄	227
$C_{11}H_{11}ClO_3$	1.1 / 1.2	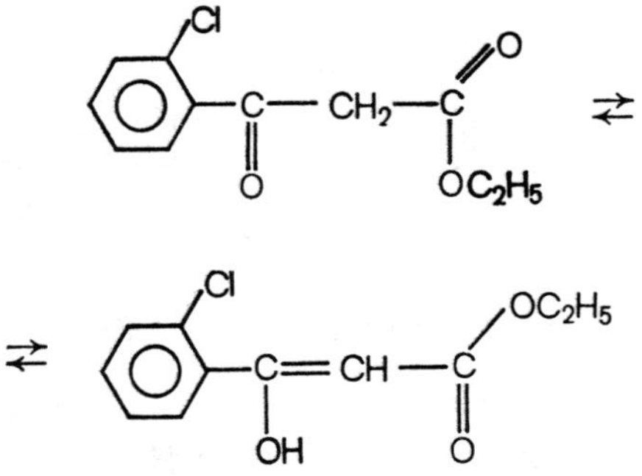 (2-Cl-C₆H₄-CO-CH₂-COOC₂H₅ ⇌ 2-Cl-C₆H₄-C(OH)=CH-COOC₂H₅)	293 (l)	−6.27 ±0.42	−16.7	SE, NMR CCl₄	227
		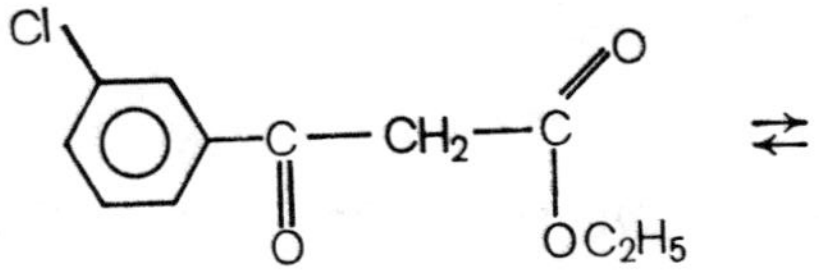 (3-Cl-C₆H₄-CO-CH₂-COOC₂H₅ ⇌ ...)	293 (l)	−5.86 ±0.42	−16.7	SE, NMR CCl₄	227

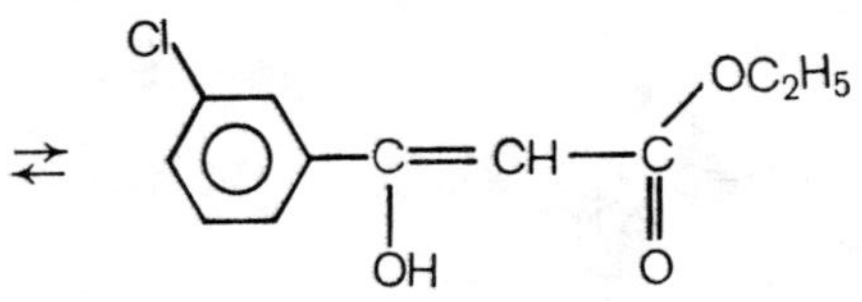

		↑↓ ethyl 3-(3-chlorophenyl)-3-hydroxyacrylate ⇌ ethyl (4-chlorobenzoyl)acetate; ↑↓ ethyl 3-(4-chlorophenyl)-3-hydroxyacrylate ⇌ ethyl (2-nitrobenzoyl)acetate ⇌ ethyl 3-(2-nitrophenyl)-3-hydroxyacrylate	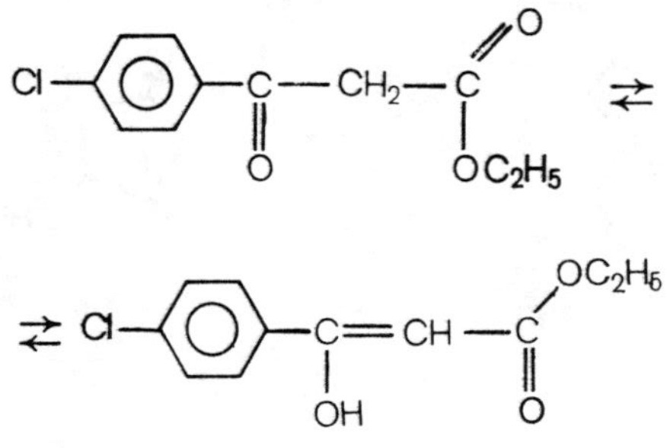 293 (l)	−5.02 ±0.42	−16.7	SE, NMR CCl$_4$	227
C$_{11}$H$_{11}$NO$_5$	1.1 1.2	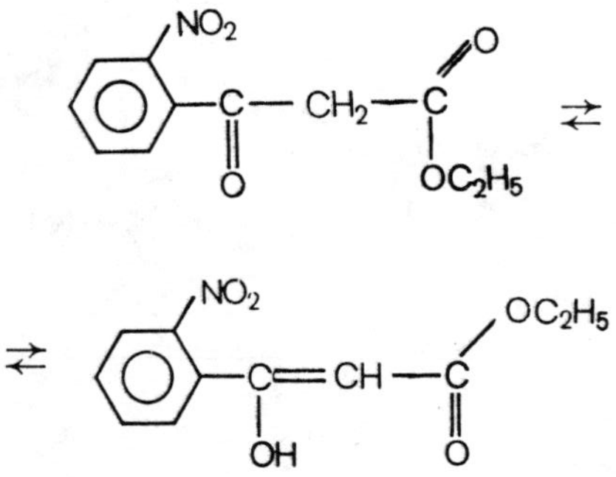	293 (l)	−2.93 ±0.84	−20.9	SE, NMR CCl$_4$	227

Table (Continued)

Formula	Type	Reaction	T, K (phase)	$\Delta_r H°_T$ ($\Delta_r G°_T$)	$\Delta_r S°_T$	Technique	Reference
$C_{11}H_{11}NO_5$	1.1 1.2	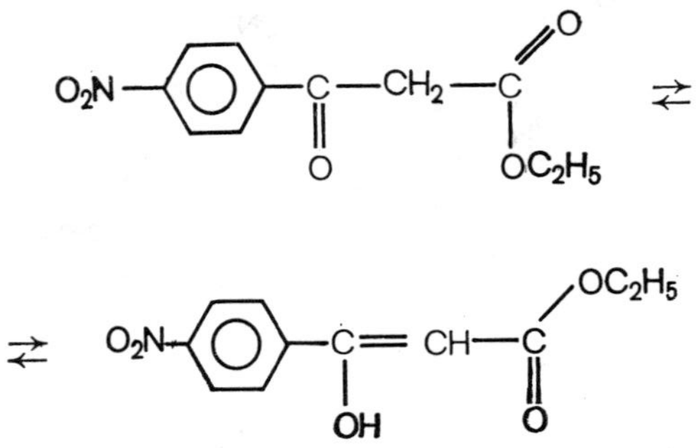 $O_2N-C_6H_4-CO-CH_2-CO-OC_2H_5 \rightleftarrows O_2N-C_6H_4-C(OH)=CH-CO-OC_2H_5$	293 (l)	-6.69 ± 0.84	-12.5	SE, NMR CCl_4	227
$C_{11}H_{12}O_2$	2.1	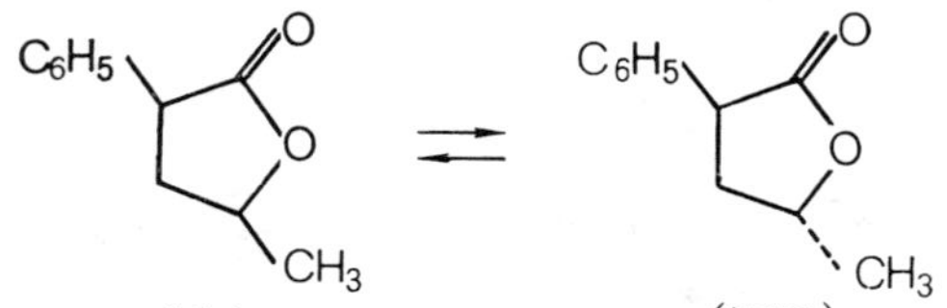 (cis-) $\rightleftarrows$ (trans-) lactone with C_6H_5 and CH_3	298 (l)	(0.79)	—	SE, CT, NMR $t-C_4H_9OH$	121
$C_{11}H_{12}O_3$	1.1 1.2	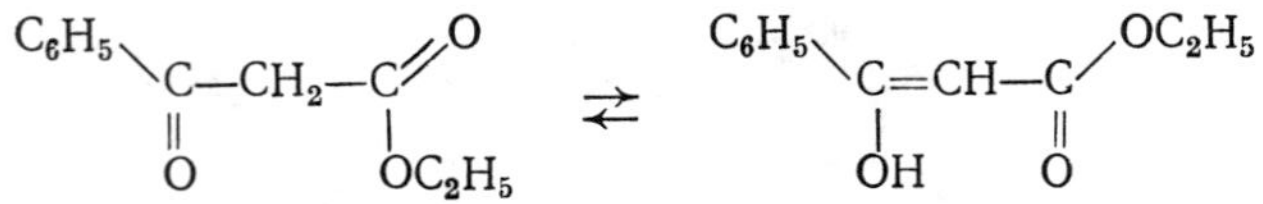 $C_6H_5-CO-CH_2-CO-OC_2H_5 \rightleftarrows C_6H_5-C(OH)=CH-CO-OC_2H_5$	293 (l)	-3.76 ± 0.84	-12.5	SE, NMR CCl_4	227

$C_{11}H_{14}$	1.3	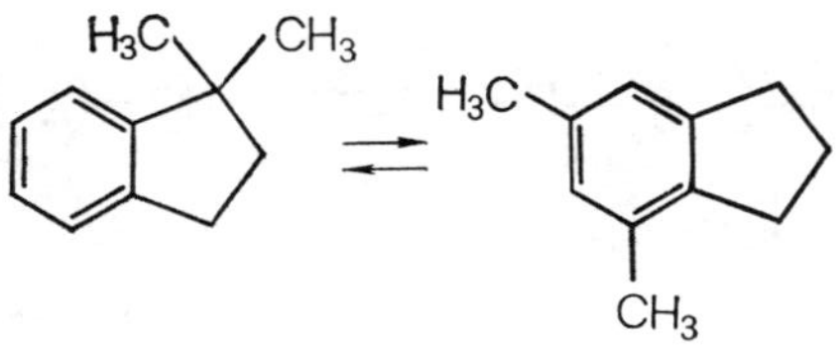 (structures: 1,1-dimethylindane ⇌ 4,6-dimethyl-1-methylindane type; and 4,7-dimethylindane ⇌ 4,6-dimethylindane)	298 (l)	−10.21 ±2.31	—	CC	228
		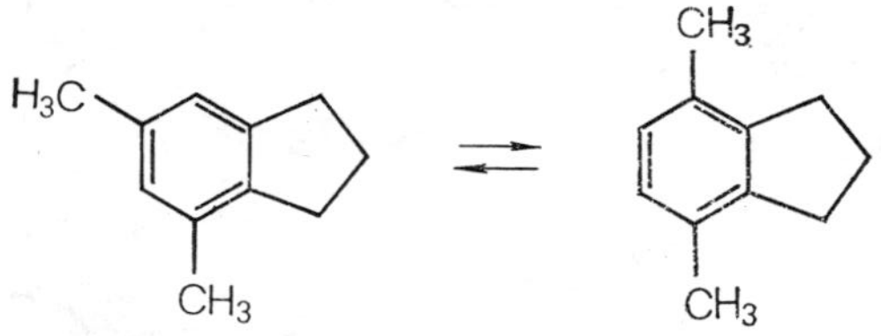	298 (l)	−1.97 ±2.12	—	CC	228
$C_{11}H_{14}O$	2.1	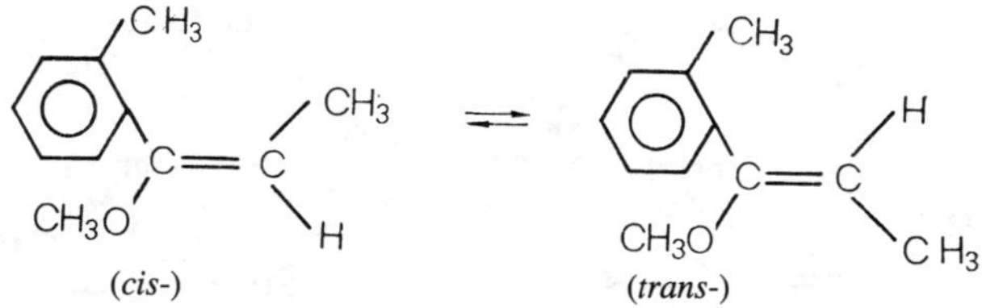 (cis-/trans- methyl 2-methylphenyl methoxymethylene ether)	298 (l)	5.35	10.12	SE, CT, NMR	173
		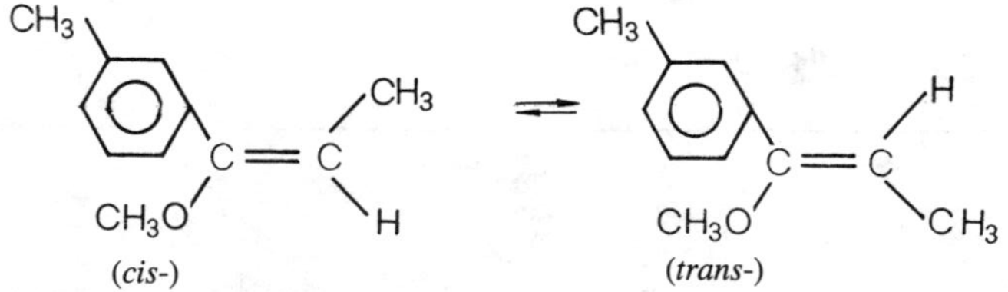 (cis-/trans- methyl 3-methylphenyl methoxymethylene ether)	298 (l)	0.13	5.59	SE, CT, NMR	173

Table (*Continued*)

Formula	Type	Reaction	T, K (phase)	$\Delta_r H_T^\circ$ ($\Delta_r G_T^\circ$)	$\Delta_r S_T^\circ$	Technique	Reference
$C_{11}H_{14}O_2$	2.1	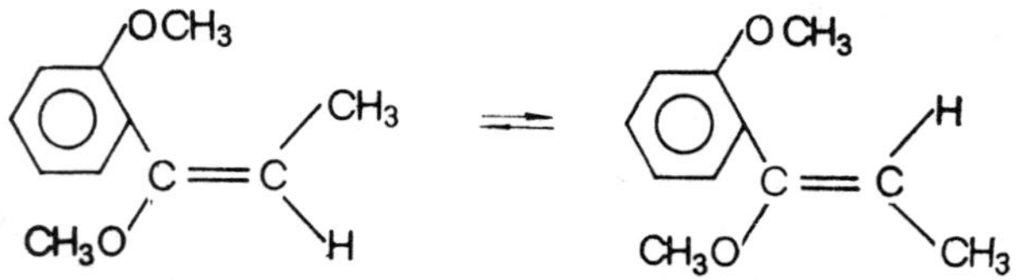 cis- ⇌ trans- (methyl methoxy methoxyphenyl enol ethers)	298 (l)	3.87	9.22	SE, CT, NMR	173
		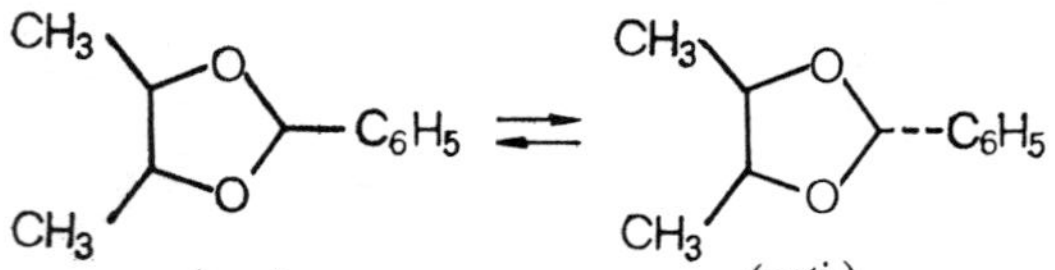 syn- ⇌ anti- (2-phenyl-4,5-dimethyl-1,3-dioxolane)	298 (l)	(1.61 ±0.03)	—	SE, CT, GLC $(C_2H_5)_2O$	93
$C_{11}H_{16}$	1.3	(1-phenylpropyl) ⇌ (1-phenylbutyl isomers)	342 (l)	−0.16 ±0.08	8.45 ±0.23	SE, CT, GLC	263
$C_{11}H_{16}O$	1.3	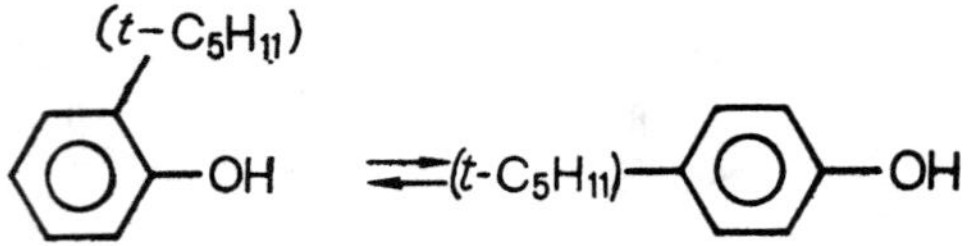 (t-C_5H_{11})-2-phenol ⇌ (t-C_5H_{11})-4-phenol	408 (l)	−16.7 ±1.8	−10.3 ±4.3	SE, CT, GLC	262

$C_{11}H_{18}O_2$	2.1	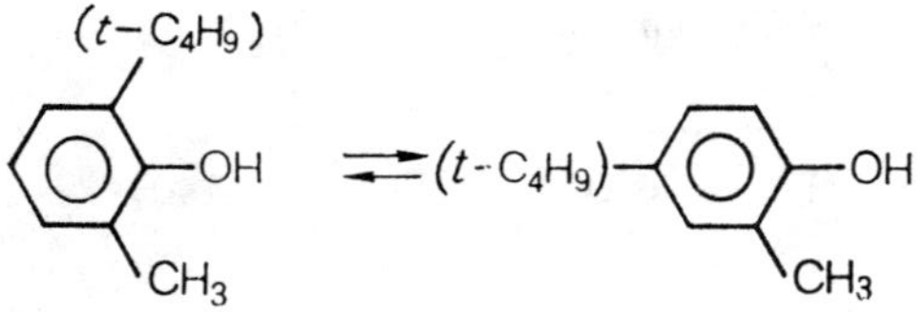 408 (l)	−17.5 ±1.2	−4.4 ±2.9	SE, CT, GLC	262
		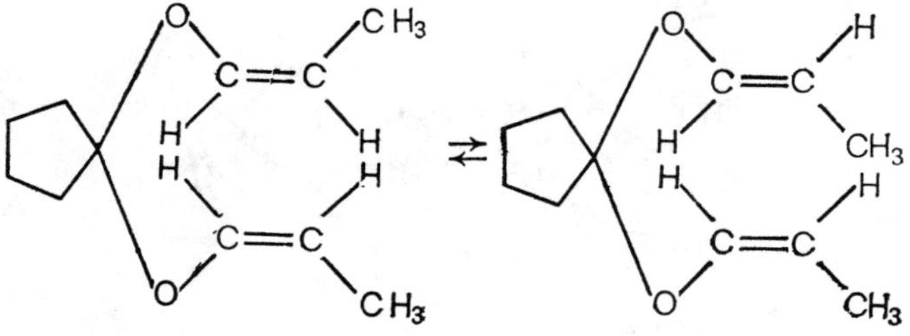 298 (l)	(0.9 ±0.1)	—	SE, CT, GLC	92
		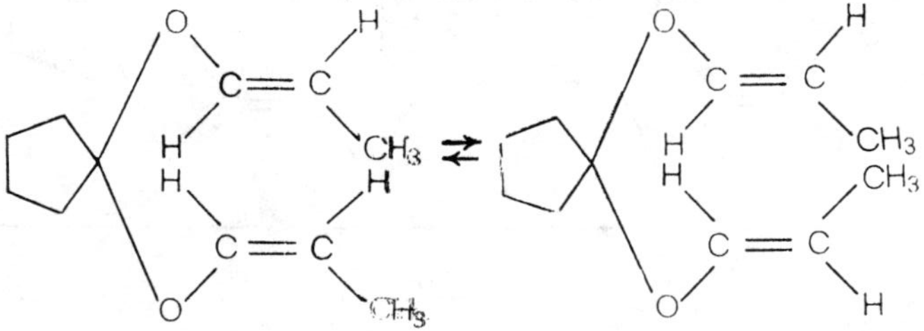 298 (l)	(4.6 ±0.2)	—	SE, CT, GLC	92

Table (Continued)

Formula	Type	Reaction	T, K (phase)	$\Delta_r H_T^\circ$ ($\Delta_r G_T^\circ$)	$\Delta_r S_T^\circ$	Technique	Reference
$C_{11}H_{18}O_2$	2.1	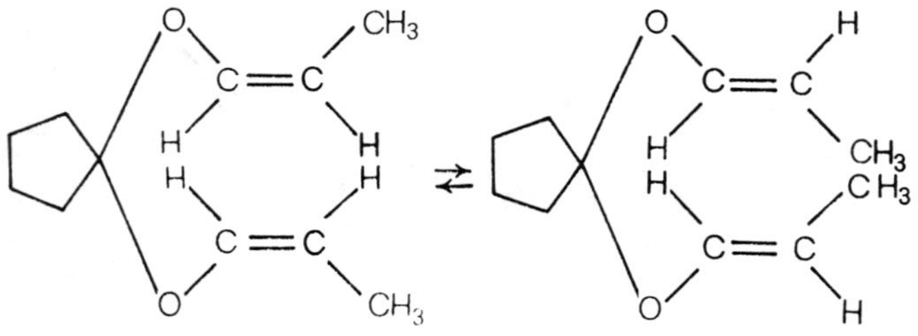 (cis-,cis-) ⇌ (trans-,trans-)	298 (l)	(5.5 ±0.2)	—	SE, CT, GLC	92
$C_{11}H_{18}O_3$	2.1	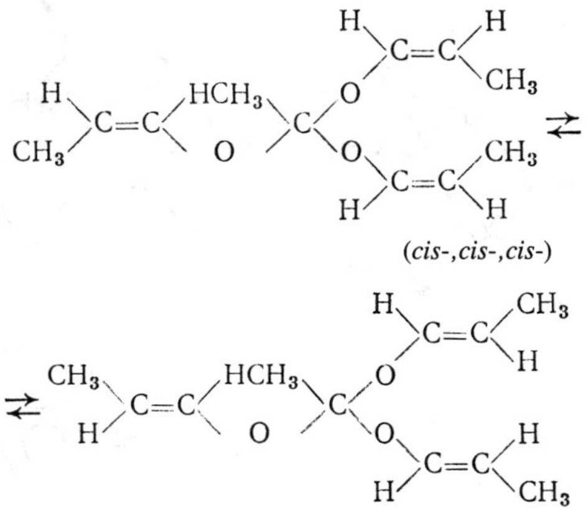 (cis-,cis-,cis-) ⇌ (trans-,trans-,trans-)	298 (l)	(8.66 ±0.13)	—	SE, CT, GLC 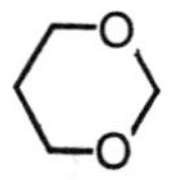	134

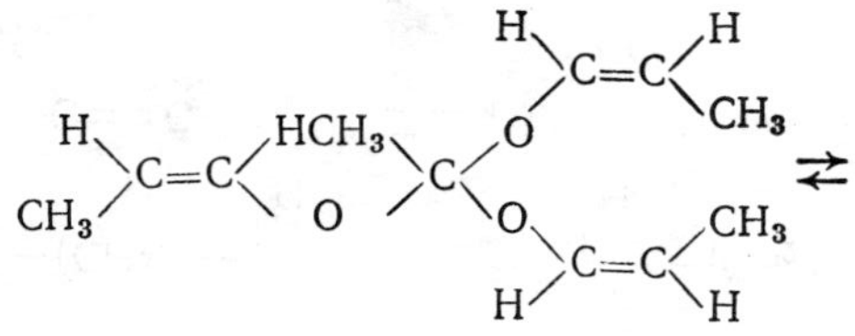

Structure	T (K)	ΔG (kcal/mol)		Method	Ref.
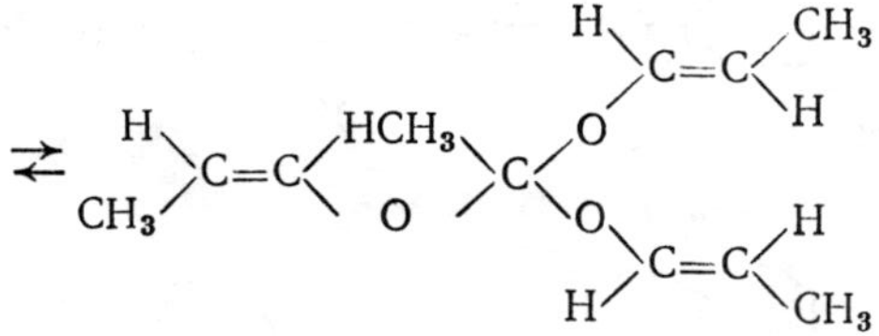 (cis-,cis-,cis-) 1,3-dioxane derivative	298 (l)	(3.25 ±0.12)	—	SE, CT, GLC 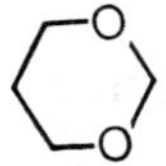	134
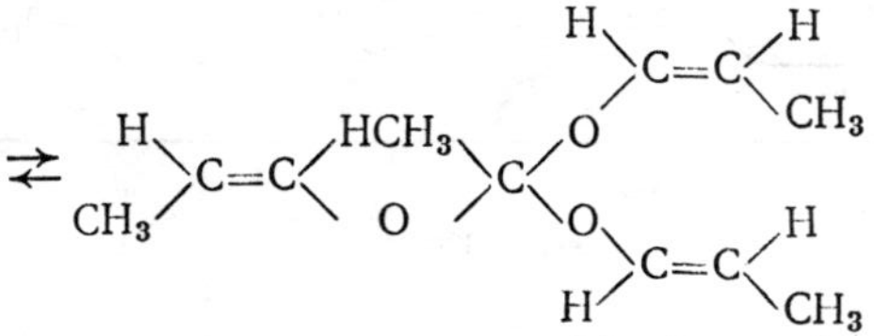 (cis-,cis-,cis-) 1,3-dioxane derivative	298 (l)	(0.52 ±0.08)	—	SE, CT, GLC 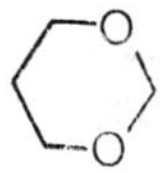	134

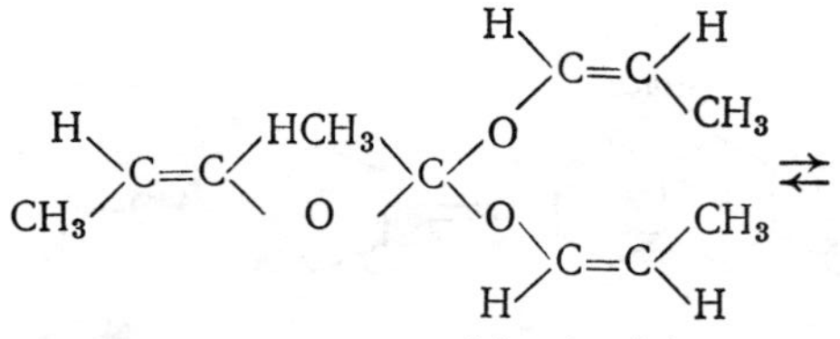

(cis-,cis-,cis-) ⇌ (cis-,trans-,trans-)

(cis-,cis-,cis-) ⇌ (cis-,cis-,trans-)

Table (Continued)

Formula	Type	Reaction	T, K (phase)	$\Delta_r H_T^\circ$ ($\Delta_r G_T^\circ$)	$\Delta_r S_T^\circ$	Technique	Reference
$C_{11}H_{19}ClO$	2.1	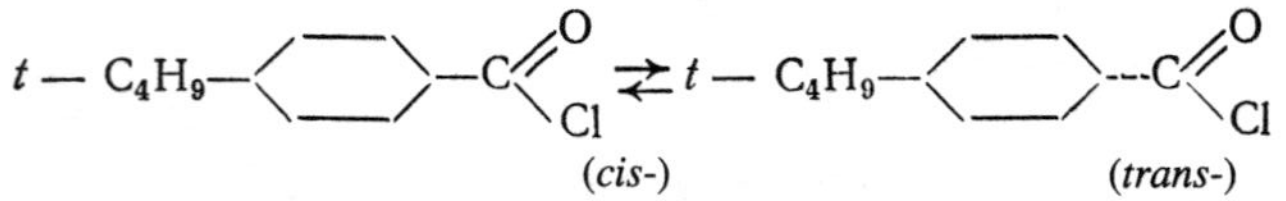 (cis-) $t-C_4H_9$–⌬–COCl ⇌ $t-C_4H_9$–⌬–COCl (trans-)	491 (l)	−5.81 ±0.21	−1.34 ±0.46	SE, GLC $C_{12}H_{26}$	229
$C_{11}H_{20}$	1.2	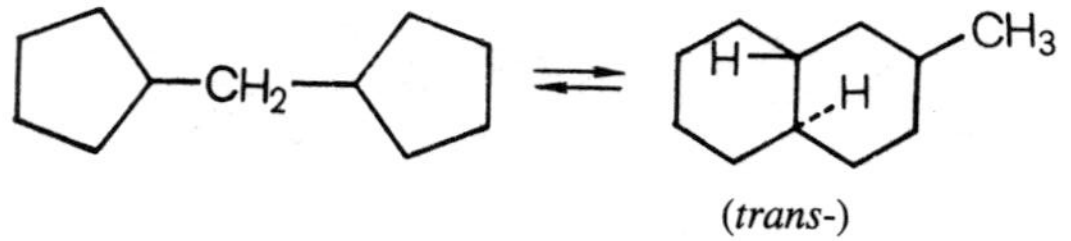 cyclopentyl-CH₂-cyclopentane ⇌ (trans-) 2-methyl decalin	298 (l)	−59.40	—	CC	222
		cyclohexyl-cyclopentane ⇌ (trans-) 2-methyl decalin	298 (l)	−34.30	—	CC	222
$C_{11}H_{20}O$	2.1	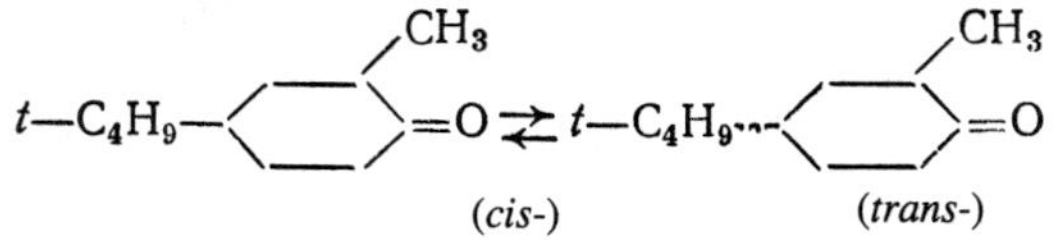 (cis-) $t-C_4H_9$–⌬(CH₃)=O ⇌ $t-C_4H_9$–⌬(CH₃)=O (trans-)	318 (l)	6.57 ±0.88	0.42 ±2.51	SE, CT, GLC $t-C_4H_9OH$	230
		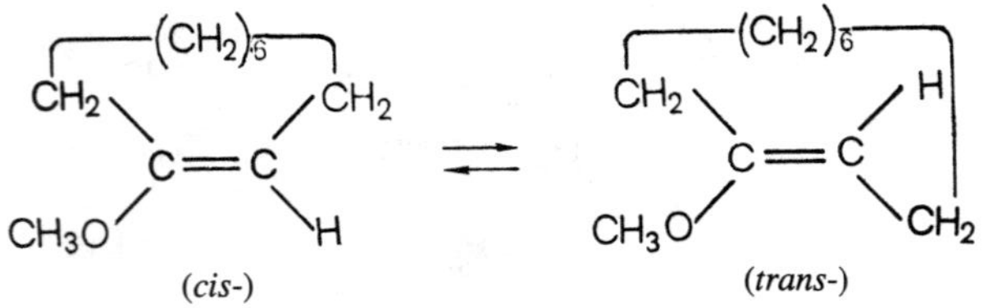 (cis-) enol ether ⇌ (trans-) enol ether	298 (l)	20.8 ±1.2	14.0 ±3.0	SE, CT, GLC	231

$C_{11}H_{20}O_2$	2.1	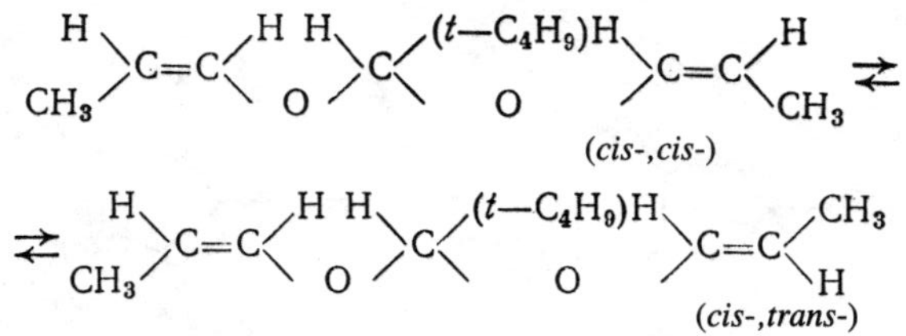 (structures: cis,cis- ⇌ cis,trans-; cis,trans- ⇌ trans,trans-; cis,cis- ⇌ trans,trans- of $CH_3CH=CHCH=CHC(t-C_4H_9)H-C(O)-O-$ esters)	298 (l)	2.3 ±0.2	3.1 ±0.7	SE, CT, GLC	92
		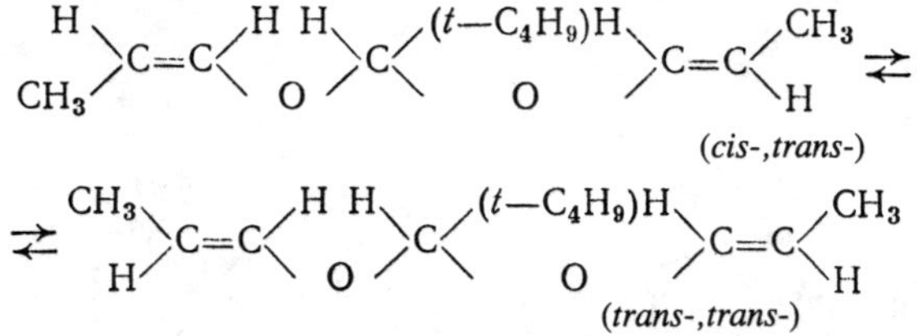	298 (l)	2.3 ±0.7	−8.7 ±2.0	SE, CT, GLC	92
		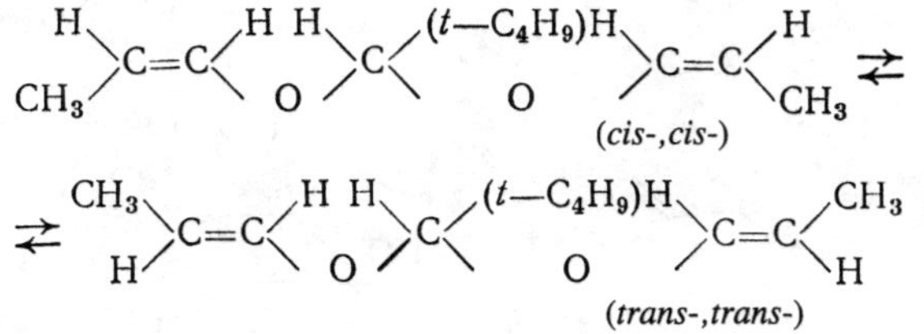	298 (l)	4.6 ±0.9	−5.5 ±2.5	SE, CT, GLC	92
		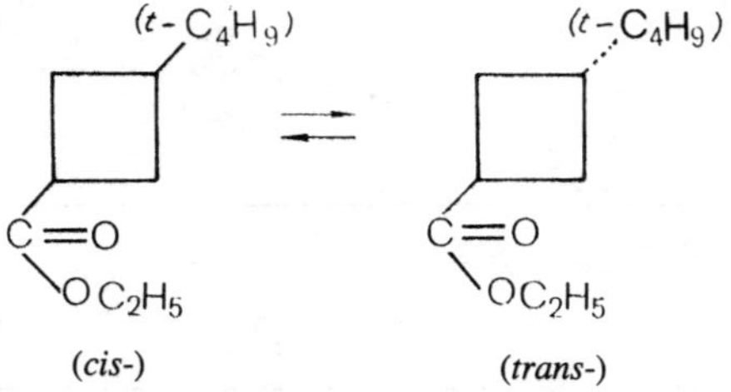 cis ⇌ trans cyclobutane $(t-C_4H_9)$–$CO_2C_2H_5$	389 (l)	3.34 ±0.84	2.93 ±2.09	SE, CT, GLC i–C_3H_7OH	184

Table (*Continued*)

Formula	Type	Reaction	T, K (phase)	$\Delta_r H_T^\circ$ ($\Delta_r G_T^\circ$)	$\Delta_r S_T^\circ$	Technique	Reference
$C_{11}H_{20}O_2$	2.1	$t{-}C_4H_9{-}\!\!\bigcirc\!\!{-}COOH$ (*cis-*) $\rightleftarrows$ $t{-}C_4H_9{-}\!\!\bigcirc\!\!{-}COOH$ (*trans-*)	546 (l)	−6.86 ±0.13	−3.55 ±0.21	SE, GLC $C_{12}H_{26}$	229
$C_{11}H_{22}O$	2.1	$t{-}C_4H_9{-}\!\!\bigcirc\!\!{-}OH$, CH_3 (*cis-,cis-*) $\rightleftarrows$ $t{-}C_4H_9{-}\!\!\bigcirc\!\!{-}OH$, CH_3 (*trans-,trans-*)	353 (l)	(4.81)	—	SE, CT, GLC $i{-}C_3H_7OH$	162
$C_{11}H_{22}O_2$	2.1	$t{-}C_4H_9{-}\!\!\bigcirc\!\!{-}(t{-}C_4H_9)$ (*cis-*) $\rightleftarrows$ $t{-}C_4H_9{-}\!\!\bigcirc\!\!{-}(t{-}C_4H_9)$ (*trans-*)	298 (l)	(1.71 ±0.08)	—	SE, CT, GLC $(C_2H_5)_2O$	93
		C_2H_5 dioxolane with $i{-}C_3H_7$, $i{-}C_3H_7$ (*syn-*) $\rightleftarrows$ (*anti-*)	323 (l)	1.64	2.09	SE, CT, NMR CCl_4	84
			298 (l)	(2.09)	—	SE, CT, NMR $(C_2H_5)_2O$	93

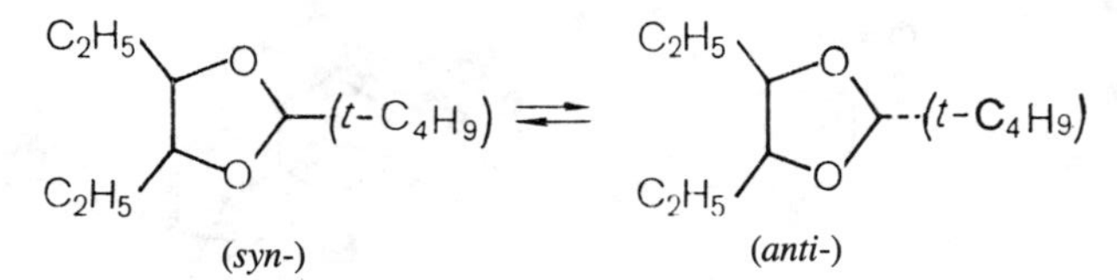 ![syn/anti dioxolane with t-C4H9 and C2H5] (syn-) ⇌ (anti-) with t-C4H9, C2H5 groups	298 (l)	(2.09)	—	SE, CT, NMR $(C_2H_5)_2O$	93
 1,3-dioxane with C6H13 and CH3 (cis-) ⇌ (trans-)	333 (l)	(−3.68 ±0.02)	—	SE, CT, GLC	165
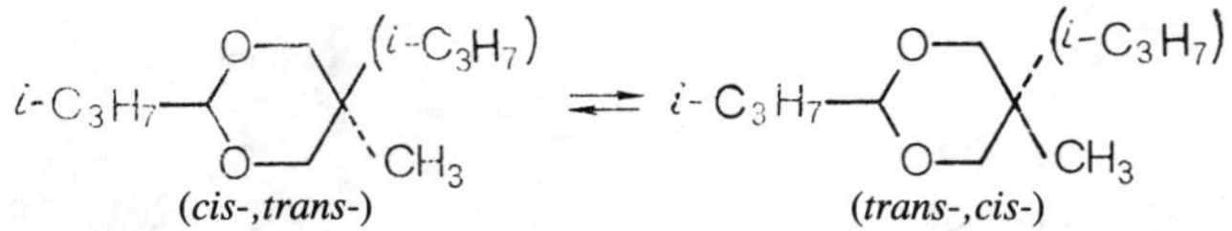 1,3-dioxane with i-C3H7, CH3 (cis-,trans-) ⇌ (trans-,cis-)	333 (l)	(−1.25 ±0.04)	—	SE, CT, GLC	133
1,3-dithiane with t-C4H9, i-C3H7 (cis-) ⇌ (trans-)	298 (l)	(−1.25)	—	SE, CT, GLC $(C_2H_5)_2O$	190
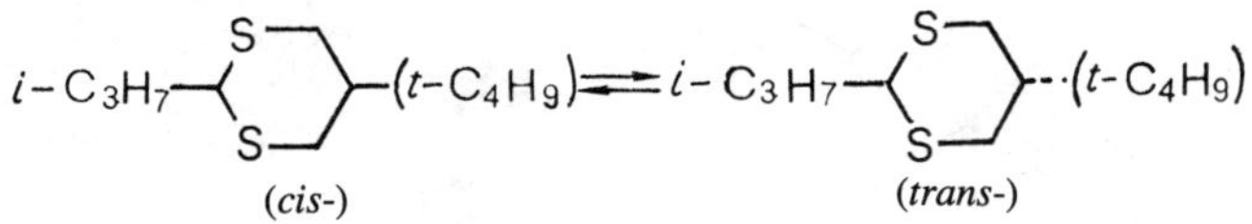 1,3-dithiane with i-C3H7, t-C4H9 (cis-) ⇌ (trans-)	298 (l)	(−6.73 ±0.04)	—	SE, CT, GLC $CHCl_3$	137
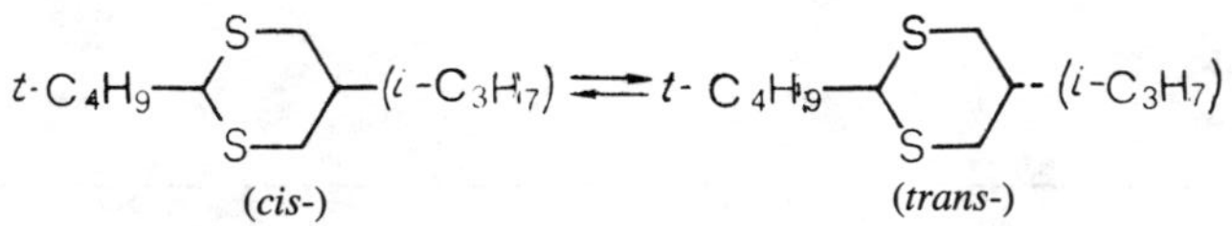 1,3-dithiane with t-C4H9, i-C3H7 (cis-) ⇌ (trans-)	342 (l)	(−3.55 ±0.04)	—	SE, CT, GLC $CHCl_3$	137

$C_{11}H_{22}S_2$ 2.1

Table (Continued)

Formula	Type	Reaction	T, K (phase)	$\Delta_r H_T^\circ$ ($\Delta_r G_T^\circ$)	$\Delta_r S_T^\circ$	Technique	Reference
$C_{12}H_{12}$	1.3	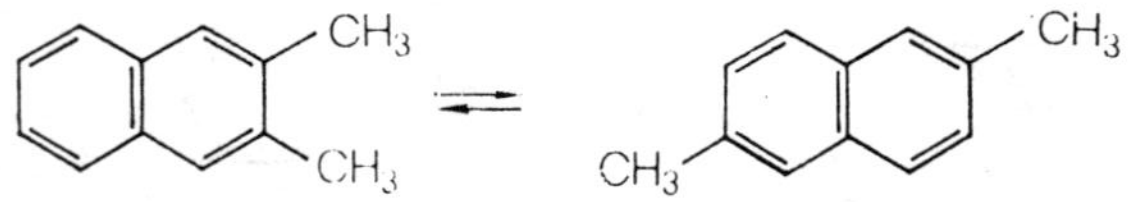 2,3-dimethylnaphthalene ⇌ 2,6-dimethylnaphthalene	298 (S)	−3.39 ±1.84	—	CC	232
		 2,6-dimethylnaphthalene ⇌ 2,7-dimethylnaphthalene	298 (S)	0.29 ±1.87	—	CC	232
		 1,8-dimethylnaphthalene ⇌ 2,3-dimethylnaphthalene	298 (S)	−28.24 ±1.53	—	CC	232
$C_{12}H_{12}O_2$	1.2	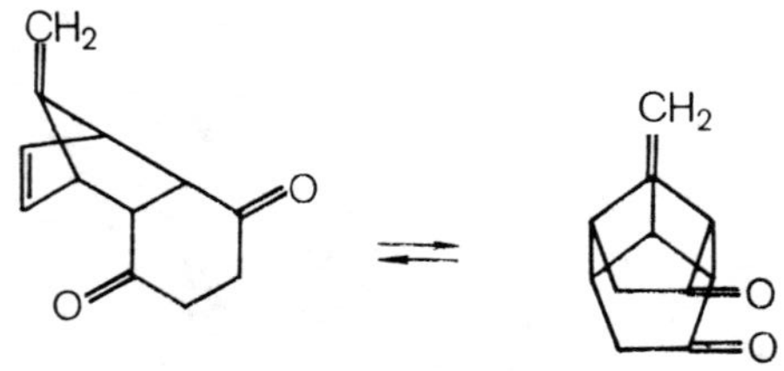	298 (g)	−17.69	—	CC	210

184

$C_{12}H_{14}O_3$	1.1 1.2	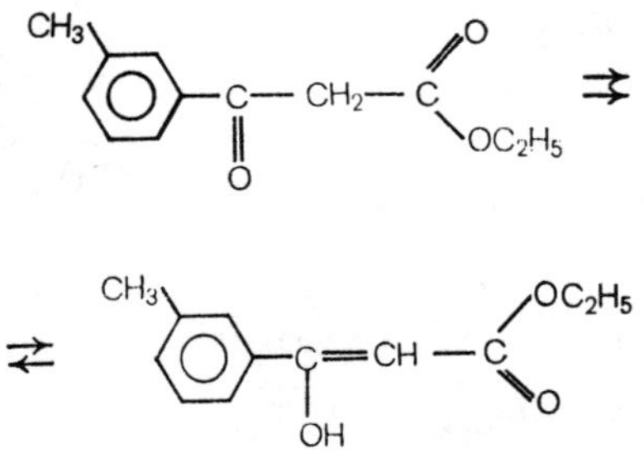	293 (l)	−5.02 ±0.84	−16.7	SE, NMR CCl_4	227
		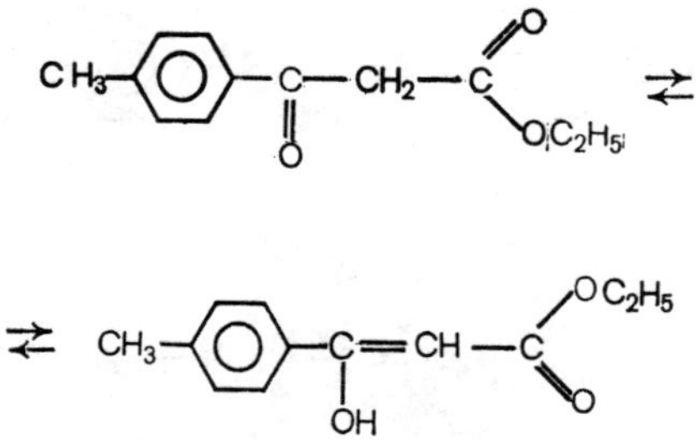	293 (l)	−4.60 ±0.84	−16.7	SE, NMR CCl_4	227
$C_{12}H_{14}O_4$	1.1 1.2	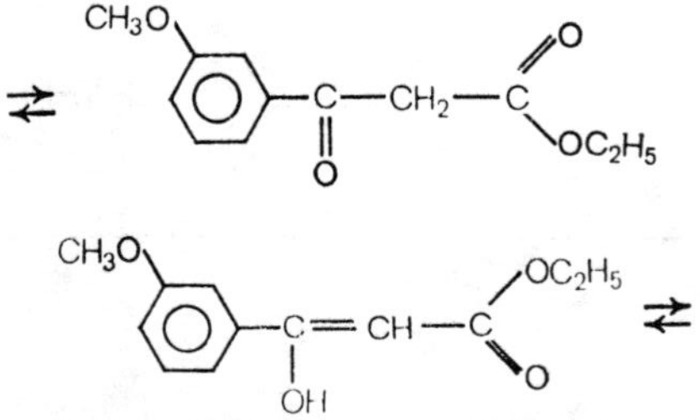	293 (l)	−5.02 ±0.84	−16.7	SE, NMR CCl_4	227

Table (Continued)

Formula	Type	Reaction	T, K (phase)	$\Delta_r H_T^\circ$ ($\Delta_r G_T^\circ$)	$\Delta_r S_T^\circ$	Technique	Reference
$C_{12}H_{14}O_4$	1,1 1,2	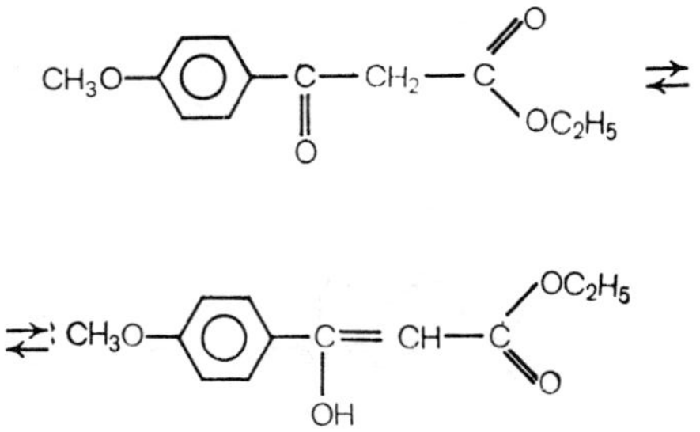 (4-methoxyphenyl-CO-CH₂-CO-OC₂H₅ ⇌ 4-methoxyphenyl-C(OH)=CH-CO-OC₂H₅)	293 (l)	−2.09 ±0.42	−16.7	SE, NMR CCl_4	227
$C_{12}H_{15}BrO_2$	2.1	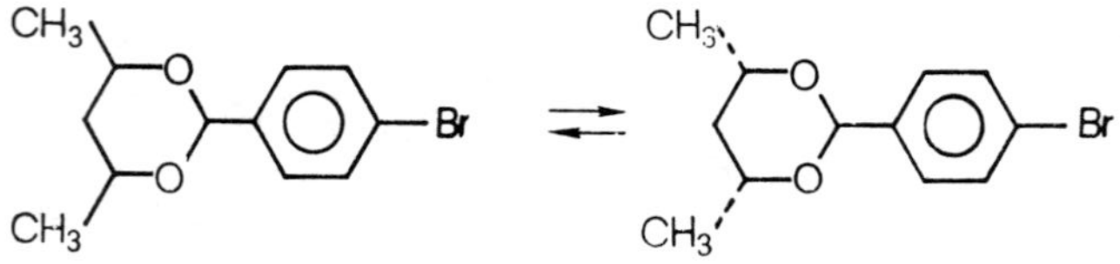 (cis,cis-) ⇌ (trans,trans-) 2-(4-bromophenyl)-4,6-dimethyl-1,3-dioxane	298 (l)	(13.30)	—	SE, CT NMR $(C_2H_5)_2O$	166
$C_{12}H_{15}FO_2$	2.1	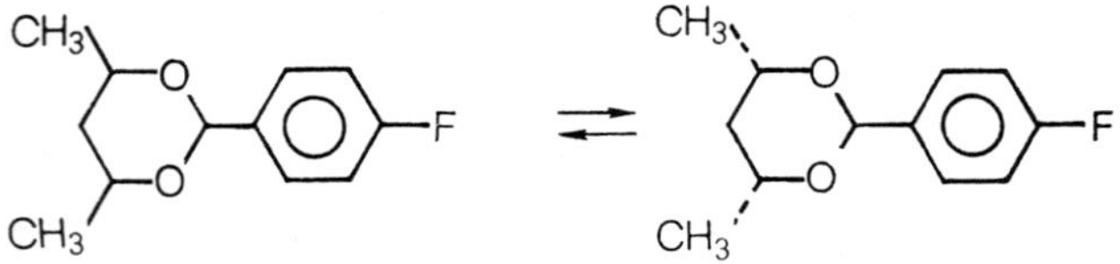 (cis,cis-) ⇌ (trans,trans-) 2-(4-fluorophenyl)-4,6-dimethyl-1,3-dioxane	298 (l)	(13.09)	—	SE, CT NMR $(C_2H_5)_2O$	166

$C_{12}H_{16}O_2$	2.1	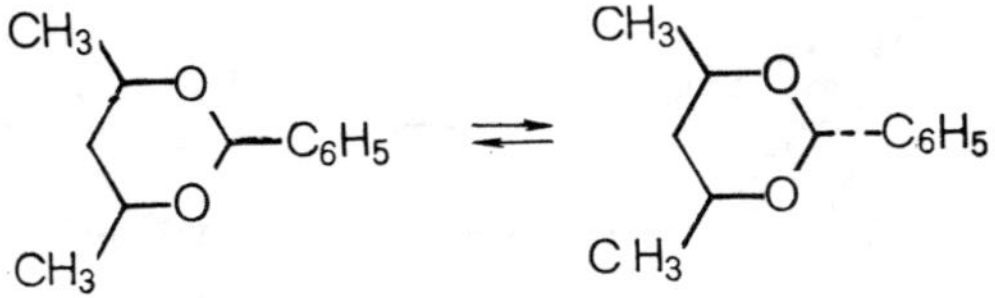 (cis-,cis-) ⇌ (trans-,trans-) [1,3-dioxane with CH₃, C₆H₅ substituents]	298 (l)	(12.97)	—	SE, CT, NMR $(C_2H_5)_2O$	166
			298 (l)	(11.96 ±0.08)	—	SE, CT, GLC C_6H_{14}	154
$C_{12}H_{18}$	1.3	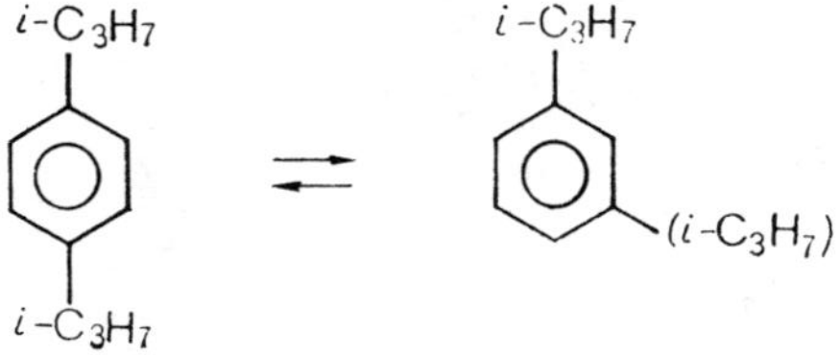 1,4-diisopropylbenzene ⇌ 1,3-diisopropylbenzene	343 (l)	0.00 ±0.54	5.19 ±0.54	SE, CT, GLC	233
		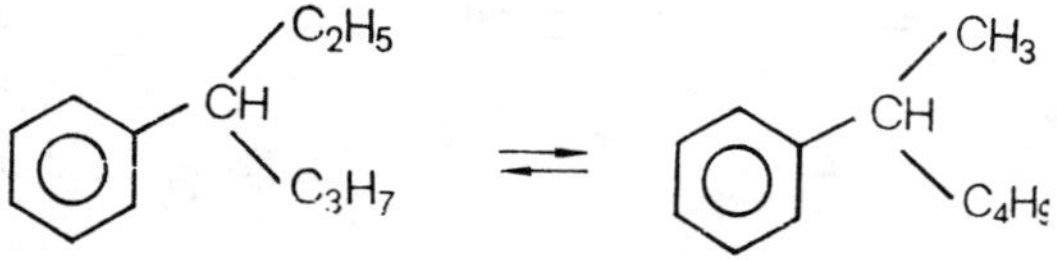 sec-butylbenzene ⇌ (1-methylpropyl)benzene	325 (l)	−0.30 ±0.07	3.85 ±0.21	SE, CT, GLC	263
$C_{12}H_{18}N_2O$	1.1 1.2	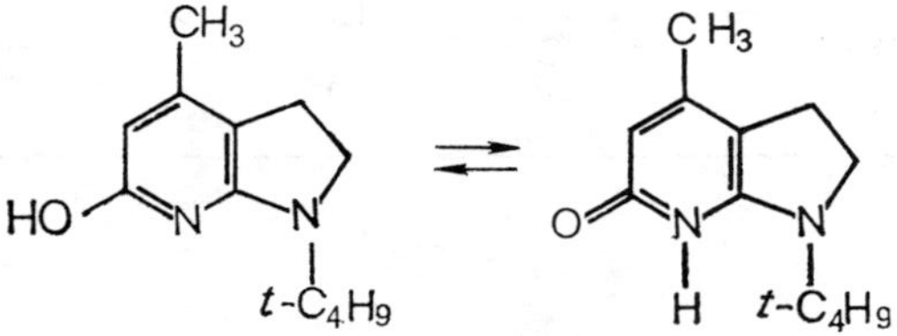 [pyrrolopyridine HO-form ⇌ oxo-form, t-C₄H₉, CH₃]	323 (l)	−12.26	−19.07	SE, UV $C_2H_5OH-H_2O$	65

Table (Continued)

Formula	Type	Reaction	T, K (phase)	$\Delta_r H_T^\circ$ ($\Delta_r G_T^\circ$)	$\Delta_r S_T^\circ$	Technique	Reference
$C_{12}H_{22}O$	2.1	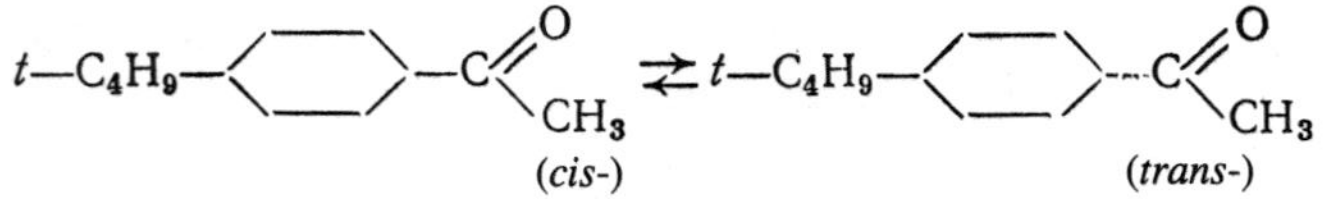 $t-C_4H_9-C_6H_4-CO-CH_3$ (cis-) ⇌ $t-C_4H_9-C_6H_4-CO-CH_3$ (trans-)	360 (l)	−4.89 ±0.04	4.85 ±0.13	SE, GLC CH_3OH	229
		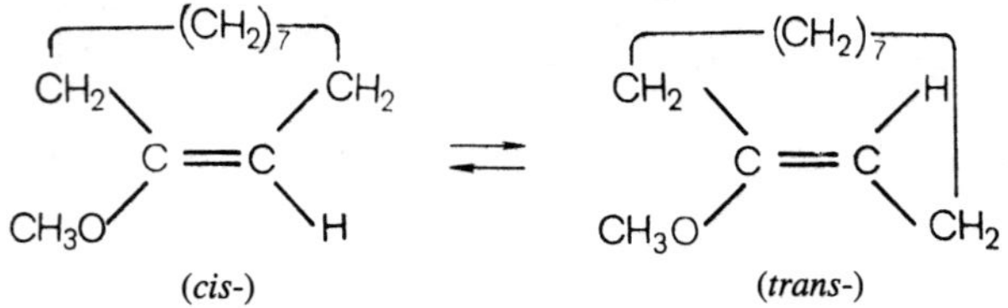 (CH₂)₇ methyl oleate cis- ⇌ trans-	298 (l)	8.8 ±0.1	8.2 ±0.3	SE, CT, GLC C_6H_6	231
$C_{12}H_{22}O_2$	2.1	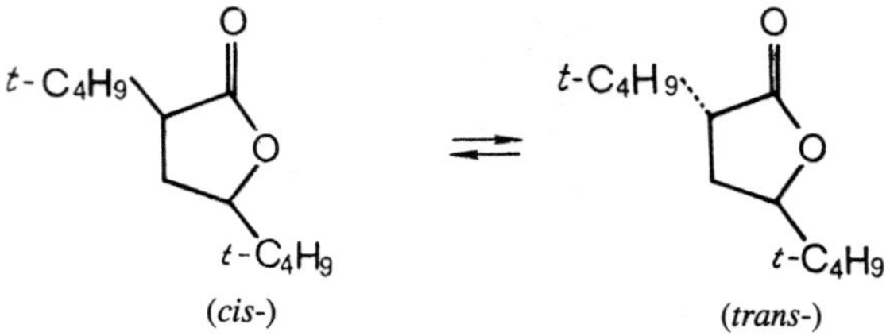 t-C₄H₉ lactone (cis-) ⇌ (trans-)	298 (l)	(5.02)	—	SE, CT, GLC $t-C_4H_9OH$	121
		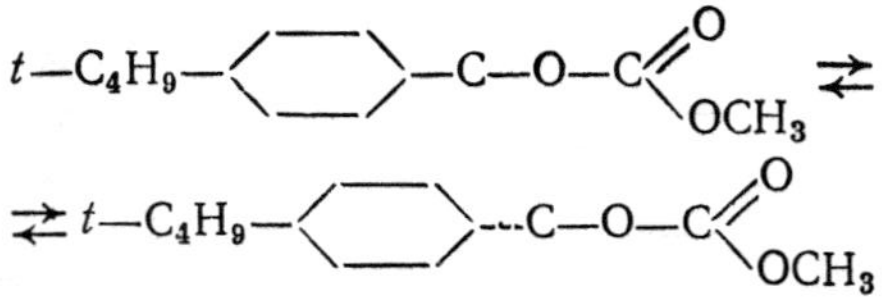 $t-C_4H_9-C_6H_4-CO-O-CH_3$ (cis-) ⇌ (trans-)	391 (l)	−4.68 ±0.17	2.09 ±0.46	SE, GLC CH_3OH	229

188

$C_{12}H_{24}$	1.3	$t-C_4H_9-CH_2$ $t-C_4H_9-CH_2$ $\Big\rangle C=CH_2 \rightleftarrows t-C_4H_9-CH=C\Big\langle {CH_3 \atop CH_2(t-C_4H_9)}$ (cis- + trans-)	348 (l)	6.27	—	SE, RSS	234
$C_{12}H_{24}O_2$	2.1	t-C₄H₉ ⟶ CH₃ on dioxolane ⇌ t-C₄H₉ ⟶ CH₃ on dioxolane (syn-) (anti-)	298 (l)	(2.09)	—	SE, CT, GLC $(C_2H_5)_2O$	93
		i-C₃H₇ dioxolane ⇌ i-C₃H₇ dioxolane (syn-) (anti-)	298 (l)	(1.78 ±0.13)	—	SE, CT, GLC $(C_2H_5)_2O$	93
		(t-C₄H₉)(CH₃) dioxane ⇌ (t-C₄H₉)(CH₃) dioxane (cis-,trans-) (trans-,cis-)	298 (l)	(−3.39)	—	SE, CT, GLC $(C_2H_5)_2O$	190
		(i-C₃H₇)(C₂H₅) dioxane ⇌ (i-C₃H₇)(CH₃) dioxane (cis-,trans-) (trans-,cis-)	298 (l)	(−1.34)	—	SE, CT, GLC $(C_2H_5)_2O$	190

Table (Continued)

Formula	Type	Reaction	T, K (phase)	$\Delta_r H_T^\circ$ ($\Delta_r G_T^\circ$)	$\Delta_r S_T^\circ$	Technique	Reference
$C_{12}H_{24}S_2$	2.1	$t\text{-}C_4H_9$—[dithiane-$t\text{-}C_4H_9$] (cis-) ⇌ $t\text{-}C_4H_9$—[dithiane-($t\text{-}C_4H_9$)] (trans-)	298 (l)	(−7.73 ±0.04)	—	SE, CT, GLC $CHCl_3$	137
$C_{12}H_{26}$	1.3	$CH_3-(CH_2)_{10}-CH_3 \rightleftharpoons CH_3-\underset{CH_3}{\overset{CH_3}{C}}-CH_2-\underset{CH_3}{CH}-CH_2-\underset{CH_3}{\overset{CH_3}{C}}-CH_3$	361 (g)	−14.32 ±0.05	−22.30 ±0.05	SE, CT, GLC $(C_2H_5)_2O-CCl_4$	136
			298 (g)	−24.52 ±1.41	—	CC	235
$C_{13}H_{15}F_3O_2$	2.1	[4-CF$_3$-phenyl-dioxane with CH$_3$ groups] ⇌ [cis-,cis- isomer]	298 (l)	(13.22)	—	SE, CT, NMR $(C_2H_5)_2O$	166
			298 (l)	(12.88 ±0.08)	—	SE, CT, NMR C_6H_{14}	154

Formula		Structures	T (K)	ΔH	ΔS	Method	Ref.
$C_{13}H_{18}$	1.3	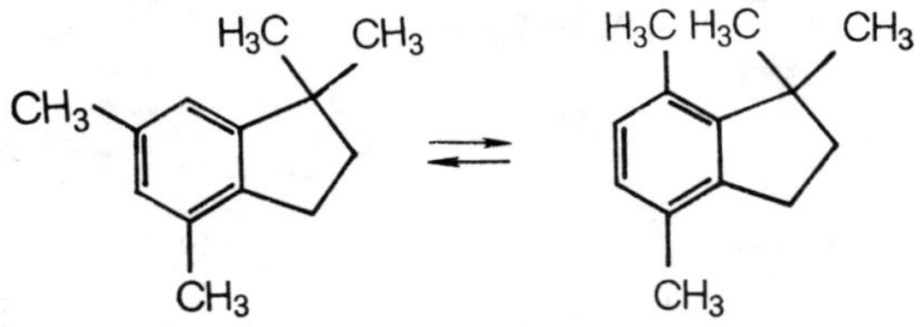 (indane structures)	298 (l)	7.95 ±3.10	—	CC	228
$C_{13}H_{19}Cl$	1.3	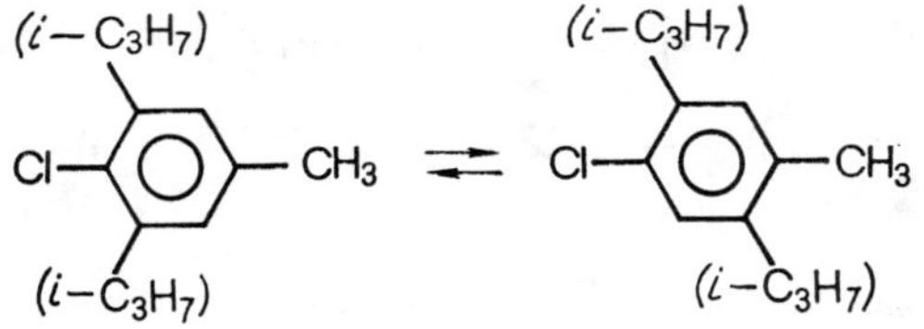 (trisubstituted chlorobenzene structures)	353 (l)	0.43 ±0.17	2.68 ±0.25	SE, CT, GLC	261
		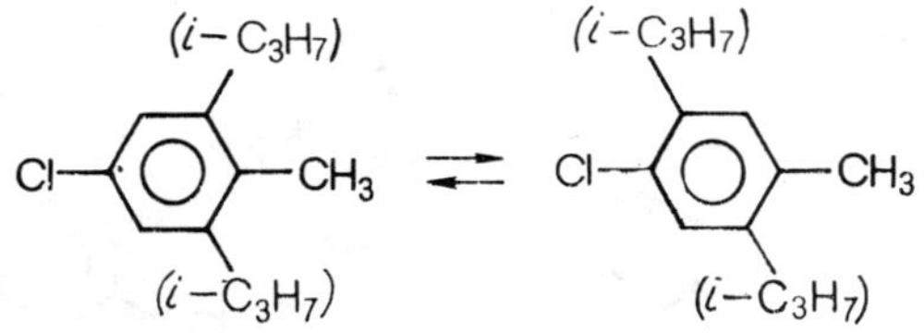 (trisubstituted chlorobenzene structures)	325 (l)	−10.08 ±4.86	5.40 ±13.14	SE, CT, GLC	261
$C_{13}H_{20}$	1.3	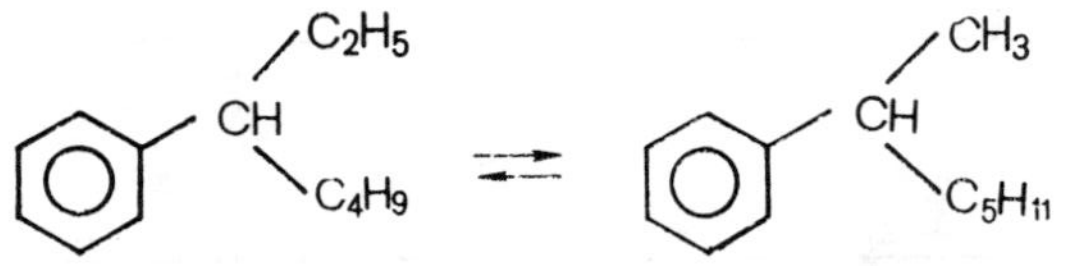 (PhCH(C₂H₅)C₄H₉ ⇌ PhCH(CH₃)C₅H₁₁)	319 (l)	−0.21 ±0.07	3.52 ±0.22	SE, CT, GLC	263

Table (Continued)

Formula	Type	Reaction	T, K (phase)	$\Delta_r H_T^\circ$ ($\Delta_r G_T^\circ$)	$\Delta_r S_T^\circ$	Technique	Reference
$C_{12}H_{20}$	1.3	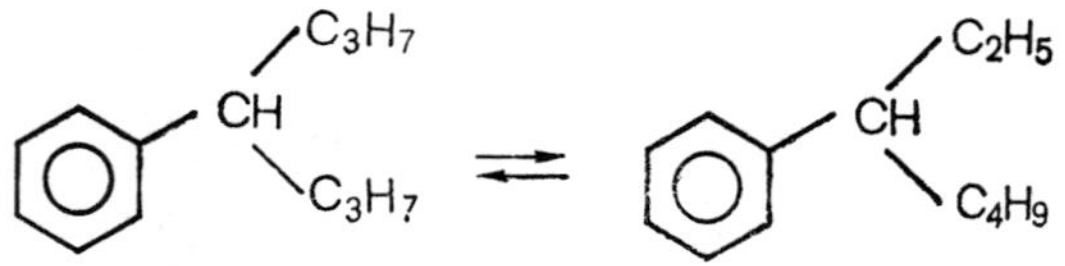 PhCH(C_3H_7)(C_3H_7) ⇌ PhCH(C_2H_5)(C_4H_9)	321 (l)	0.02 ±0.41	7.57 ±1.29	SE, CT, GLC	263
$C_{13}H_{24}O$	2.1	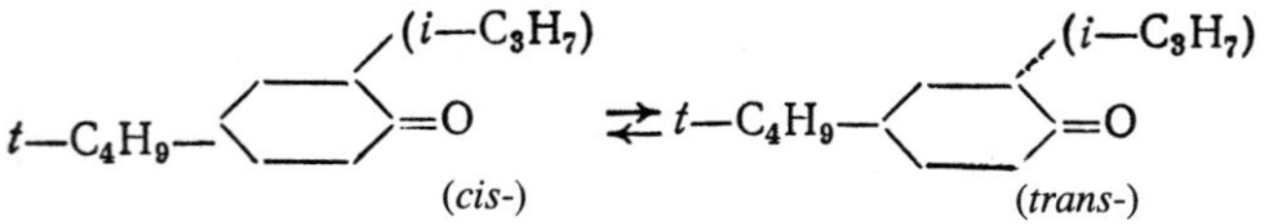 $t-C_4H_9$—(cis-, $i-C_3H_7$, =O) ⇌ $t-C_4H_9$—(trans-, $i-C_3H_7$, =O)	318 (l)	1.84 ±0.50	−2.09 ±1.25	SE, CT, GLC $t-C_4H_9OH$	230
	2.1	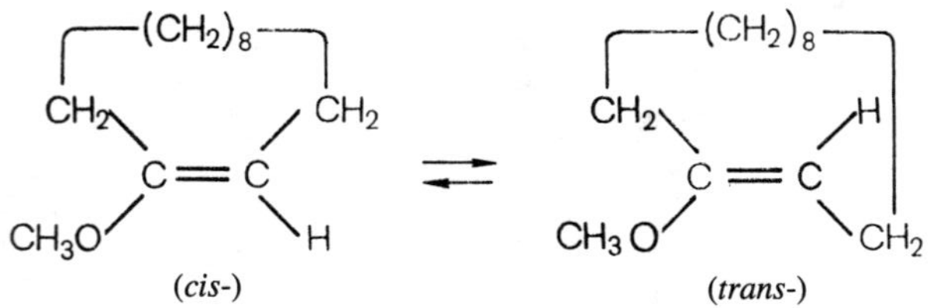 (cis-) ⇌ (trans-) cyclic structure with CH_3O	298 (l)	9.5 ±0.3	9.6 ±0.7	SE, CT, GLC	231
$C_{13}H_{24}O_2$	2.1	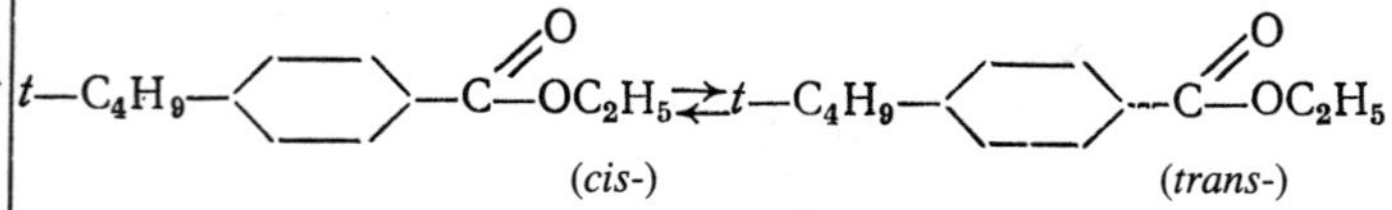 $t-C_4H_9$—C$_6$H$_4$—C(=O)—OC$_2$H$_5$ (cis-) ⇌ $t-C_4H_9$—C$_6$H$_4$—C(=O)—OC$_2$H$_5$ (trans-)	298 (l)	−4.60	1.55	SE, GLC C_2H_5OH	142

$C_{13}H_{26}O_2$	2.1	(structures: syn- ⇌ anti- 2-ethyl-4,5-di-t-butyl-1,3-dioxolane)	298 (l)	−4.56	1.67	SE, GLC C_2H_5OH	237
		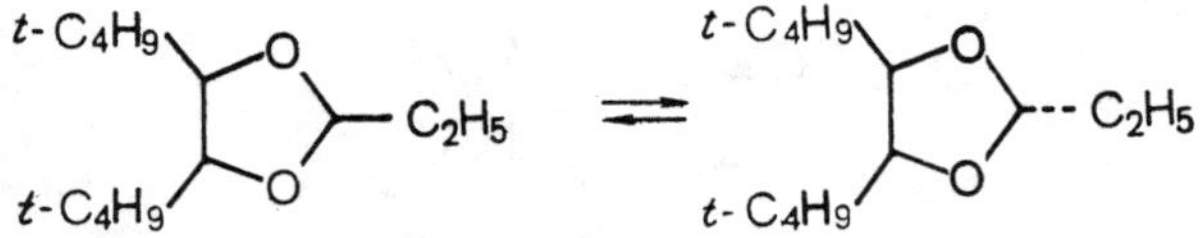 (structures: syn- ⇌ anti- 2-t-butyl-4,5-di-i-propyl-1,3-dioxolane)	298 (l)	(0.75 ±0.04)	—	SE, CT, GLC $(C_2H_5)_2O$	93
		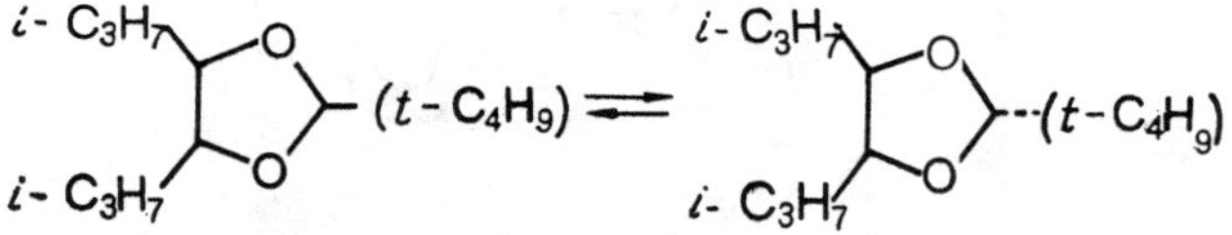 (structures: syn- ⇌ anti- 2-t-butyl-4,5-di-i-propyl-1,3-dioxolane)	298 (l)	(−0.93 ±0.06)	—	SE, CT, GLC $(C_2H_5)_2O$	93
$C_{14}H_8F_4$	2.1	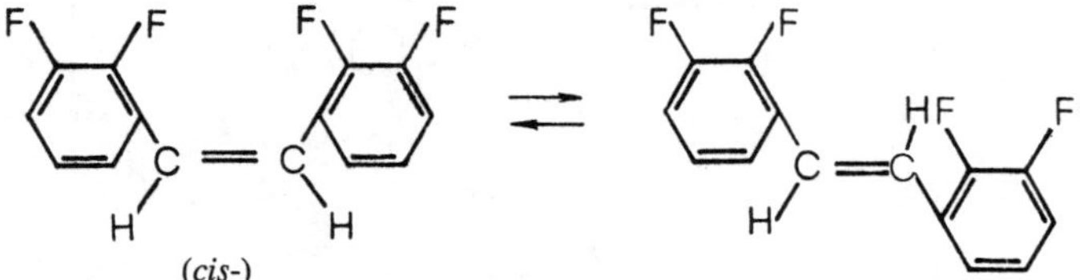 (cis- ⇌ trans- bis(2,3-difluorophenyl) ether/enol)	298 (l)	(−6.78 ±0.13)	—	SE, GLC	238
$C_{14}H_{10}$	1.3	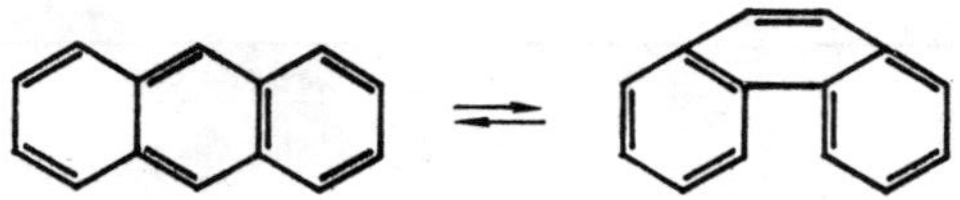 (anthracene ⇌ phenanthrene)	298 (g)	−23.88 ±6.27	—	CC	239

Table (Continued)

Formula	Type	Reaction	T, K (phase)	$\Delta_r H_T^\circ$ ($\Delta_r G_T^\circ$)	$\Delta_r S_T^\circ$	Technique	Reference
$C_{14}H_{10}Br_2$	2.1	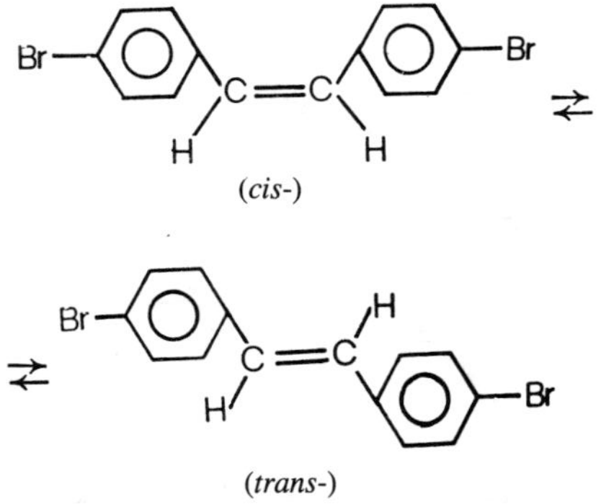 (4-Br-C₆H₄)CH=CH(4-Br-C₆H₄) (cis-) ⇌ (trans-)	298 (l)	(−15.06 ±0.42)	—	SE, GLC	238
$C_{14}H_{10}N_2O_4$	2.1	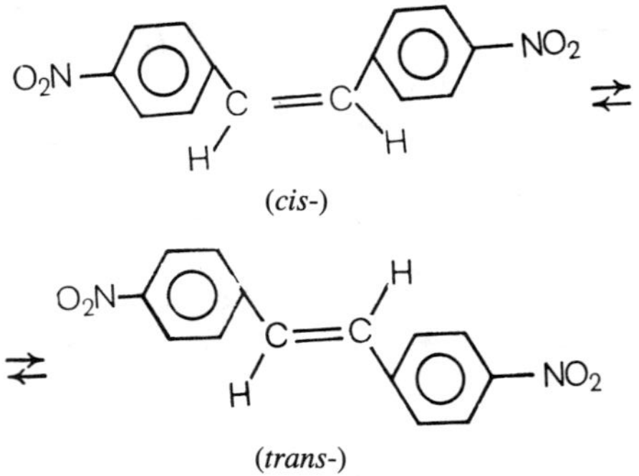 (4-NO₂-C₆H₄)CH=CH(4-NO₂-C₆H₄) (cis-) ⇌ (trans-)	298 (l)	(−17.57 ±0.42)	—	SE, GLC	238

Formula		Structures	T (K)	ΔH	ΔG	Method / Solvent	Ref
$C_{14}H_{10}O$	1.1 / 1.2	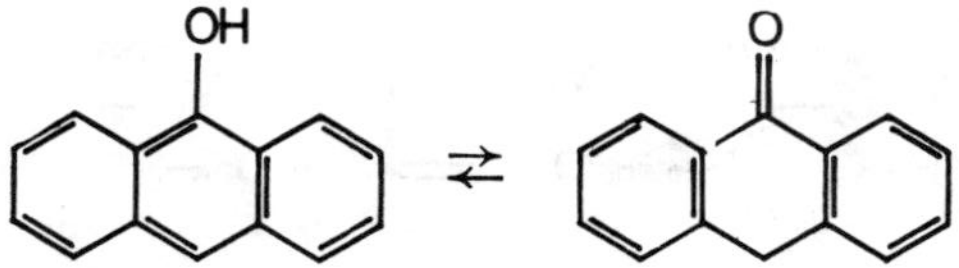 9-hydroxyanthracene ⇌ anthrone	293 (l)	10.88	—	SE, UV  / benzene	240
$C_{14}H_{12}$	2.1	 cis-stilbene ⇌ trans-stilbene	298 (l)	(−15.47 ±0.42)	—	SE, GLC	238
$C_{14}H_{14}N_2O$	1.1 / 1.2	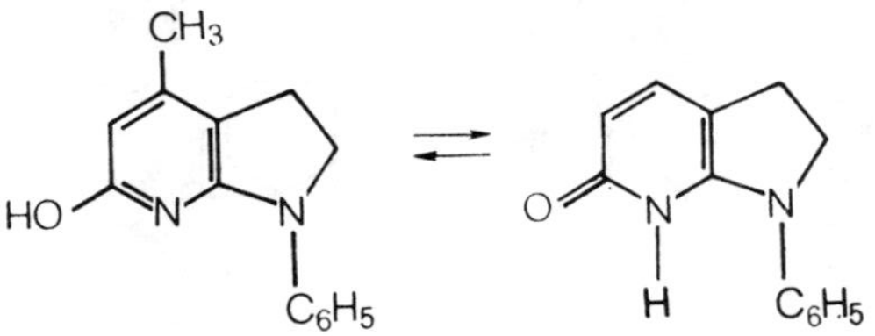 enol ⇌ keto (pyrrolopyridine)	323 (l)	−2.97	−4.18	SE, UV / $C_2H_5OH-H_2O$	240
$C_{14}H_{16}O_2$	2.1	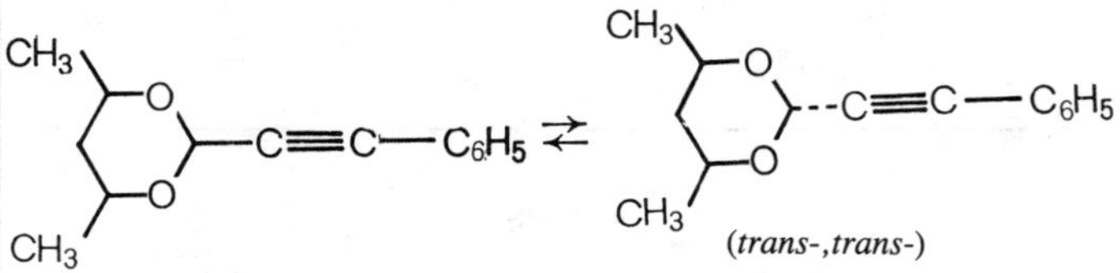 (cis-,cis-) ⇌ (trans-,trans-) dioxane-phenylacetylene	298 (l)	(−1.32 ±0.13)	—	SE, CT, UV / CCl_4	154

Table (Continued)

Formula	Type	Reaction	T, K (phase)	$\Delta_r H°_T$ ($\Delta_r G°_T$)	$\Delta_r S°_T$	Technique	Reference
$C_{14}H_{20}O$	2.1	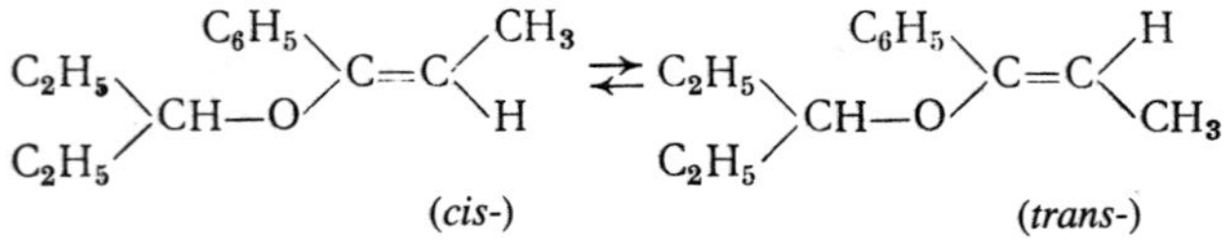 (structure: cis- ⇌ trans- stilbene ether with C₂H₅, C₆H₅, CH—O, CH₃ groups)	298 (l)	−12.35	−22.02	SE, CT, NMR (C₆H₆)	173
	1.3	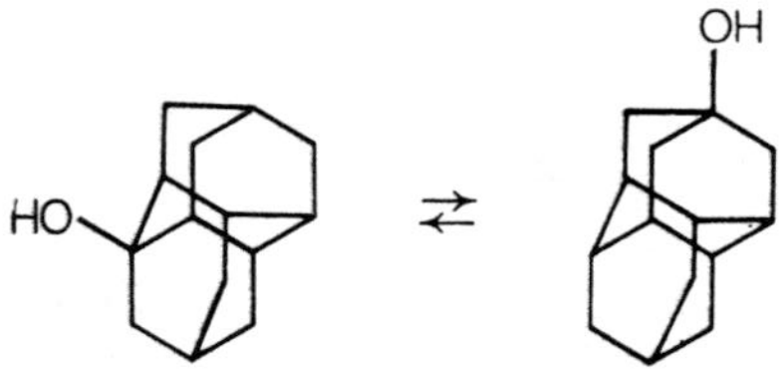 (structure: adamantanol isomers ⇌)	373 (l)	−4.60 ±0.42	−14.22 ±1.25	SE, GLC H_2SO_4	241
$C_{14}H_{20}O_2$	2.1	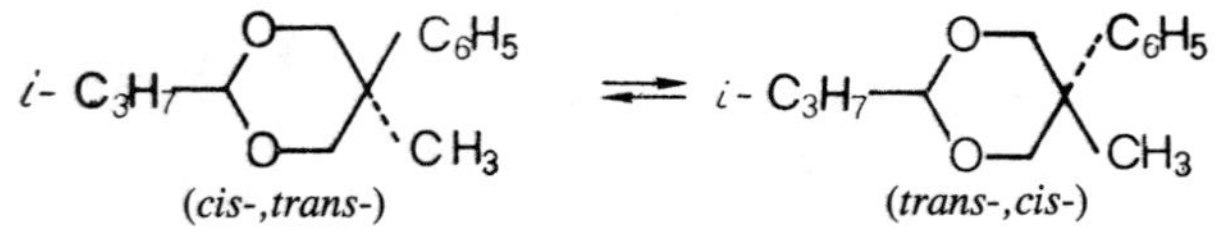 (1,3-dioxane: cis-,trans- ⇌ trans-,cis- with i-C_3H_7, C_6H_5, CH_3)	298 (l)	(−2.26)	—	SE, CT, GLC $(C_2H_5)_2O$	190
$C_{14}H_{20}S_2$	2.1	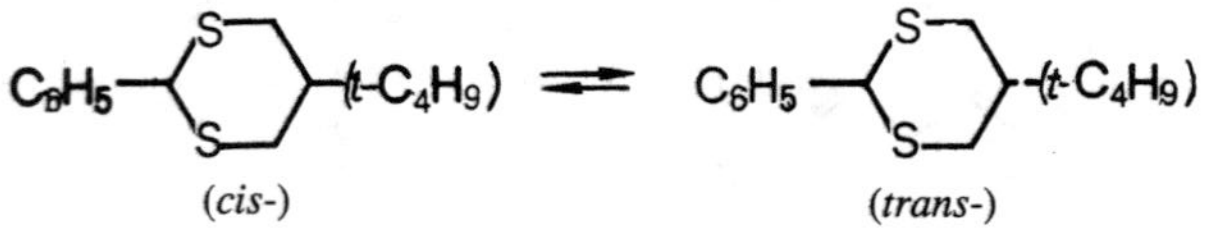 (1,3-dithiane: cis- ⇌ trans- with t-C_4H_9, C_6H_5)	298 (l)	(−8.12 ±0.08)	—	SE, CT, GLC $CHCl_3$	137

$C_{14}H_{22}$	1.3	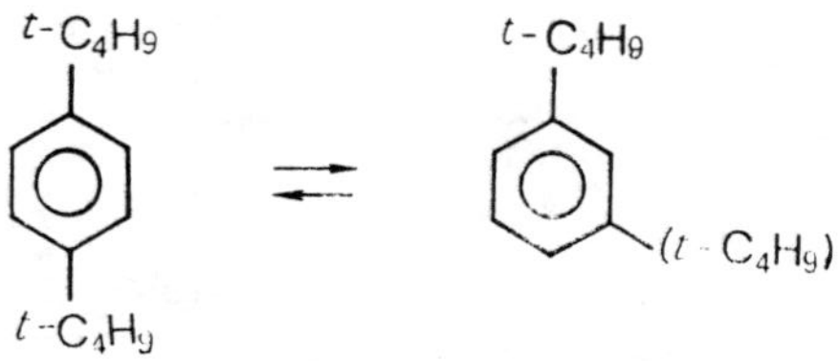 t-C4H9—⟨benzene⟩—t-C4H9 ⇌ t-C4H9—⟨benzene(t-C4H9)⟩	313 (l)	1.48 ±0.33	6.26 ±0.73	SE, CT, GLC	242
$C_{14}H_{22}O$	1.3	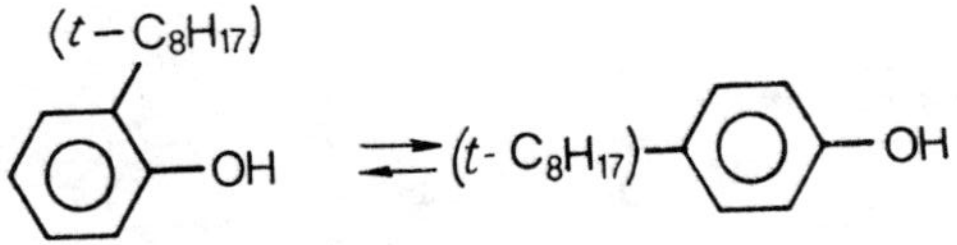 (t-C8H17)—⟨phenol⟩ ⇌ ⟨phenol(t-C8H17)⟩—OH	408 (l)	−14.2 ±1.6	−4.5 ±3.9	SE, CT, GLC	262
$C_{14}H_{26}O$	2.1	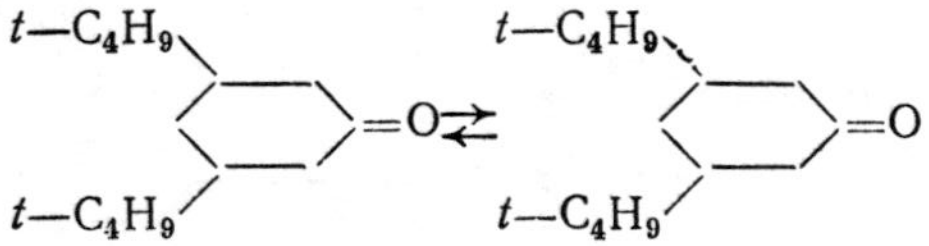 cis/trans cyclohexanone di-t-C4H9	525 (l)	16.36 ±0.42	14.89 ±0.42	SE, CT, GLC t—C4H9OH	243
		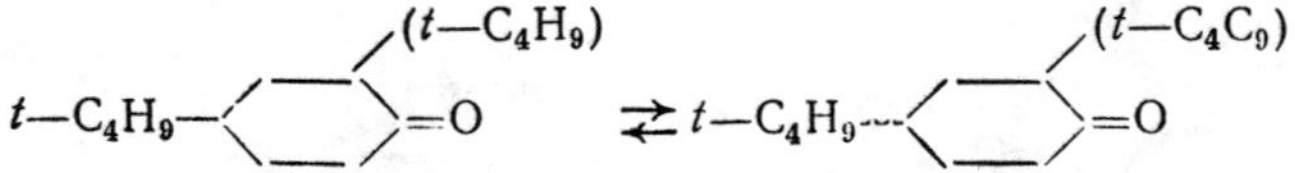 cis/trans cyclohexanone t-C4H9	343 (l)	10.00 ±1.59	10.46 ±4.18	SE, CT, GLC t—C4H9OH	230
		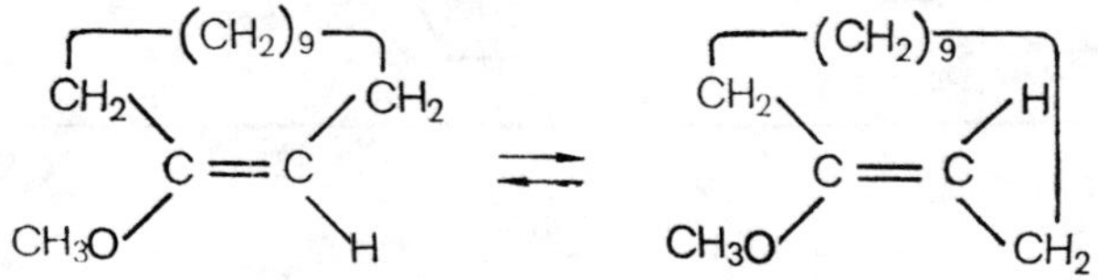 cis/trans methoxy alkene	298 (l)	7.2 ±0.3	8.3 ±0.8	SE, CT, GLC	231

Table (*Continued*)

Formula	Type	Reaction	T, K (phase)	$\Delta_r H_T^\circ$ ($\Delta_r G_T^\circ$)	$\Delta_r S_T^\circ$	Technique	Reference
$C_{14}H_{26}O_2$	2.1	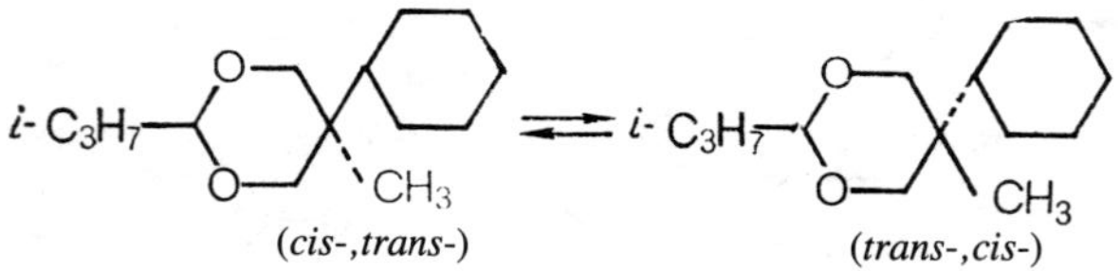 (cis-,trans-) ⇌ i-C_3H_7 (trans-,cis-)	298 (l)	(−1.17 ±0.04)	—	SE, CT, GLC $(C_2H_5)_2O$	167
$C_{14}H_{28}O$	2.1	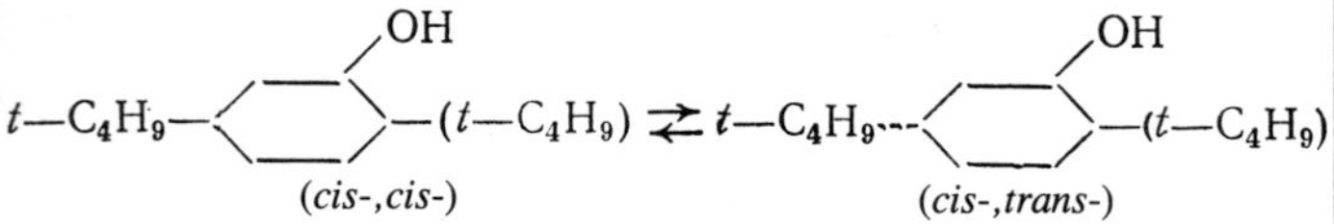 $t-C_4H_9$—(cis-,cis-) ⇌ $t-C_4H_9$—(cis-,trans-)	422 (l)	−27.86 ±1.51	−29.24 ±3.56	SE, CT, GLC	224
		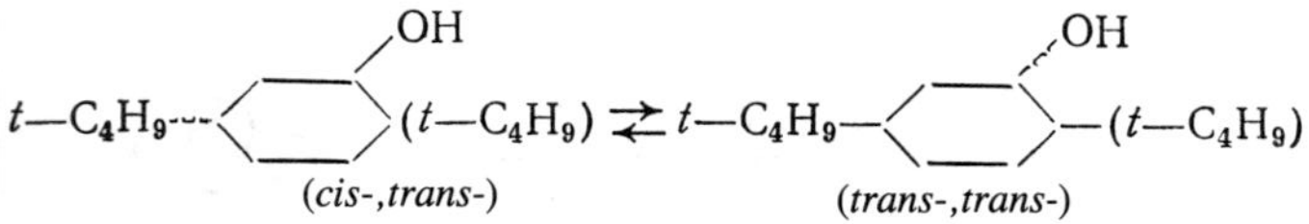 $t-C_4H_9$—(cis-,trans-) ⇌ $t-C_4H_9$—(trans-,trans-)	422 (l)	19.28 ±0.40	19.95 ±2.51	SE, CT, GLC 	224
		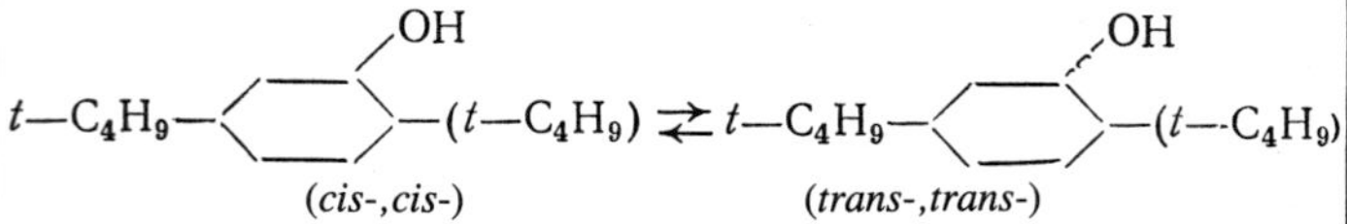 $t-C_4H_9$—(cis-,cis-) ⇌ $t-C_4H_9$—(trans-,trans-)	422 (l)	−8.58 ±0.47	−9.29 ±1.13	SE, CT, GLC	224
		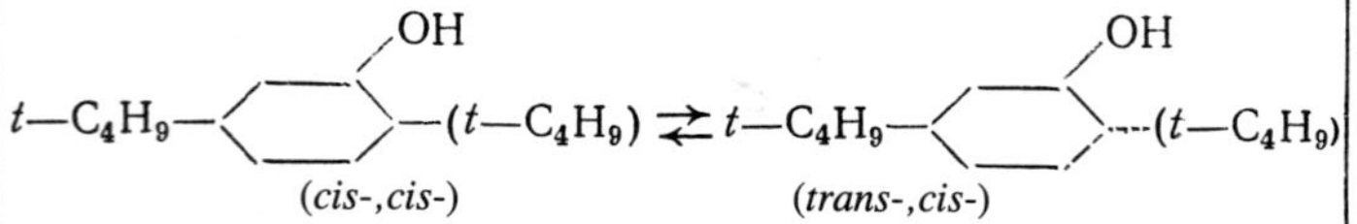 $t-C_4H_9$—(cis-,cis-) ⇌ $t-C_4H_9$—(trans-,cis-)	422 (l)	−28.28 ±1.13	−32.29 ±2.59	SE, CT, GLC	224

$C_{14}H_{28}O_2$	2.1	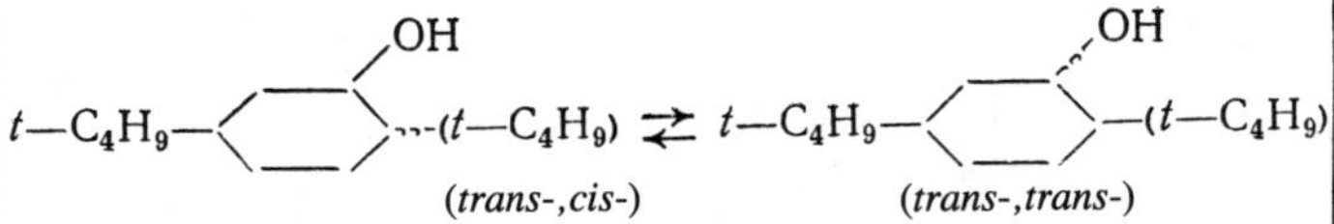 $t-C_4H_9$—⌬—(t-C_4H_9) ⇌ t-C_4H_9—⌬—(t-C_4H_9) (trans-,cis-) / (trans-,trans-) OH / OH	422 (l)	17.44 ±1.09	17.74 ±2.59	SE, CT, GLC  ⌬	224
		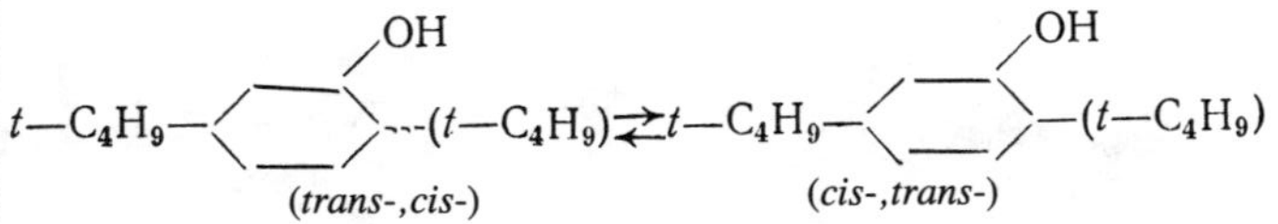 $t-C_4H_9$—⌬—(t-C_4H_9) ⇌ t-C_4H_9—⌬—(t-C_4H_9) (trans-,cis-) / (cis-,trans-) OH / OH	422 (l)	−1.51 ±0.04	1.38 ±0.17	SE, CT, GLC ⌬	224
		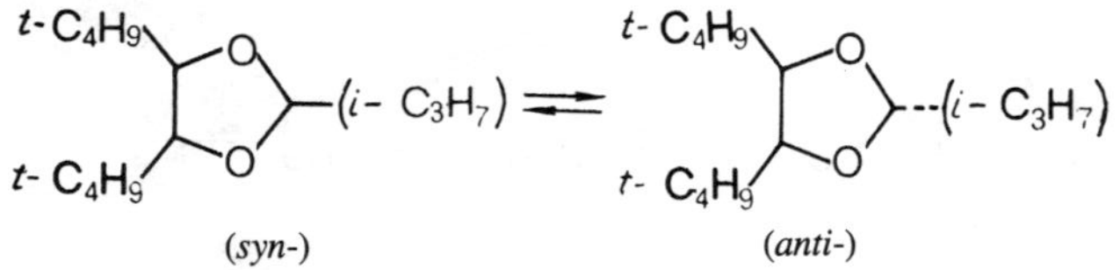 t-C_4H_9—[dioxolane-(i-C_3H_7)]—t-C_4H_9 (syn-) ⇌ (anti-)	298 (l)	(−0.05 ±0.01)	—	SE, CT, GLC $(C_2H_5)_2O$	93
$C_{15}H_{14}ClN$	1.1 1.2	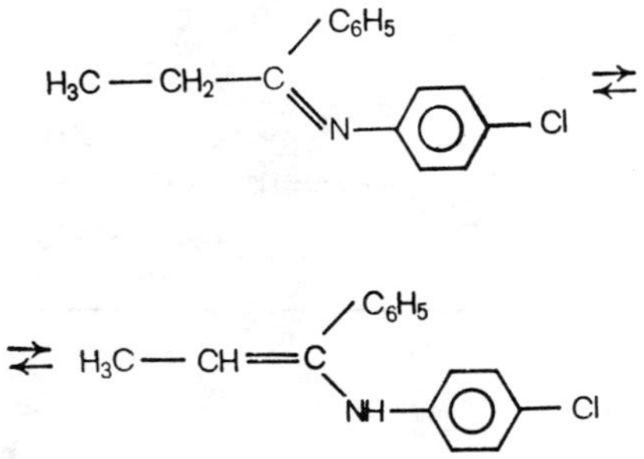 H_3C—CH_2—C(C_6H_5)=N—⌬—Cl ⇌ H_3C—CH=C(C_6H_5)—NH—⌬—Cl	333 (l)	10.0	3.3	SE, NMR  ⌬	213

Table (Continued)

Formula	Type	Reaction	T, K (phase)	$\Delta_r H_T^\circ$ ($\Delta_r G_T^\circ$)	$\Delta_r S_T^\circ$	Technique	Reference
$C_{15}H_{14}FN$	1.1 1.2	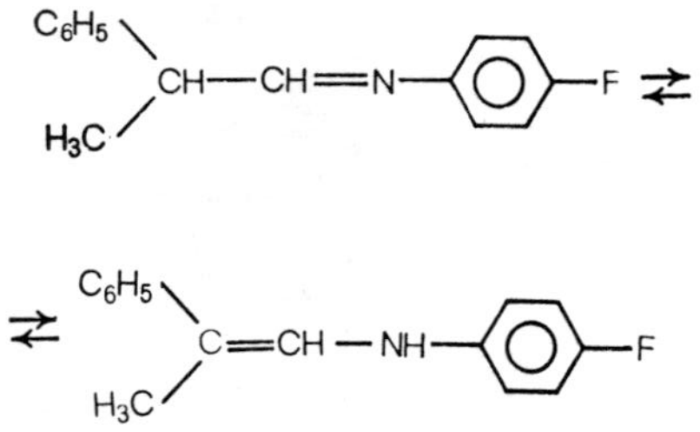	333 (l)	−9.2	−15.9	SE, NMR 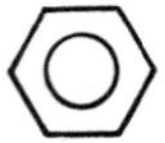	213
$C_{15}H_{15}N$	1.2 1.1		333 (l)	−10.9	−14.6	SE, NMR 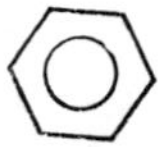	213
$C_{15}H_{18}BrNO_4S$	1.1 1.2	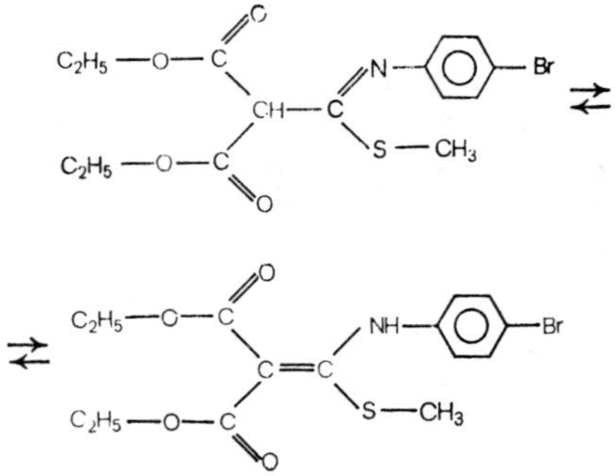	326 (l)	−15.02	−39.32	SE, NMR $CDBr_3$	224

$C_{15}H_{22}O_2$	2.1	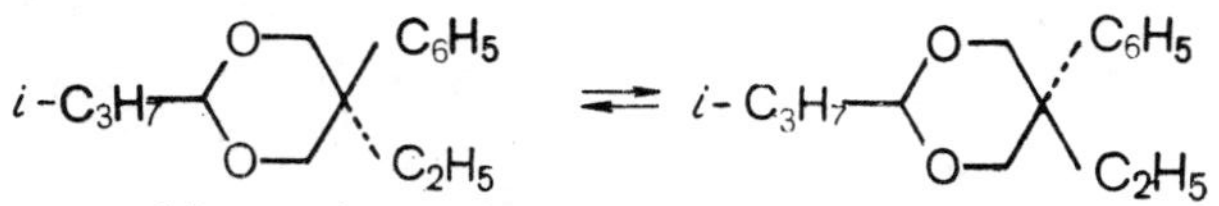 (cis-,trans-) ⇌ (trans-,cis-)	298 (l)	(−2.13)	—	SE, CT, GLC $(C_2H_5)_2O$	190
$C_{15}H_{30}O_2$	2.1	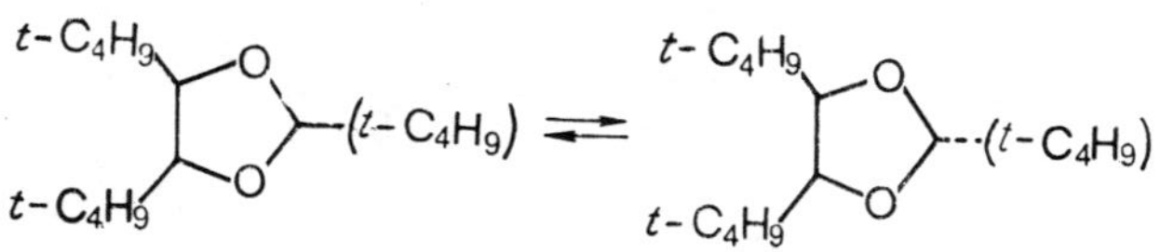 (syn-) ⇌ (anti-)	298 (l)	(−3.84 ±0.35)	—	SE, CT, GLC $(C_2H_5)_2O$	93
$C_{16}H_{14}O_2$	2.1	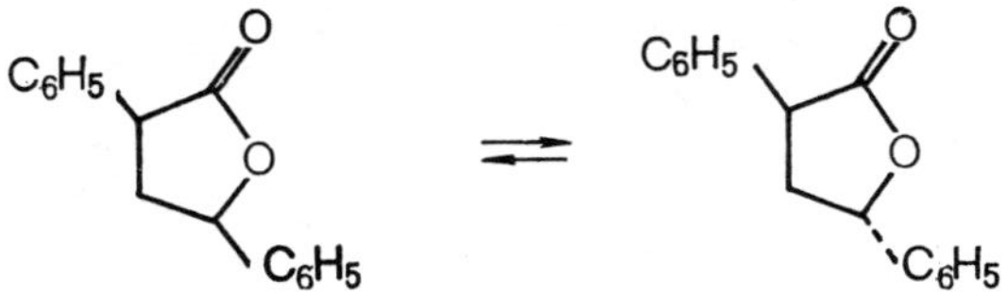 (cis-) ⇌ (trans-)	298 (l)	(1.00)	—	SE, NMR $(C_2H_5)_3N$—CCl_4	121
$C_{16}H_{16}$	2.1	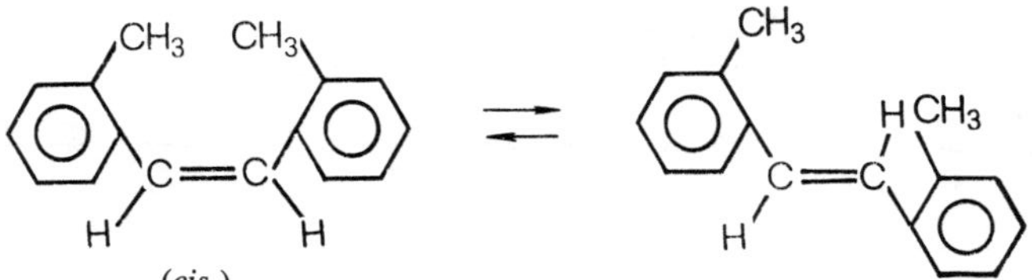 (cis-) ⇌ (trans-)	298 (l)	(−3.47 ±0.08)	—	SE, GLC	238

Table (Continued)

Formula	Type	Reaction	T, K (phase)	$\Delta_r H_T^\circ$ ($\Delta_r G_T^\circ$)	$\Delta_r S_T^\circ$	Technique	Reference
$C_{16}H_{16}$	2.1	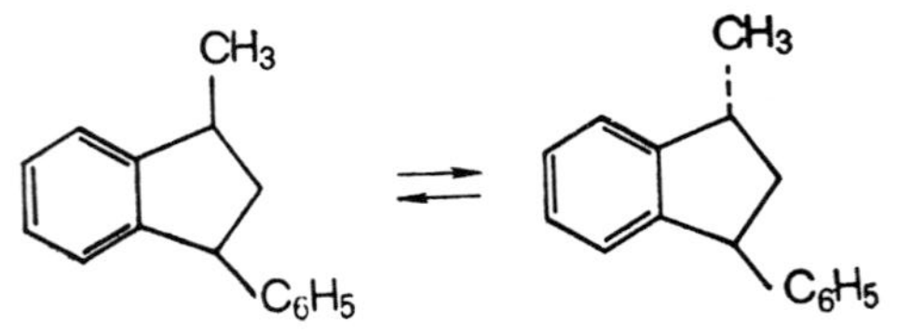 (cis-) ⇌ (trans-)	323 (l)	2.84 ±0.67	5.23 ±2.05	SE, CT, GLC	245
$C_{16}H_{16}O_2$	2.1	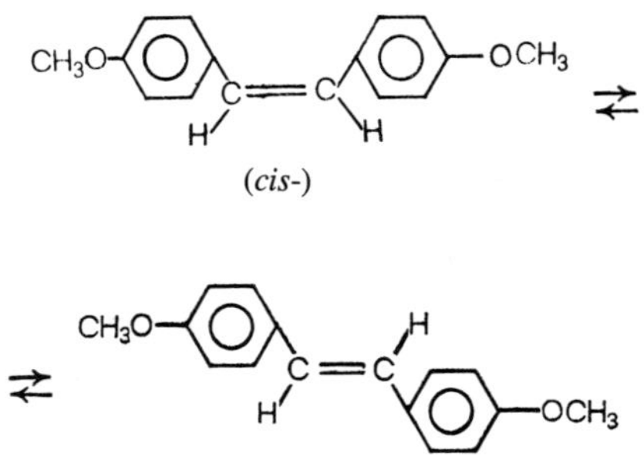 (cis-) ⇌ (trans-)	298 (l)	(−15.06 ±0.42)	—	SE, GLC	238
$C_{16}H_{17}N$	1.1 1.2	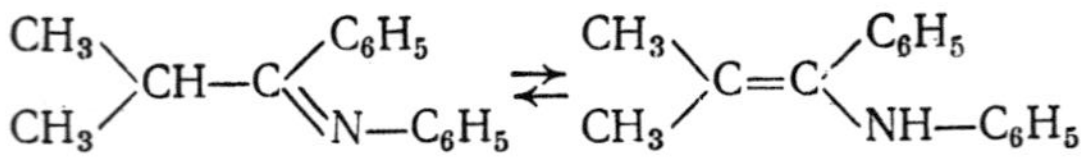 ⇌	333 (l)	−2.9	−17.2	SE, NMR 	213

Formula		Structure			Method	Ref.
$C_{16}H_{17}NO$	1.1 1.2	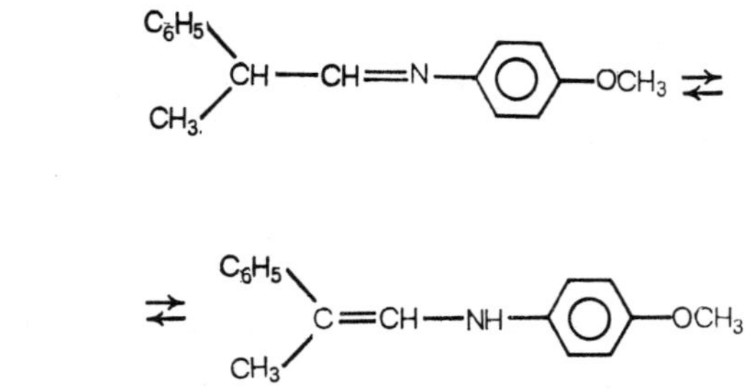 C₆H₅–CH(CH₃)–CH=N–C₆H₄–OCH₃ ⇌	333 (l)	−6.7	−13.4	SE, NMR 	213
	2.1	C₆H₅–C(CH₃)=CH–NH–C₆H₄–OCH₃ ⇌					
$C_{16}H_{24}$	1.3	t-C₄H₉–C₆H₄–C₆H₅ (trans-) ⇌ C₆H₅–C₆H₄–t-C₄H₉ (cis-)	365 (l)	−15.10 ±0.67	−8.74 ±1.88	SE, CT, GLC	246
		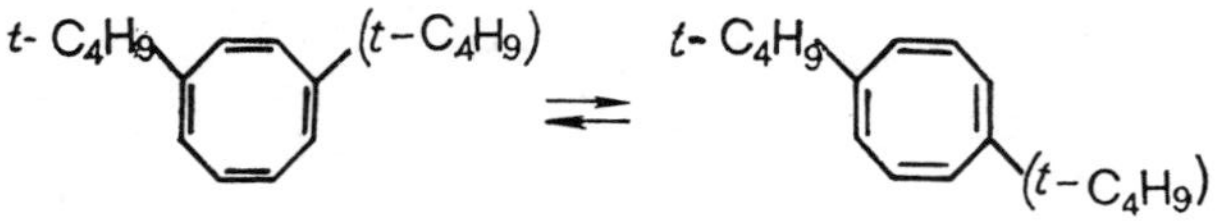 t-C₄H₉ / t-C₄H₉ (cycloheptatriene) ⇌ t-C₄H₉ / t-C₄H₉	311 (l)	2.01	3.85	SE, CT, GLC CDCl₃	247
$C_{16}H_{26}$	1.3	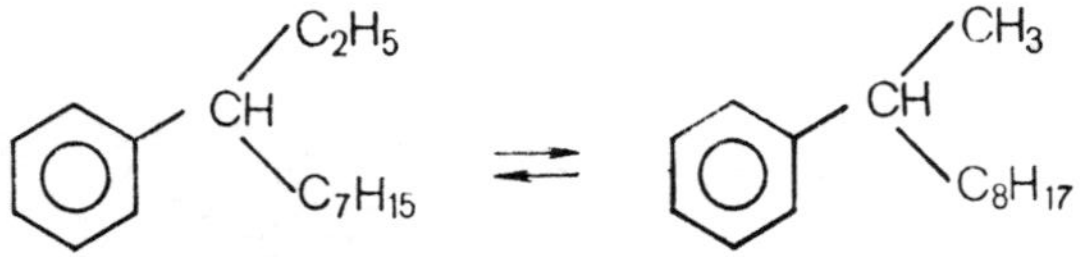 C₆H₅–CH(CH₃)–C₈H₁₇ ⇌ C₆H₅–CH(C₂H₅)–C₇H₁₅	314 (l)	−0.23 ±0.14	3.51 ±0.43	SE, CT, GLC	263
		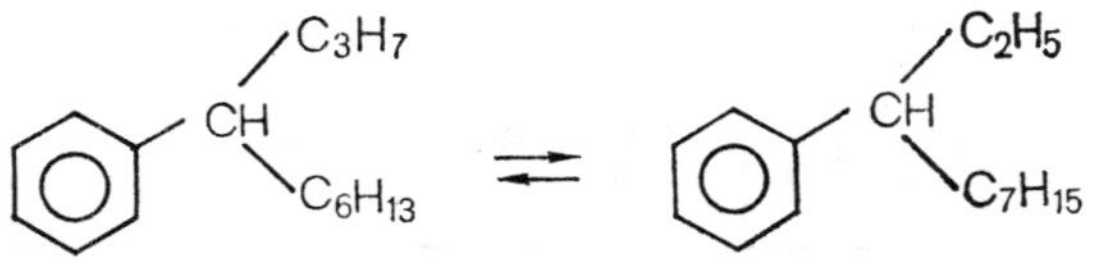 C₆H₅–CH(C₂H₅)–C₇H₁₅ ⇌ C₆H₅–CH(C₃H₇)–C₆H₁₃	322 (l)	0.09 ±0.41	1.69 ±1.28	SE, CT, GLC	263

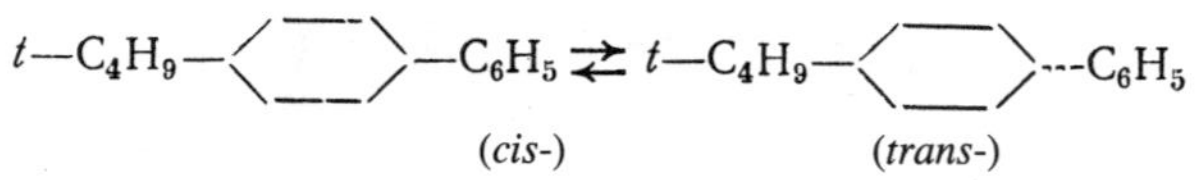

Table (Continued)

Formula	Type	Reaction	T, K (phase)	$\Delta_r H_T^\circ$ ($\Delta_r G_T^\circ$)	$\Delta_r S_T^\circ$	Technique	Reference
$C_{16}H_{26}$	1.3	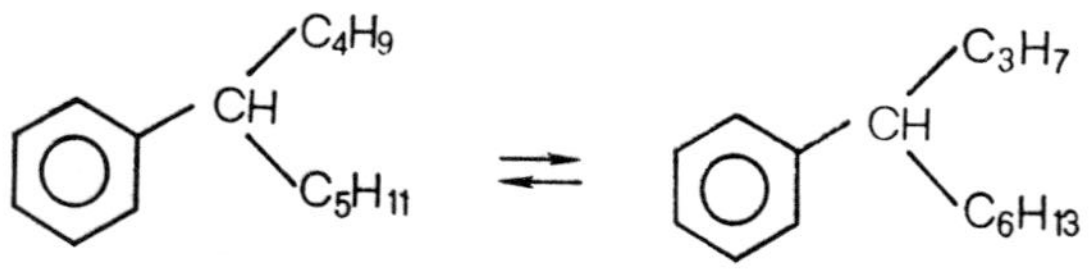 [PhCH(C₄H₉)(C₅H₁₁)] ⇌ [PhCH(C₃H₇)(C₆H₁₃)]	342 (l)	−0.01 ±0.09	0.18 ±0.25	SE, CT, GLC	263
		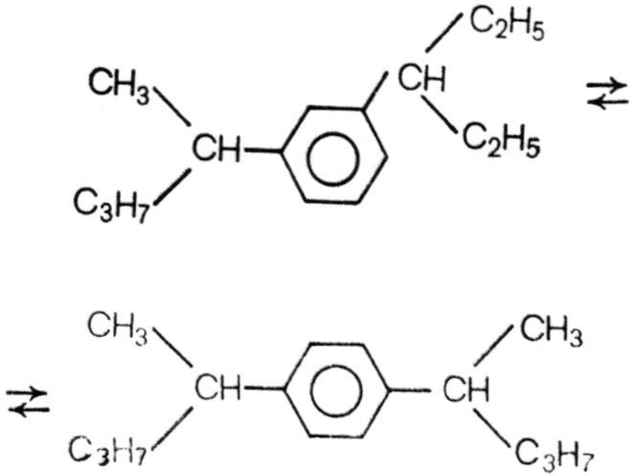 [m-(CH₃)(C₃H₇)CH-C₆H₄-CH(C₂H₅)₂] ⇌ [m-(C₂H₅)(C₂H₅)CH-C₆H₄-CH(CH₃)(C₃H₇)]	368 (l)	−0.42 ±0.28	1.76 ±0.75		263
		[p-(CH₃)(C₃H₇)CH-C₆H₄-CH(CH₃)(C₃H₇)] ⇌ [p-(CH₃)(C₃H₇)CH-C₆H₄-CH(C₂H₅)(C₂H₅)]					
		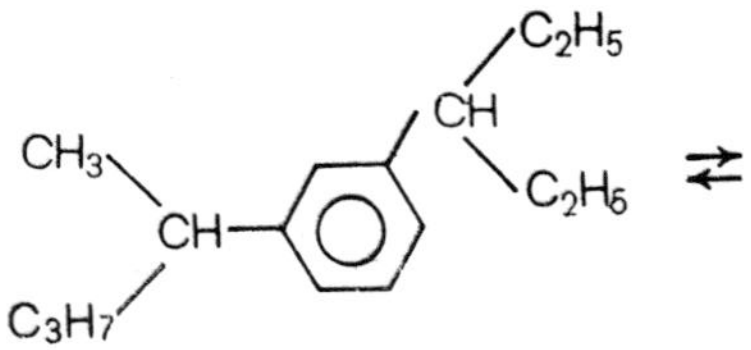 [m-(CH₃)(C₃H₇)CH-C₆H₄-CH(C₂H₅)(C₂H₆)] ⇌ ...	344 (l)	−0.38 ±0.41	1.89 ±1.19	SE, CT, GLC	263

	327 (I)	−0.30 ±0.26	4.68 ±0.80	SE, CT, GLC	263
	336 (I)	−0.23 ±1.12	4.85 ±3.36	SE, CT, GLC	263

$C_{16}H_{26}$ 1.3

Table (Continued)

Formula	Type	Reaction	T, K (phase)	$\Delta_r H_T^\circ$ ($\Delta_r G_T^\circ$)	$\Delta_r S_T^\circ$	Technique	Reference
$C_{16}H_{26}$	1.3	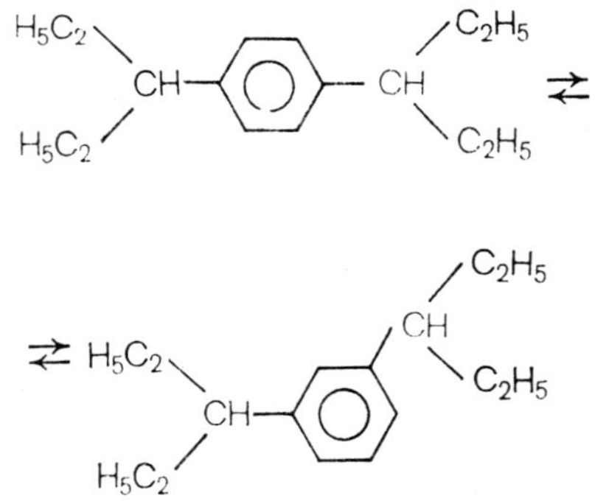	342 (l)	−0.25 ±0.27	4.77 ±0.80	SE, CT, GLC	263
$C_{16}H_{30}O$	2.1	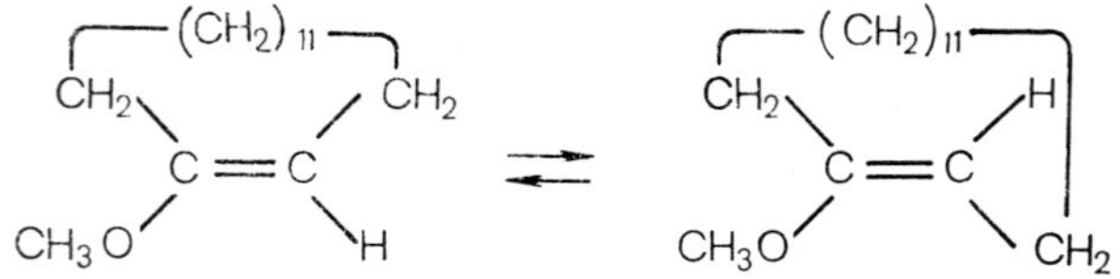	298 (l)	2.7 ±0.5	−3.7 ±1.3	SE, CT, NMR	231
$C_{17}H_{13}NO$	1.1	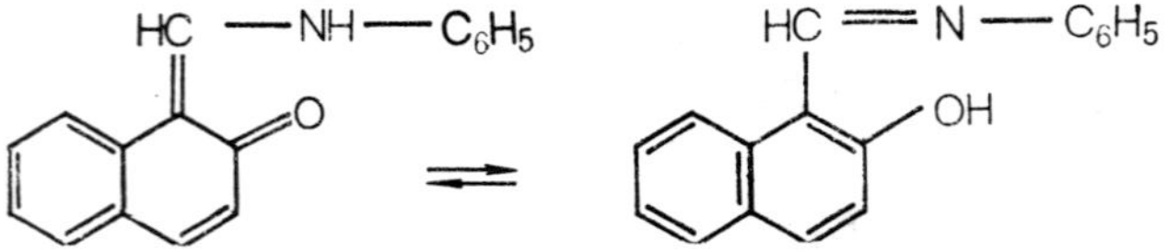	303 (l)	−6.27	—	SE, UV $CDCl_3$	248

$C_{17}H_{18}$	2.1	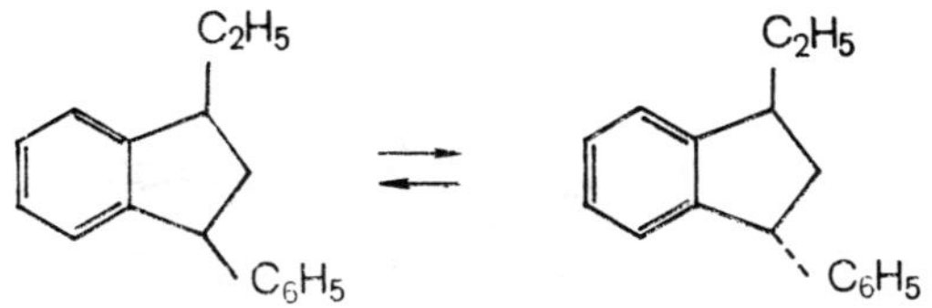 (cis-) ⇌ (trans-)	313 (l)	1.96 ±0.20	4.2 ±0.4	SE, CT, GLC	245
		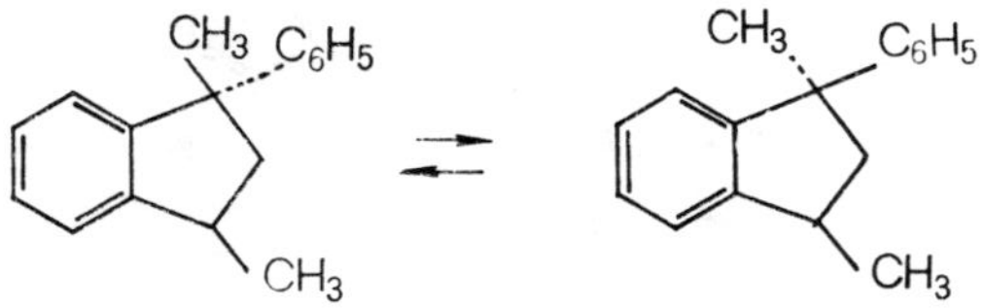 (cis-,trans-) ⇌ (trans-,cis-)	338 (l)	−2.31 ±0.71	−3.60 ±1.55	SE, CT, GLC	245
$C_{17}H_{20}N_2$	1.1 1.2	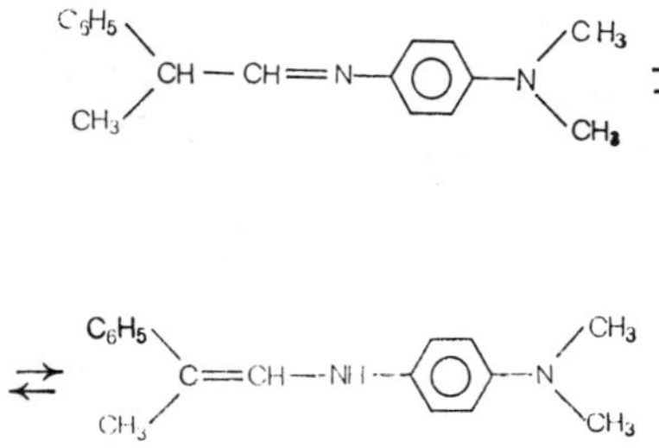	333 (l)	−4.2	−9.6	SE, NMR 	213

Table (Continued)

Formula	Type	Reaction	T, K (phase)	$\Delta_r H_T^\circ$ ($\Delta_r G_T^\circ$)	$\Delta_r S_T^\circ$	Technique	Reference
$C_{18}F_{30}$	1.2	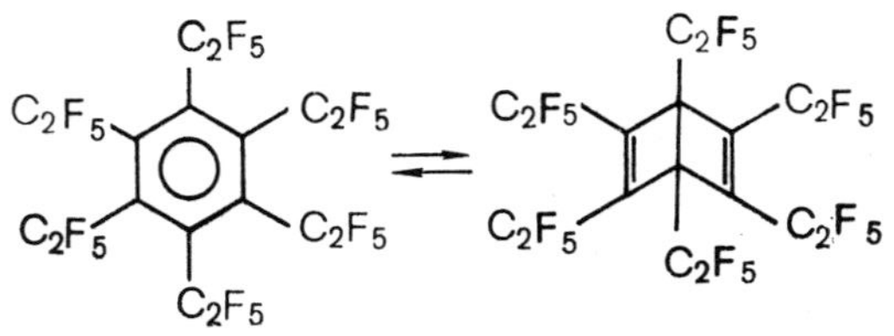 (hexakis(pentafluoroethyl)Dewar benzene ⇌ hexakis(pentafluoroethyl)benzene)	528 (g)	34.9 ±1.9	62.6 ±3.8	FE, GLC	249
$C_{18}H_{15}NO$	1.1 1.2	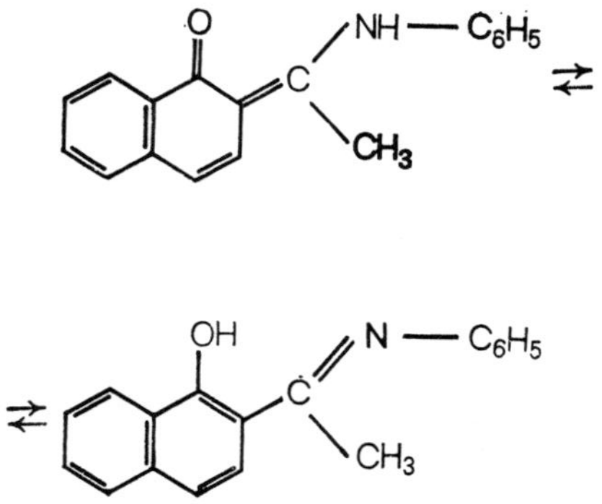 (keto-NH form ⇌ enol-N form of 1-naphthol phenylimine)	303 (l)	−4.60	—	SE, UV CDCl$_3$	248

| $C_{18}H_{20}$ | 2.1 | (structures: cis/trans 1-propyl-3-phenylindane) | 338 (l) | 1.76 ±0.43 | 3.30 ±0.29 | SE, CT, GLC | 245 |
| $C_{20}H_{24}$ | 2.1 | (structures: cis/trans dimesityl ketone-like) | 298 (l) | (−6.44 ±0.13) | — | SE, CT, GLC | 238 |

REFERENCES

1. Armstrong, G. T., Marantz, S. Heats of formation of two isomers of difluorodiazine. *J. Chem. Phys.*, vol. 38, no. 1, pp. 169–172, 1963.
2. Pankratov, A. V., Sokolov, O. M. Vzaimnoye prevrashcheniye izomerov diftordiazina (Interconversion of difluorodiazine isomers). *Zh. Neorgan. Khim.*, vol. 11, no. 8, pp. 1761–1764, 1966.
3. Kuzyakov, Yu. Ya., Moskvitina, E. N. Struktura, chastoty kolebaniya i termodinamicheskiye funktsii izomerov diftordiazina (The structure, vibration frequencies and thermodynamic functions of difluorodiazine isomers). *Vestn. Mosk. Ser. Khim.*, vol. 12, no. 2, pp. 164–167, 1971.
4. Gaines, D. F., Walsh, J. L. Kinetics of the isomerization of 1-chloro- and 2-chloropentaborane. *Inorg. Chem.*, vol. 17, no. 4, pp. 806–809, 1978.
5. Geiseler, G., Rätzsch, M. Über das thermische Gleichgewicht der Umlagerungsreaktion methylnitrit ⇌ nitromethan. *Z. Phys. Chem. (BRD)*, vol. 26, no. 10, pp. 131–137, 1960.
6. Craig, N. C., et al. Vibrational assignments and thermodynamic functions for *cis*- and *trans*-1,2-difluoro-*l*-chloroethylenes. *J. Phys. Chem.*, vol. 74, no. 26, pp. 4520–4527, 1970.
7. Levanova, S. V., et al. Issledovaniye izomerizatsii dibrometilenov (Investigation of dibromoethylene isomerization). *Izv. Vuzov. Ser. Khim. Khim. Tekhnol.*, vol. 15, no. 12, pp. 1821–1823, 1972.
8. Karaseva, S. Ya., Andreevskii, D. N. Ravnovesiye izomerizatsii vtorichnykh monokhlorpentanov i degidrokhlorirovaniya 2-khlorpentana (Isomerization equilibrium of secondary

monochloropentanes and 2-chloropentane dehydrochlorination equilibrium). *Zh. Fiz. Khim.*, vol. 43, no. 9, pp. 2203–2207, 1969.
9. Rozhnov, A. M., et al. Ravnovesiye izomerizatsii dichloretilenov (Isomerization equilibrium of dichloroethylenes). *Zh. Prikl. Khim.*, vol. 47, no. 3, pp. 661–662, 1974.
10. Craig, N. C., Entemann, E. A. Thermodynamics of *cis*-, *trans*-isomerisations: The 1,2-difluoroethylenes. *J. Am. Chem. Soc.*, vol. 83, no. 14, pp. 3047–3050, 1961.
11. Furujama, S., Golden, D. M., Benson, S. W. The thermochemistry of the gas-phase, equilibria *trans*-1,2-diiodoethylene ⇌ acetylene + I_2 and *trans*-1,2-diiodoethylene ⇌ *cis*-1,2-diiodoethylene. *J. Phys. Chem.*, vol. 72, no. 9, pp. 3204–3208, 1968.
12. Levanova, S. V., et al. Issledovaniye zhidkofaznovo gidrokhlorirovaniya simmetrichnykh dikhloretilenov (An investigation of liquid-phase hydrochlorination of symmetric dichloroethylenes). *Zh. Prikl. Khim.*, vol. 48, no. 7, pp. 1574–1577, 1975.
13. Baghal-Vayjooce, M. H., Callester, J. L., Pritchard, H. O. The enthalpy of isomerisation of methyl isocyanide. *Can. J. Chem.*, vol. 55, no. 14, pp. 2634–2636, 1977.
14. Levanova, S. V., Rozhnov, A. M., Sedov, S. M., et al. Ravnovesnye prevrashcheniya dibrometana (Equilibrium conversions of dibromoethane). *Izv. Vuzov. Ser. Khim. Khim. Tekhnol.*, vol. 13, no. 1, pp. 62–65, 1970.
15. Rozhnov, A. M. Ravnovesiye izomerizatsii dikhloretana (Isomerization equilibrium of dichloroethane). *Neftekhimiya*, vol. 8, no. 3, pp. 431–434, 1968.
16. Crump, T. W. The *cis*-, *trans*-izomerization of some simple ethylene derivatives. *J. Org. Chem.*, vol. 28, no. 4, pp. 953–956, 1973.
17. Dolbier, W. R., Medinger, K. S. The thermodynamic effect of fluorine as a substituent vinylic CF_2 and CFH and allylic CF_2C. *Tetrahedron*, vol. 38, no. 15, pp. 2415–2420, 1982.
18. Abell, P. I., Adolf, P. K. HBr catalyzed photoisomerization of allyl haileds. *J. Chem. Thermodyn.*, vol. 1, no. 4, pp. 333–338, 1969.
19. Sharonov, K. G., Rozhnov, A. M. K voprosu ob izomerizatsii 1-brom-1-propenes (On isomerization of 1-bromo-1-propenes). *Izv. Vuzov. Ser. Khim. Khim. Tekhnol.*, vol. 14, no. 4, pp. 557–560, 1971.
20. Alfassi, Z. B., Golden, D. M., Benson, S. W. The thermochemistry isomerization of 3-halopropenes (allyl halides) to 1-halopropenes; entropy and enthalpy of formation contribution of the C_d-(H) (X)-group. *J. Chem. Thermodyn.*, vol. 5, no. 3, pp. 411–420, 1973.
21. Shevtsova, L. A., Andreevskii, D. N. Cis-, *trans*-isomerizatsiya 1-khlorpropena-1 (*Cis*-, *trans*-isomerization of 1-chloropropene-1). *Izv. Vuzov. Ser. Khim. Khim. Tekhnol.*, vol. 14, pp. 718–720, 1971.
22. Sharonov, K. G., Rozhnov, A. M., Cherkasova, R. I. Issledovaniye ravnovesiya dikhlorpropanov (Investigation of equilibrium of dichloropropanes). *Izv. Vuzov. Ser. Khim. Khim. Tekhnol.*, vol. 15, no. 3, pp. 375–377, 1972.
23. Izmaylov, V. D. Issledovaniye degidrokhlorirovaniya i izomerizatsii 1,1-dikhlorpropana (An investigation of 1,1-dichloropropane dehydrochlorination and isomerization). *Izv. Vuzov. Ser. Khim. Khim. Tekhnol.*, vol. 17, no. 5, pp. 706–709, 1974.
24. Hine, J., Arata, K. Keto-enol tautomerism. 2. The calorimetrical determination of the equilibrium constants for ketoenol tautomerism for cyclohexanoul and acetone. *Bull. Chem. Soc. Japan*, vol. 49, no. 11, pp. 3089–3092, 1976.
25. Rozhnov, A. M., Andreevskii, D. N. Izomerizatsiya brompropana (Bromopropane isomerization). *Neftekhimiya*, vol. 4, no. 1, pp. 111–118, 1964.
26. Kabo, G. Ya., Andreevskii, D. N. Termodinamika izomerizatsii monokhlorpropanov (Thermodynamics of isomerization of monocloropropanes). *Neftekhimiya*, vol. 5, no. 1, pp. 132–135, 1965.
27. Andreevskii, D. N., Rozhnov, A. M. K voprosu termodinamiki galoidnykh alkilov: Izomerizatsiya iodpropana (On the question of thermodynamics of alkyl halides: Iodopropane isomerization). *Neftekhimiya*, vol. 2, no. 3, pp. 778–783, 1962.
28. Yogev, A., Benmair, R. M. T. Photochemistry in the electronic ground state: Isotope separa-

tion by infrared multiphoton isomerization of complex molecules. *Chem. Phys. Lett.*, vol. 63, no. 3, pp. 558-561, 1979.
29. Schlag, E. W., Kaiser, E. W. The thermal *trans-*, *cis-*isomerization of octafluorobutene-2. *J. Am. Chem. Soc.*, vol. 87, no. 15, pp. 3311-3314, 1965.
30. Dolbier, W. R., Fielder, T. H. Thermal isomerization of 2,2-difluoromethylenecyclopropane. *J. Am. Chem. Soc.*, vol. 100, no. 17, pp. 5577-5578, 1978.
31. Shainman, B. A., Mirskova, A. N. Stroyeniye i reaktsionnaya sposobnost' polifunktsional'nykh amidov. 7. Kinetika priosoyedineniya spirtov k N-atsilkhloral'iminam (Structure and reactivity of polyfunctional amides. 7. Kinetics of alcohol addition to N-acylchloralimines). *Zh. Org. Khim.*, vol. 12, no. 6, pp. 1188-1191, 1976.
32. Wiberg, K. B., Fenoglio, R. A. Heats of formation of C_4H_6 hydrocarbons. *J. Am. Chem. Soc.*, vol. 90, no. 13, pp. 3395-3397, 1968.
33. Taskinen, E. Thermodynamics of vinyl ethers. 5. The relative stabilities of 4-methylene-1,3-dioxolane and 4-methyl-1,3-dioxole. *J. Chem. Thermodyn.*, vol. 6, no. 11, pp. 1021-1025, 1974.
34. Sharonov, K. G., et al. Ravnovesiye izomerizatsii geometricheskikh form 1-brombutena-1 i 2-brombutena-2 (Isomerization equilibrium of geometrical forms of 1-bromobut-1-ene and 2-bromobut-2-ene). *Izv. Vuzov. Ser. Khim. Khim. Tekhnol.*, vol. 18, no. 5, pp. 736-738, 1975.
35. Taskinen, E. Sainio thermodynamics of vinyl ethers. 15. Halogen-containing vinyl ethers. *Tetrahedron*, vol. 42, no. 5, 593-595, 1976.
36. Polyakov, V. M., Danov, S. M., Bodrikov, I. V. Izomerizatsiya sodineniy allil'novo tipa. 1. Gasofaznaya kataliticheskaya izomerizatsiya β-metallylkhlorida (Isomerization of allyl-type compounds. 1. Gas-phase catalytic isomerization of β-methallyl chloride). *Izv. Vuzov. Ser. Khim. Khim. Tekhnol.*, vol. 12, no. 2, pp. 208-210, 1969.
37. Levanova, S. V., et al. Issledovaniye ravnovesnykh prevrashcheniy 2,2-dikhlorbutenov (Investigation of equilibrium conversions of 2,2-dichlorobutenes). *Zh. Fiz. Khim.*, vol. 47, no. 4, p. 1064, 1973.
38. Levanova, S. V., et al. Issledovaniye izomerizatsii khlorbutenov (Investigation of isomerization of chlorobutenes). *Zh. Fiz. Khim.*, vol. 49, no. 1, pp. 76-80, 1975.
39. Rodova, R. M., et al. Issledovaniye degidrokhlorirovaniya 1,1-dikhlorbutana and izomerizatsii khlorbutenov (Investigation of 1,1-dichlorobutane dehydrochlorination and isomerization of chlorobutanes). *Izv. Vuzov. Ser. Khim. Khim. Tekhnol.*, vol. 17, no. 3, pp. 379-381, 1974.
40. Okujama, T., Fueno, T., Furukawa, J. Structure and reactivity of α,β-unsaturated ethers. 9. The *cis-*, *trans-*isomerization equilibria in the liquid phase. *Tetrahedron*, vol. 25, no. 22, pp. 5409-5414, 1969.
41. Levanova, S. V., Andreevskii, D. N. Ravnovesiye reaktsii degidrokhlorirovaniya 2-khlorbutana (Equilibrium of 2-chlorobutane dehydrochlorination). *Neftekhimiya*, vol. 4, no. 2, pp. 329-336, 1964.
42. Benson, S. W., Bose, A. N. The iodine-catalyzed positional isomerization of olefins: A new tool for the precise measurement of thermodynamic data. *J. Am. Chem. Soc.*, vol. 85, no. 10, pp. 1385-1387, 1963.
43. Golden, D. M., Egger, K. W., Benson, S. W. Iodine-catalyzed isomerization of olefins. 1. Thermodynamic data from equilibrium studies of positional and geometrical isomerization. *J. Am. Chem. Soc.*, vol. 86, no. 24, pp. 5416-5420, 1964.
44. Maccoll, A., Ross, R. A. The hydrogen bromide catalyzed isomerization of *n*-butenes. 1. Equilibrium values. *J. Am. Chem. Soc.*, vol. 87, no. 6, pp. 1169-1170, 1965.
45. Meyer, E. F., Stroz, D. G. Thermodynamics of *n*-butene isomerization. *J. Am. Chem. Soc.*, vol. 94, no. 18, pp. 6344-6347, 1972.
46. Kaptein, F., Van der Steen, A. J., Mol, J. C. Thermodynamics of the geometrical isomerization of 2-butene and 2-pentene. *J. Chem. Thermodyn.*, vol. 15, no. 2, pp. 137-146, 1983.
47. Akimoto, H., Sprung, J. L., Pitts, J. N. Nitrogen dioxide catalyzed isomerization of 2-butene

and 2-pentene: A precise method for determining the enthalpies and entropies of geometric isomerizations. *J. Am. Chem. Soc.*, vol. 94, no. 14, pp. 4850–4855, 1972.
48. Abell, P. I. Bromine atom catalyzed isomerization of terminal olefins. *J. Am. Chem. Soc.*, vol. 88, no. 7, pp. 1346–1348, 1966.
49. Nesterova, T. N., Rozhnov, A. M., Kovaleva, T. V. Issledovaniye ravnovesiya izomerizatsii 1,2- 1,3-, 2,3-dibrombutanov (A study of isomerization equilibrium of 1,2-, 1,3-, and 2,3-dibromobutanes). *Zh. Fiz. Khim.*, vol. 47, no. 5, p. 1330, 1973.
50. Rozhnov, A. M., Nesterova, T. N., Kovaleva, T. V. Issledovaniye izomerizatsii dibrombutanov. 1. Izomerizatsiya 1,2-, 1,3-, 1,4-dibrombutanov (A study of isomerization equilibrium of dibromobutanes. 1. Isomerization of 1,2-, 1,3-, and 1,4-dibromobutanes). *Zh. Org. Khim.*, vol. 8, no. 8, pp. 1560–1564, 1972.
51. Nesterova, T. N., Rozhnov, A. M., Borodenko, N. A. Issledovaniye ravnovesiya izomerizatsii 2,2- i 2,3-dibrombutanov (A study of isomerization equilibrium of 2,2- and 2,3-dibromobutanes). *Izv. Vuzov. Ser. Khim. Khim. Tekhnol.*, vol. 16, no. 8, pp. 1216–1218, 1973.
52. Rozhnov, A. M., Nesterova, T. N., Mukovnina, G. S. Issledovaniye ravnovesiya izomerizatsii dibromalkanov. 2. Izomerizatsiya dibrombutanov izostroeniya (A study of isomerization equilibrium of dibromoalkanes. 2. Isomerization of branched-chain dibromobutanes). *Zh. Fiz. Khim.*, vol. 49, no. 11, pp. 2842–2846, 1975.
53. Nesterova, T. N., Rozhnov, A. M., Migacheva, E. G. Ravnovesiya izomerizatsii opticheskikh form 2,3-dibrombutana (Isomerization equilibrium of the optical forms of 2,3-dibromobutane). *Izv. Vuzov. Ser. Khim. Khim. Tekhnol.*, vol. 13, no. 9, pp. 1300–1302, 1970.
54. Rodova, R. M. et al. Issledovaniye ravnovesnykh prevrashcheniy 2,3-dikhlorbutana (A study of equilibrium conversions of 2,3-dichlorobutane). *Zh. Fiz. Khim.*, vol. 47, no. 4, p. 1064, 1973.
55. Pavskiy, V. I., Batarova, N. I., Rozhnov, A. M. *Meso-, d,l*-izomerizatsiya 2,3-dikhlorbutana (*Meso-, d,l*-isomerization of 2,3-dichlorobutane). *Zh. Orgo. Khim.*, vol. 7, no. 7, pp. 1325–1327, 1971.
56. Phoads, S. J., Chattopadhyay, T. K., Waali, E. E. Double-bond isomerization in unsaturated esters and enol ethers. 1. Equilibrium studies in cyclic and acyclic systems. *J. Org. Chem.*, vol. 35, no. 10, pp. 3352–3358, 1970.
57. Sváta, V., Prochazka, M., Bacos, V. Prototropic isomerization of unsaturated sulfides, sulfoxides and sulfones. *Collect. Chech. Chem. Commun.*, vol. 43, no. 10, pp. 2619–2634, 1978.
58. Lebedeva, N. D. Teploty sgoraniya, obrazovaniya i izomerizatsii ryada monokarbonovykh kislot. Termodinamichieskiye i termokhimicheskiye konstanty *(Heats of Combustion, Formation and Isomerization of Monocarboxylic Acids. Thermodynamic and Thermochemical Constants)*. Nauka, pp. 57–61, 1970.
59. Byström, K., Mansson, M. Enthalpies of formation of some cyclic 1,3- and 1,4-di- and polyesters: Thermochemical strain in the —O—C—O— and —O—C—C—O— groups. *J. Chem. Soc. Perkin Trans.*, part 2, no. 5, pp. 565–569, 1982.
60. Waldron, J. T., Snyder, W. H. Thermodynamics of *cis-, trans*-isomerizations: The relative stabilities of the 1,2-dimethoxyethylenes. *J. Am. Chem. Soc.*, vol. 95, no. 17, pp. 5491–5495, 1973.
61. Peshchenko, A. D., Andreevskii, D. N. Ravnovesiye izomerizatsii monobrombutanov normal'novo stroyeniya (Isomerization equilibrium of straight-chain monobromobutanes). *Izv. AN Bel-SSR. Ser. Khim. Nauk*, no. 2, pp. 122–123, 1968.
62. Nesterova, T. N., Rozhnov, A. M. Izucheniye reaktsii izomerizatsii monobrombutanov izostroyeniya (Isomerization studies of branched-chain monobromobutanes). *Izv. Vuzov. Ser. Khim. Khim. Tekhnol.*, vol. 17, no. 4, pp. 556–558, 1974.

63. Beak, P., Lee, J., Zeigler, J. M. Equilibrium studies: amide-imidate and thioamide-thioimidate functions. *J. Org. Chem.*, vol. 43, no. 8, pp. 1536–1538, 1978.
64. Booth, M. R., Frankiss, S. G. The constitution, vibrational spectra and proton resonance spectra of trimethylsilyl cyanide and isocyanide. *Spectrochim. Acta*, vol. A26, no. 4, pp. 859–869, 1970.
65. Sheynker, Yu. N., et al. Temperaturnaya zavisimost' i termodinamicheskiye kharakteristiki laktam-laktim tautomernovo ravnovesiya nekotorykh proizvodnykh 2-oksipiridina (Temperature dependence and thermodynamic functions of lactam-lactim tautomeric equilibrium of selected derivatives of 2-hydroxypyridine). *DAN SSSR*, vol. 192, no. 6, pp. 1295–1298, 1970.
66. Beak, P., Covington, J. B., White, J. M. Quantitative model of solvent effects on hydroxypyridine-pyridone and mercaptopyridine-thiopyridon equilibria: Correlation with reaction-field and hydrogen-bonding effects. *J. Org. Chem.*, vol. 45, no. 8, pp. 1347–1354, 1980.
67. Cook, M. J., et al. Quantitative measure of the aromatic stabilization energy of 2-pyridone and related compounds. *J. Chem. Soc. Chem. Commun.*, no. 10, pp. 510–511, 1971.
68. Beak, P., et al. Equilibration studies. Protometric equilibria of 2- and 4-hydroxypyridines, 2-hydroxypyrimidines, 2- and 4-mercaptopyridines and structurally related compounds in gas phase. *J. Am. Chem. Soc.*, vol. 98, no. 1, pp. 171–179, 1976.
69. Sheynker, Yu. N., Peresleni, E. M. O. tautomerii nekotorykh proizvodnykh geterotsiklicheskikh soedineniy. 12. Temperaturnaya zavisimost' i nekotorye termodinanicheskiye kharakteristiki aminoimino tautomernovo ravnovesiya atsilirovannykh geterotsiklicheskikh aminov (On tautomerism of some derivatives of heterocyclic compounds. 12. Temperature dependence and selected thermodynamic functions of amine-imine tautomeric equilibrium of acylated heterocyclic amines). *Zh. Fiz. Khim.*, vol. 35, no. 11, pp. 2623–2627, 1961.
70. Rull, M., Vilarrasa, J. 6-pivaloil-1,2,3,3a,6-pentaazapentalene: Steric effects on the 2-azidoimidazole/imidazo-1,2-*d*-tetrazole equilibrium. *Tetrahedron Lett.*, no. 46, pp. 4175–4176, 1976.
71. Bickerton, J., Pilcher, G., Al-Takhin, G. Enthalpies of combustion of the three aminopyridines and the three cyanopyridines. *J. Chem. Thermodyn.*, vol. 16, no. 4, pp. 373–378, 1984.
72. Egger, K. W., Benson, S. W. Nitric oxide and iodine catalyzed isomerization of olefins. 4. Thermodynamic data from equilibrium studies of the geometrical isomerization of 1,3-pentadione. *J. Am. Chem. Soc.*, vol. 87, no. 15, pp. 3311–3314, 1965.
73. Herling, J., Shabtai, J., Gill-Av, E. Relative stability of cyclic olefins. 1. Equilibrium isomerization of monocyclic olefins containing four- to six-member substituted divinyl ethers. *J. Org. Chem.*, vol. 42, no. 8, pp. 1443–1449, 1977.
74. Taskinen, E., Virtanen, R. Thermodynamics of vinyl ethers. 19. Alkyl-substituted divinyl ethers. *J. Org. Chem.*, vol. 42, no. 8, pp. 1443–1449, 1977.
75. Cocks, A. T., Egger, K. W. Gas-phase thermal unimolecular isomerizations of acetylcyclopropane. 1. The equilibrium between 2,3-dihydro-5-methylfuran and acetylcyclopropane. *J. Chem. Soc. Perkin Trans.*, part 2, no. 2, pp. 197–199, 1973.
76. Hine, J., Arata, K. Keto-enol tautomerism. 1. The calorimetrical determination of the equilibrium constant for keto-enol tautomerism for cyclopentanone. *Bull. Chem. Soc. Japan*, vol. 49, no. 11, pp. 3085–3088, 1976.
77. Mines, G. W., Thompson, H. Infrared and photoelectron spectra and keto-enol tautomerism of acetylacetones and acetoacetic esters. *Proc. R. Soc. London Ser. A*, vol. A342, no. 1630, pp. 327–339, 1975.
78. Nakanishi, H., Morita, H., Nagakura, S. Electronic structures and spectra of the keto- and enol-forms of acetylacetone. *Bull. Chem. Soc. Japan*, vol. 50, no. 9, pp. 2255–2261, 1977.
79. Schweig, A., Vermeer, H., Weidner, U. A photoelectron spectroskopic study of keto-enol tautomerism in acetylacetones: A new application of photoelectron spectroscopy. *Chem. Phys. Lett.*, vol. 26, no. 2, pp. 229–233, 1974.

80. Thompson, D. W., Allred, A. L. Keto-enol equilibria in 2,4-pentanedione and 3,3-dideuterio-2,4-pentanedione. *J. Phys. Chem.*, vol. 75, no. 3, pp. 433–435, 1971.
81. Taskinen, E. Thermodynamics of vinyl ethers. 17. Thermodynamic stability of 1,2-dialkoxyethylenes. *Tetrahedron*, vol. 32, no. 19, pp. 2327–2330, 1976.
82. Taskinen, E. Thermodynamics of vinyl ethers. 9. The relative stabilities of 4-methylene-1,3-dioxane and 4-methyl-1,3-dioxene-(4). *Acta Chem. Scand.*, vol. B28, no. 10, pp. 1234–1236, 1974.
83. Salomaa, P., Hautoniemi, L. Relative thermodynamic stabilities of 4-methylene-1,3-dioxane and the isomeric 4-methyl-1,3-dioxene-(4) and the kinetics of their hydrolytic. *Acta Chem. Scand.*, vol. 23, no. 2, pp. 709–711, 1969.
84. Rommelaere, Y., Anteunis, M. NMR experiments on acetals. 18. *Cis-*, *trans*-equilibria in some 2,4-disubstituted 1,3-dioxolanes. *Bull. Soc. Chim. Belg.*, vol. 79, no. 1–2, pp. 11–14, 1970.
85. Radyuk, Z. A., Kabo, G. Ya., Andreevskii, D. N. Ravnovesiye izomerizatsii i termodinamicheskiye svoystva metilbutenov (Isomerization equilibrium and thermodynamic properties of methylbutenes). *Neftekhimiya*, vol. 13, no. 3, pp. 356–360, 1973.
86. Good, W. D., Smith, N. K. The enthalpies of combustion of the isomeric pentenes in the liquid state: A warning to combustion calorimetrists about sample drying. *J. Chem. Thermodyn.*, vol. 11, no. 2, pp. 111–118, 1979.
87. Karaseva, S. Ya., Andreevskii, D. N. Ravnovesiya izomerizatsii pentenov (Isomerization equilibrium of pentenes). *Neftekhimiya*, vol. 8, no. 3, pp. 431–434, 1968.
88. Rozhnov, A. M., et al. Issledovaniye ravnovesiya izomerizatsii dibromalkanov. 1. Isomerizatsiya 1,2-, 1,3-, 1,4-, 1,5-dibrompentanov. (A study of isomerization equilibrium of dibromoalkanes. 1. Isomerization of 1,2-, 1,3-, 1,4-, and 1,5-dibromopentanes). *Zh. Fiz. Khim.*, vol. 48, no. 8, p. 2136, 1974.
89. Rozhnov, A. M., et al. Issledovaniye ravnovesiya izomerizatsii dibromalkanov. 3. Isomerizatsiya 1,4-, 2,3- i 2,4-dibrompentanov (A study of isomerization equilibrium of dibromoalkanes. 1. Isomerization of 1,4-, 2,3-, and 2,4-dibromopentanes). *Zh. Fiz. Khim.*, vol. 48, no. 7, p. 1876, 1974.
90. Cherkasova, R. I., Rozhnov, A. M., Sharonov, K. K. Ravnovesiye izomerizatsii opticheskikh form 2,3- i 2,4-dikhlorpentanov (Isomerization equilibrium of optical forms of 2,3- and 2,4-dichloropentanes). *Izv. Vuzov. Ser. Khim. Khim. Tekhnol.*, vol. 19, no. 1, pp. 15–17, 1976.
91. Taskinen, E. Thermodynamics of vinyl ethers. 3. On the relative stabilities of 2-methoxy-1-butene, (E)-2-methoxy-2-butene and (Z)-2-methoxy-2-butene and related isomeric compounds. *J. Chem. Thermodyn.*, vol. 6, no. 4, pp. 345–353, 1974.
92. Taskinen, E., Länteenmäki, H. Thermodynamics of vinyl ethers. 18. Unsaturated acetals. *Tetrahedron*, vol. 32, no. 19, pp. 2331–2333, 1976.
93. Willy, W. E., Binsch, G., Eliel, E. L. Conformational analysis. 23. 1,3-Dioxolanes. *J. Am. Chem. Soc.*, vol. 92, no. 18, pp. 5394–5402, 1970.
94. Good, W. D. Enthalpies of combustion of 18 organic sulfur compounds related to petroleum. *Chem. Eng. Data*, vol. 17, no. 2, pp. 158–162, 1972.
95. Roganov, G. N., Kabo, G. Ya. Ravnovesiye izomerizatsii nekotorykh vtorichnykh n-bromalkanov (Isomerization equilibrium of selected secondary n-bromoalkanes). *Izv. AN Bel-SSR. Ser. Khim. Nauk*, no. 4, pp. 116–118, 1973.
96. Roganov, G. N., Kabo, G. Ya., Andreevskii, D. N. Additivnost' termokhimicheskikh svoystv izomernykh vtorichnyk khloridov n-alkanov (C_5–C_{10}) (Additivity of the thermochemical properties of isomeric secondary C_5–C_{10} n-alkane chlorides). *Zh. Org. Khim.*, vol. 5, no. 12, pp. 2097–2102, 1969.
97. Esipenok, G. E., Kabo, G. Ya., Andreevskii, D. N. Ravnovesiye i termodinamika izomerizatsii metilkhlorbutanov (Isomerization equilibrium and thermodynamics of methylchlorobutanes). *Zh. Fiz. Khim.*, vol. 47, no. 3, p. 739, 1973.

98. Pilcher, G., Chadwick, J. D. M. Measurements of heats of combustion by flame calorimetry. 4. n-Pentane, *iso*-pentane, *neo*-pentane. *Trans. Faraday Soc.*, vol. 62, no. 4, pp. 821-827, 1966.
99. Good, W. D. The enthalpies of combustion and formation of the isomeric pentanes. *J. Chem. Thermodyn.*, vol. 3, no. 1, pp. 97-103, 1971.
100. Belavin, I. Yu., et al. O- i C-elementoorganicheskiye soedineniya. 10. Germanotropnoye ravnovesiye (O- and C-organoelemental compounds. 10. Germanotropic equilibrium). *Zh. Obshch. Khim.*, no. 5, pp. 1065-1075, 1970.
101. Chesick, J. P. Kinetics and thermodynamics of the thermal interconversion of decafluoro-1,2-dimethylcyclobutene and decafluoro-2,3-dimethylbutadiene-1,3. *J. Chem. Soc.*, vol. 88, no. 21, pp. 4800-4803, 1966.
102. Butler, J. B., Liemezs, J. Thermodynamic functions for halogenated benzenes. *J. Chem. Eng. Data*, vol. 14, no. 3, pp. 335-341, 1969.
103. Godnev, I. N., Sverdlin, A. S. O ravnovesii izomerov dikhlorbenzolov (On the equilibrium of dichlorobenzene isomers). *Zh. Fiz. Khim.*, vol. 35, no. 2, pp. 474-475, 1961.
104. Desai, P. D., Wilhoit, R. C., Zwolinski, B. J. Heat of combustion of resorcinol and enthalpies of isomerization of dihydroxybenzenes. *J. Chem. Eng. Data*, vol. 13, no. 3, pp. 334-335, 1968.
105. Camps, F., et al. NMR study of the keto-enol equilibrium of ethyl γ,γ,γ-trifluoroacetoacetate and its reaction with water and alcohols. *Tetrahedron*, vol. 33, no. 13, pp. 1637-1640, 1977.
106. Beak, P., Bonham, J., Lee, J. T. Equilibrium studies: The energy differences for some six-membered heterocyclic methyl amide-amidate isomer pairs. *J. Am. Chem. Soc.*, vol. 90, no. 6, pp. 1569-1582, 1968.
107. Beak, P., Mueller, D. S., Lee, J. Equilibrium studies. Determination of the enthalpy difference between methyltropic isomers from heats of methylation. *J. Am. Chem. Soc.*, vol. 96, no. 12, pp. 3867-3874, 1974.
108. Taskinen, E., Mukkala, V.-M. Thermodynamics of vinyl ethers. 27. Thermodynamic stability of β-methoxy-substituted α,β-unsaturated ketones and the corresponding carboxylic esters. *Tetrahedron*, vol. 38, no. 5, pp. 613-616, 1982.
109. Petrova-Kuminskaya, S. V., Roganov, G. N., Kabo, G. Ya. Termodinamica izomerizatsii geksadienov (Thermodynamics of hexadiene isomerization). *Neftekhimiya*, vol. 23, no. 4, pp. 489-494, 1983.
110. Döring, C.-E., Hauthal, H. G. Gaschromatographische Bestimmung von Reaktionsgleichgewichtes der Hexadiene-(2,4). *J. Prakt. Chem.*, vol. 24, no. 1-2, pp. 27-37, 1964.
111. Becker, J. Y. Isomerization of mono- and diacetyl hydrocarbons. *Tetrahedron*, vol. 32, no. 24, pp. 3041-3043, 1976.
112. Rogers, D. W., Dagdagan, O. A., Allinger, N. L. Heats of hydrogenation and formation of linear alkynes and a molecular mechanics interpretation. *J. Am. Chem. Soc.*, vol. 101, no. 3, pp. 671-676, 1979.
113. Yursha, I. A., Kabo, G. Ya. Termodinamika izomerizatsii metiltsiklopentenov (Thermodynamics of methylcyclopentene isomerization). *Zh. Fiz. Khim.*, vol. 49, no. 5, pp. 1302-1303, 1975.
114. Kabbouf, A., Rossini, F. D. Heats of combustion, formation and hydrogenation of 14 selected cyclomonoölefin hydrocarbons. *J. Phys. Chem.*, vol. 65, no. 3, pp. 476-480, 1961.
115. Fuchs, R., Peucock, L. A. Heats of vaporization and gaseous heats of formation of some five- and six-membered ring alkenes. *Can. J. Chem.*, vol. 57, no. 17, pp. 2302-2304, 1979.
116. Coppens, P., et al. Relative stability of cyclic olefins. 2. Calculations on 1-, 3- and 4-substituted cyclenes. *J. Am. Chem. Soc.*, vol. 87, no. 18, pp. 4111-4113, 1965.
117. Taskinen, E., Kuusisto, M. Thermodynamics of vinyl ethers. 25. Relative stabilities of some methoxy derivatives of 1,3- and 1,4-pentadiene. *Acta. Chem. Scand.*, vol. B34, no. 8, pp. 571-574, 1980.
118. Taskinen, E. Thermodynamics of vinyl ethers. 22. Effect of unsaturated substituents on the

relative stabilities of the geometrical isomers of α-substituted methyl propenyl ethers. *Finn. Chem. Lett.*, no. 8, pp. 276–278, 1978.
119. Taskinen, E. Thermodynamics of vinyl ethers. 12. The relative stabilities of 2-alkylidenetetrahydrofurans and 5-alkyl-2,3-dihydrofurans. *Acta. Chem. Scand.*, vol. B29, no. 2, pp. 245–248, 1975.
120. Young, W. G., Green, H. E., Diaz, A. F. The acid-catalyzed isomerization of the butenyl acetates. *J. Am. Chem. Soc.*, vol. 93, no. 19, pp. 4782–4787, 1971.
121. Hussain, S. A. M., et al. Conformational equilibria in 2,4-disubstituted γ-butyrolactones. *J. Chem. Soc. Chem. Commun.*, no. 21, pp. 873–874, 1974.
122. Král, V., Lešetický, L. The effect of solvent, base strength and isotopic substitution on the kinetics of nitro group rearrangement. *Coll. Czech. Chem. Commun.*, vol. 40, no. 9, pp. 2816–2825, 1975.
123. Bartolo, H. F., Rossini, F. D. Heats of isomerization of the seventeen isomeric hexenes. *J. Phys. Chem.*, vol. 64, no. 11, pp. 1865–1869, 1960.
124. Yursha, I. A., Kabo, G. Ya. Ravnovesiye izomerizatsii i termodinamicheskiye svoystva 2-metilpentenov (Isomerization equilibrium and thermodynamic properties of 2-methylpentenes). *Zh. Fiz. Khim.*, vol. 50, no. 2, pp. 558–559, 1976.
125. Maurel, R., Guisnet, M., Bove, L. Isomerisation catalytique des hydrocarbures éthyléniques. 2. Equilibres d'isomérisation dex hexénes. *Bull. Soc. Chem. Fr.*, no. 6, pp. 1975–1981, 1969.
126. Radyuk, Z. A., Kabo, G. Ya., Andreevskii, D. N. Ravnovesiye i termodinamika izomerizatsii nekotorykh izomerov geksena (Isomerization equilibrium and thermodynamics of selected hexene isomers). *Neftekhimiya*, vol. 12, no. 5, pp. 679–686, 1972.
127. Maurel, R., et al. Etude de l'isomérisation des oléfins sur les catalyseurs méttaliques. *Bull. Soc. Chim. Fr.*, no. 10, pp. 3082–3085, 1966.
128. Rodgers, A. S., Wu, M.-C. R. Thermochemistry of the gas-phase iodine catalyzed isomerization: 2,3-dimethyl-1-butene–2,3-dimethyl-2-butene. *J. Chem. Thermodyn.*, vol. 3, no. 5, pp. 591–597, 1971.
129. Kabo, G. Ya., Andreevskii, D. N., Radyuk, Z. A. Termodinamika izomerizatsii n-hexenov (Thermodynamics of isomerisation of *n*-hexenes). *Neftekhimiya*, vol. 10, no. 3, pp. 330–334, 1979.
130. Rogers, D. W., Papadimetriou, P. M., Siddique, N. A. An improved hydrogen microcalorimeter for use with large molecules. *Microchim. Acta*, vol. 2, no. 4–5, pp. 389–400, 1975.
131. Konnecke, H.-G., Schmier, H., Gawalec, G. Zur Isomerisierung von alicyclischen Kohlenwasserstoffen. *Z. Phys. Chem. (DDR)*, vol. 218, no. 3–4, pp. 233–249, 1961.
132. Kabo, G. Ya., Andreevskii, D. N. Termodinamicheskiye kharacteristiki reaktsii tsiklogeksan-metiltsiklopentan (Thermodynamic functions of the cyclohexane-methylcyclopentane reaction). *Zh. Fiz. Khim.*, vol. 47, no. 1, pp. 272–273, 1973.
133. Mamontov, V. P. O zavisimosti konstanty ravnovesiya knofiguratsionnov izomerizatsii 1,3-dioksanov ot knotsentratsii katalizatora (On the dependence of the equilibrium constant of configurational isomerization of 1,3-dioxanes on catalyst concentration). *Vopr. Stereokhim.*, issue 3, pp. 57–59, 1953.
134. Taskinen, E., Lanteenmäki, H. Thermodynamics of vinyl ethers. 20. α,β-unsaturated orthoesters. *Finn. Chem. Lett.*, no. 1, pp. 47–50, 1978.
135. Eliel, E. L., Giza, C. Conformational analysis. 17. 2-alkoxy- and 2-alkylthiotetrahydropyrans and 2-alkoxy-1,3-dioxane. The anomeric effect. *J. Org. Chem.*, vol. 33, no. 10, pp. 3754–3758, 1968.
136. Pihlaja, K. Conformational analysis. 11. Chemical equilibrium of diastereomeric alkyl-1,3-dithians. Conformational preference of alkyl substituents and the chair-boat energy difference. A revision of the chair-boat energy difference of cyclohexane. *J. Chem. Soc. Perkin Trans.*, no. 8, pp. 890–896, 1974.
137. Eliel, E. L., Hutchins, R. O. Conformational analysis. 18. 1,3-dithianes. Conformational

preferences of alkyl substituents and the chair-boat energy difference. *J. Am. Chem. Soc.*, vol. 91, no. 10, pp. 2703–2715, 1969.
138. Andreevskii, D. N., Kabo, G. Ya., Esipenok, G. E. Ravnovesiye i termodinamika reaktsiy izomerizatsii metilkhlorpentanov (Equilibrium and thermodynamics of the isomerizations of methylchloropentanes). *Zh. Fiz. Khim.*, vol. 48, no. 6, p. 1614, 1974.
139. Roganov, G. N., Kabo, G. Ya., Andreevskii, D. N. Termodinamika izomerizatsii metilpentanov i metilgeptanov (Thermodynamics of the isomerization of methylpentanes and methylheptanes). *Neftekhimiya*, vol. 12, no. 4, pp. 495–500, 1972.
140. Potekhin, A. A., Shevchenko, S. M. Kol'chato-tsepnaya tautomeriya zameshchennykh gidrazonov: Sintez 2-gidrazino-1-propantriola i struktura evo alkilidenovykh proizvodnykh (Ring-chain tautomerism of substituted hydrazones: Synthesis of 2-hydrazino-1-propanetriol and the structure of its alkylidene derivatives). *Khim. Geterotsikl. Soed.*, no. 10, pp. 1355–1362, 1981.
141. Sheynker, Yu. N., et al. O ravnovesii tetrazol'noy i azidnoy form u benzthiazolotetrazola (On equilibrium between the tetrazole and the azide forms in benzthiazolotetrazole). *DAN SSSR*, vol. 141, no. 6, pp. 1388–1390, 1961.
142. Kabo, G. J., Frenkel, M. L. Thermodynamics of diastereomeric transformations of alcohols with different carbon-skeleton structures. *J. Chem. Thermodyn.*, vol. 15, no. 4, pp. 377–381, 1983.
143. Wiberg, K. B., Cannon, H. A. Enthalpy of the metal catalyzed isomerizations of quadricyclane and the thicyclo [4.1.0.0] heptane. *J. Am. Chem. Soc.*, vol. 98, no. 17, pp. 5411–5412, 1976.
144. Steele, W. V. The standard enthalpies of formation of a series of C_7 bridged-ring hydrocarbons: Norbornane, norbornene, nortricyclene, norbornadiene and quadricyclane. *J. Chem. Thermodyn.*, vol. 10, no. 10, pp. 919–927, 1978.
145. Egger, K. W., James, T. L. Thermodynamic data from equilibrium studies of the nitric oxide-catalyzed isomerization of 1,3,5-heptatriene in the gas phase. *J. Chem. Soc.*, no. 2, pp. 348–350, 1971.
146. Skuratov, S. M., et al. Teploty sgoraniya, obrazovaniya, izomerizatsii i gidrogenizatsii 2-metilenbitsiklo(2,2,1)-geptana i smesi ekzo- i endo-5-metilbitsiklo(2,2,1)-geptana-2 (Heats of combustion, formation isomerization and hydrogenation of 2-methylenebicyclo(2,2,1)-heptane and a mixture of exo- and endo-5-methylbicylo(2,2,1)-heptane-2). *DAN SSSR*, vol. 187, no. 2, pp. 343–346, 1969.
147. Kozina, M. P., et al. Entalpii obrazovaniya nortritsiclena i norbornena (The enthalpies of formation of nortricyclene and norbornene). *DAN SSSR*, vol. 226, no. 5, pp. 1105–1108, 1976.
148. Frey, H. M., Lamont, A. M., Walsh, R. The [1,5]-hydrogene transfer and *cis-, trans*-isomerization of *cis*-2,3-dimethylpenta-1,3-diene: The *cis-, trans*-isomerization of penta-1,3-diene. Kinetics and equilibrium measurements. *J. Chem. Soc.*, no. 16, pp. 2642–2646, 1971.
149. Guisnet, M., Canesson, P., Maurel, R. Isomerisation catalytique des hydrocarbures éthyléniques. 4. Utilisation d'un réacteur chromatographique pour déterminer les équilibres. *Bull. Soc. Chim. Fr.*, no. 10, pp. 3566–3571, 1990.
150. Yursha, I. A., Kabo, G. Ya., Andreevskii, D. N. Ravnovesiya i termodinamika izomerizatsii metiltsiklogeksenov (Izomerization equilibria and thermodynamics of methylcycloxenes). *Neftekhimiya*, vol. 14, no. 5, pp. 688–693, 1974.
151. Hine, J., Linden, S.-M. Equilibration of 5-methyl-3-hexen-2-one and 5-methyl-4-hexen-2-one and of $XCH_2CH=CYZ/XCH=CHCHYZ$ pairs in general. *J. Org. Chem.*, vol. 48, no. 4, pp. 584–587, 1983.
152. Taskinen, E. Thermodynamics of vinyl ethers: Relative stabilities of the six isomeric methyl enol ethers of acetylacetone. *Finn. Chem. Lett.*, no. 5, pp. 154–157, 1979.
153. Eliel, E. L., Kaloustian, M. K. Configuration preference of 5-heterosubstituents in 2-isopropyl-1,3-dioxanes. *J. Chem. Soc. Chem. Commun.*, no. 5, pp. 290–291, 1970.

154. Bailey, W. F., Eliel, E. L. Conformation analysis. 29. 2-substituted and 2,2-disubstituted 1,3-dioxanes: The generalyzed and reverse anomeric effects. *J. Am. Chem. Soc.*, vol. 96, no. 6, pp. 1798–1806, 1974.
155. Hammaker, R. M., Gugler, B. A. An NMR study of hindered internal rotation in N,N-dialkylamides. *J. Mol. Spectrosc.*, vol. 17, no. 2, pp. 356–364, 1965.
156. Rockenfeller, J. D., Rossini, F. D. Heats of combustion, isomerization and formation of selected C_7, C_8, and C_{10} monoölefin hydrocarbons. *J. Phys. Chem.*, vol. 65, no. 2, pp. 267–272, 1961.
157. Kabo, G. Ya., Andreevskii, D. N., Savinetskaya, G. A. Ravnovesie izomerizatsii vtorichnykh n-monokhlorgeptanov i n-geptenov (Isomerization equilibrium of secondary n-monocholoroheptanes and n-heptenes). *Neftekhimiya*, vol. 7, no. 3, pp. 364–368, 1967.
158. Good, W. D. The enthalpies of formation of five isomeric heptenes. *J. Chem. Thermodyn.*, vol. 8, no. 1, pp. 67–71, 1976.
159. Zakharenko, V. A., Delone, I. O., Petrov, A. A. Termodinamicheskaya ustoychivost' C_7–C_8 tsiklopentanov i tsiklohekcanov (Thermodynamic stability of C_7–C_8 cyclopentanes and cyclohexanes). *Neftekhimiya*, vol. 8, no. 5, pp. 675–680, 1968.
160. Taskinen, E. Thermodynamics of vinyl ethers. 2. Configurational assignment of the geometric isomers of 3-methoxy-4-methyl-2-pentene by means of thermodynamic data of isomerization. *J. Chem. Thermodyn.*, vol. 6, no. 3, pp. 271–280, 1974.
161. Chiurdoglu, G., Masschelein, W. Etudes conformationnales. 8. Equilibration des 2-, 3- et 4-methylcyclohexanols et des 3,3,5-trimethylcyclohexanols stereoisomeres. *Bull. Soc. Chim. Belg.*, vol. 70, no. 11–12, pp. 782–793, 1961.
162. Eliel, E. L., et al. Conformational analysis. 11. Configurational equilibria and chromic acid oxidation rates of alkylcyclohexanols. Deformation effect. *J. Am. Chem. Soc.*, vol. 88, no. 14, pp. 3327–3333, 1966.
163. Frenkel', M. L., Kabo, G. Ya. Termodimanika stereoizomernykh prevracheniy metiltsiklogeksanolov (Thermodynamics of the stereoisomeric conversions of methylcyclohexanols). *Termodinamika Org. Soed.*, issue 8, pp. 104–106, 1979.
164. Zaykin, I. D., et al. Standartnye teploty obrazovaniya nekotorykh epoxisoedineniy (Standard heats of formation of some epoxides). *Izv. Vuzov. Ser. Khim. Khim. Tekhnol.*, vol. 15, no. 8, pp. 1193–1195, 1972.
165. Gren', A. I. Izucheniye predpochtitel' nosti konformatsii geteroanalogov tsiklogeksana metodom konfiguratsionnogo ravnovesiya (A study of conformation preference of heteroanalogues of cyclohexane by the configurational equilibrium method). *Vopr. Stereokhimii*, issue 3, pp. 60–65, 1973.
166. Nader, F. W., Eliel, E. L. Conformational analysis. 22. Conformational equilibria in 2-substituted 1,3-dioxanes. *J. Am. Chem. Soc.*, vol. 92, no. 10, pp. 3050–3055, 1970.
167. Eliel, E. L., Enanoza, R. M. Conformational analysis. 26. Conformational equilibrium in 5,5-disubstituted 1,3-dioxanes. *J. Am. Chem. Soc.*, vol. 94, no. 23, pp. 8072–8081, 1972.
168. Esipenok, G. E., Kabo, G. Yu., Andreevskii, D. N. Ravnovesiye i termodinamika izomerizatsii tretichnykh 2,3-dimetilkhlorpentanov (Isomerization equilibrium and thermodynamics of tert. 2,3-dimethylchloropentanes). *Zh. Fiz. Khim.*, vol. 49, no. 11, pp. 16–21, 1975.
169. Roganov, G. N., et al. Termodinamika izomerizatsii metilgeksanov (Thermodynamics of isomerization of methylhexanes). *Neftekhimiya*, vol. 10, no. 1, pp. 16–21, 1970.
170. Kramer, G. M., Schriesheim, A. Heptane isomerization. *J. Phys. Chem.*, vol. 64, no. 7, pp. 849–851, 1960.
171. Paleta, O., et al. Anionic *cis*-, *trans*-isomerization of dimethyl perfluoro(4-methyl-2-pentene)dioate. *Collect. Czech. Chem. Commun.*, vol. 45, no. 12, pp. 3360–3369, 1980.
172. Merdzhanov, V. R., et al. Issledovaniye ravnovesnykh prevrashcheniy etilbrombenzolov (A study of equilibrium conversions of ethylbromobenzenes). *Izv. Vuzov. Ser. Khim. Khim. Tekhnol.*, vol. 25, no. 9, pp. 1047–1049, 1982.
173. Taskinen, E., Mustonen, E. Thermodynamics of vinyl ethers. 14. Effect of aromatic and

heteroaromatic α-substituents on the relative stabilities of geometric isomers. *Acta Chem. Scand.*, vol. 30, no. 1, pp. 1–4, 1976.
174. Kozina, M. P., et al. Toploty sgoraniya i svobodnye energii izomerizatsii nekotorykh proizvodnykh bitsiklo(2,2,1)-geptana (The heats of combustion and the free energies of izomerization of some bicyclo(2,2,1)-heptane derivatives). V knige "Termodinamicheskiye i termokhimicheskiye konstanty. In the book: "Thermodynamic and thermochemical constants. M. Nauka, pp. 158–162, 1970.
175. Egger, K. W. The thermal intramolecular rearrangement of methyl 1,3,5-cycloheptanes in the gas phase. 2. Thermodynamic data from equilibrium studies of the positional isomers. *J. Am. Chem. Soc.*, vol. 90, no. 1, pp. 1–5, 1968.
176. Staley, S. W. On the thermodynamic significance of delocalization in dienes. Thermodynamics of the vinylcyclopropyl system. *J. Am. Chem. Soc.*, vol. 89, no. 6, 1532–1533, 1967.
177. Kozina, M. P., et al. The enthalpies of combustion of some bicyclic compounds. *J. Chem. Thermodyn.*, vol. 3, no. 4, pp. 563–570, 1971.
178. Van Hoboken, N. J., Wiering, P. G., Steinberg, H. The effect of pressure on the base-catalyzed isomerization of gem-dimethyl- and isopropyl-cyclohexenes. *Rec. Trav. Chim.*, vol. 94, no. 11, pp. 243–245, 1975.
179. Pons, A., Chapat, J. P. Effects de substituants en serie dialkyl-1,2-cyclohexanique. *Tetrahedron*, vol. 36, no. 15, pp. 2219–2224, 1980.
180. Siroký, M., Prochazka, M. Prototropic isomerization of isopentene derivatives. *Collect. Czech. Chem. Commun.*, vol. 43, no. 10, pp. 2635–2642, 1978.
181. Good, W. D. The enthalpies of combustion and formation of *n*-propylcyclopentane and five methylethylcyclopentanes. *J. Chem. Thermodyn.*, vol. 3, no. 1, pp. 97–103, 1971.
182. Anfilogova, S. N., Balenkova, E. S., Dmitriyev, A. B. Otnositel' naya ustoychivost' *cis*- i *trans*-1,2-dimetilsiklogeptanov i 1,2-dimetilsiklooktanov (Relative stability of *cis*- and *trans*-1,2-dimethylcycloheptanes and 1,2-dimethylcyclooctanes). *Neftekhimiya*, vol. 14, no. 5, pp. 673–676, 1974.
183. Taskinen, E., Antilla, M. Thermodynamics of vinyl ethers. 21. Evaluation of some *cis* interaction energies. *Tetrahedron*, vol. 33, no. 18, pp. 2423–2427, 1977.
184. Lampman, G. M., Hager, G. D., Couchman, G. L. Enthalpy, entropy and free-energy changes in the equilibration of *cis*- and *trans*-ethyl-3-*t*-butylcyclobutanecarboxylate and 3-*t*-butylcyclobutanol. *J. Org. Chem.*, vol. 35, no. 7, pp. 2398–2402, 1970.
185. Eliel, E. L., Schroeter, S. H. Conformational analysis. 9. Equilibrations with Raney nickel: The conformational energy of the hydroxyl group as a function of solvent. *J. Am. Chem. Soc.*, vol. 87, no. 22, pp. 5031–5038, 1965.
186. Eliel, E. L., Knoeber, M. C. Conformational analysis. 16. 1,3-Dioxanes. *J. Am. Chem. Soc.*, vol. 90, no. 13, pp. 3444–3458, 1968.
187. Eliel, E. L., Evans, S. A. An unusually strong intramolecular interaction between the sulfone or sulfoxide and the alkoxide functions. *J. Am. Chem. Soc.*, vol. 94, no. 24, pp. 8587–8589, 1972.
188. Eliel, E. L., Juaristi, E. Conformational analysis. 37. Gaucherepulsive interactions in 5-methoxy- and 5-methoxythio-1,3-dithianes. *J. Am. Chem. Soc.*, vol. 100, no. 19, pp. 6114–6119, 1978.
189. Eliel, E. L., Hofer, O. Conformational analysis. 27. Solvent effects in conformational equilibria of heterosubstituted 1,3-dioxanes. *J. Am. Chem. Soc.*, vol. 95, no. 24, pp. 8041–8045, 1973.
190. Eliel, E. L. Insights gained from conformational analysis in heterocyclic systems. *Pure Appl. Chem.*, vol. 25, no. 3, pp. 509–525, 1971.
191. Good, W. D. The enthalpies of combustion and formation of *n*-octane and 2,2,3,3-methylbutane. *J. Chem. Thermodyn.*, vol. 4, no. 5, pp. 709–714, 1972.
192. Klimova, N. N., et al. Termodinamicheskiye parametry keto-enol'nogo ravnoesiya yaderno-

zameshchennykh benzoilacetic acids (The thermodynamic parameters of keto-enol equilibrium of ring-substituted benzoylacetic acids). *Zh. Org. Khim.*, vol. 10, no. 2, p. 405, 1974.
193. Taskinen, E., Ylivainio, P. Thermodynamics of vinyl ethers. 10. The stabilizing effect of α-phenyl substitution of α,β-unsaturated ethers. *Acta. Chem. Scand.*, vol. b29, no. 1, pp. 1-6, 1975.
194. Merdzhanov, V. R., et al. Opredeleniye ental'pii obrazovoaniya i entropii isopropilbrombenzolov po dannym o ravnovesii reaktsiy (Determination of the enthalpy of formation and entropy of isopropylbromobenzenes from the data on reaction equilibria). *Termodin. Org. Soed.*, no. 9, pp. 69-72, 1981.
195. Merdzhanov, V. R., et al. Issledovanie ravnoveslya v sistemakh ftorbenzol-izopropylftorbenzoly (An equilibrium study in the fluorobenzene-isopropylfluorobenzenes systems). *Termodin. Org. Soed.*, no. 10, pp. 65-66, 1982.
196. Rennekamp, M. E., Pankstelis, J. V., Cooks, R. G. An investigation into the mechanism of gas-phase tautomerism using mass spectrometry. *Tetrahedron*, vol. 27, no. 19, pp. 4407-4415, 1971.
197. Jochems, R., et al. Enthalpies of formation of bicyclo[3.3.1]-non-2-ene, bicyclo[3.2.2]-non-6-ene, bicyclo[4.2.1]-non-3-ene. *J. Chem. Thermodyn.*, vol. 15, no. 1, pp. 95-99, 1983.
198. Kozina, M. P., et al. Ental'pii sgoraniya i obrazovaniya 2-metilenbistsiklo(2,2,2)oktana i 2-metilbitsiklo(2,2,2)oktena-2 (The enthalpies of combustion and formation of 2-methylenebicyclo(2,2,2)octane and 2-methylbicyclo(2,2,2)octene-2). *Zh. Fiz. Khim.*, vol. 48, no. 8, pp. 2075-2077, 1974.
199. Walter, W., Meyer, H.-W. Lösungmittel-und Temperaturabhangigkeit der Thioimidsänreeker/keten-*S,N*-acetal-tautomerie. *Lieb. Ann. Chem.*, no. 1, pp. 35-40, 1975.
200. Browne, C. C., Rossini, F. D. Heats of combustion, formation and isomerization of the *cis*- and *trans*-isomers of hexahydroindan. *J. Phys. Chem.*, vol. 64, no. 7, pp. 927-931, 1960.
201. Allinger, N. L., Coke, J. L. The relative stabilities of *cis* and *trans* isomers. 7. The hydrindanes. *J. Am. Chem. Soc.*, vol. 82, no. 10, pp. 2553-2556, 1960.
202. Blanchard, K. R., Schleyer, P. R. Quantitative study of the interconversion of hydrindane isomers by aluminum bromide. *J. Org. Chem.*, vol. 28, no. 1, pp. 247-248, 1963.
203. Frye, C. G., Weitkamp, A. W. Equilibrium hydrogenations of multi-ring aromatics. *J. Chem. Eng. Data*, vol. 14, no. 3, pp. 372-376, 1969.
204. Finke, H. L., et al. *Cis*- and *trans*-hexahydroindan: Chemical thermodynamic properties and isomerization equilibrium. *J. Chem. Thermodyn.*, vol. 4, no. 3, pp. 477-494, 1972.
205. Ferrao, M. L. C. C. H., et al. Enthalpies of combustion of four methyl-substituted heptane-3,5-diones and benzoylacetone. *J. Chem. Thermodyn.*, vol. 13, no. 6, pp. 567-571, 1981.
206. Mager, S., Eliel, E. L. Synthesis and configurational equilibrium of 2-isopropyl-5-acetyl-1,3-dioxane. *Rev. Roum. Chim.*, vol. 18, no. 8, 1379-1387, 1973.
207. Good, W. D. The enthalpies of combustion and formation of *non*-propylcyclohexane and six methylethylcyclohexanes. *J. Chem. Thermodyn.*, vol. 2, no. 3, pp. 399-405, 1970.
208. Eliel, E. L., Gilbert, E. C. Conformational analysis. 19. The conformational enthalpy and entropy of the hydroxyl group in various solvents. Conformation energy of methoxyl. *J. Am. Chem. Soc.*, vol. 91, no. 20, pp. 5487-5495, 1969.
209. Eliel, E. L., Raileanu, D. I. C. Effect of solvent on the position of equilibrium of the methyl-5-*t*-butyl-1,3-dioxanes and on their nuclear magnetic resonance spectra. *J. Chem. Soc. Chem. Commun.*, no. 5, p. 291-292, 1970.
210. Cookson, R. C., et al. Photochemical cyclisation of Diels-Alder adducts. *J. Chem. Soc.*, no. 9, pp. 3062-3075, 1963.
211. Pimenova, S. M. et al. Teploty sgoraniya *cis*- i *trans*-izomerov 1-fenyltsiklopropankarbonovoy-2 kisloty (The heats of combustion of *cis*- and *trans*- isomers of 1-phenylcyclopropanecarboxylic-2 acid). *Zh. Obshchey Khim.*, vol. 40, no. 9, pp. 2117-2120, 1970.
212. Kovzel', et al. Issledovaniye ravnovesiya v sisteme khlorbenzol-*t*-butylkhlorbenzoly (An

equilibrium study in the chlorobenzene-*t*-butylchlorobenzenes system). *Termodin. Org. Soed.*, no. 9, pp. 65–68, 1981.
213. Ahlbrecht, H., Kalas, R.-D. Vinylamine. 22. Über die Lösungsmittelabhangigkeit der Imin-Enamin-Tautomerie. *Lieb. Ann. Chem.*, no. 1, pp. 102–120, 1979.
214. Good, W. D. The enthalpies of combustion and formation of *n*-butylbenzene, the diethylbenzenes, the methyl-*n*-propyl-benzene and the methyl-*iso*-propylbenzenes. *J. Chem. Thermodyn.*, vol. 5, no. 5, pp. 707–714, 1973.
215. Good, W. D. The standard enthalpies of combustion and formation of *n*-butylbenzene, the dimethylethylbenzenes and tetramethylbenzenes in the condensed state. *J. Chem. Thermodyn.*, vol. 7, no. 1, pp. 49–59, 1975.
216. Pil'shchikov, V. A., Nesterova, T. I., Rozhnov, A. M. Issledovaniye ravnovesiya v sisteme fenol–*t*-butulfenoly (An equilibrium study in the phenol–*t*-butylphenols system). *Zh. Prikl. Khim.*, vol. 54, no. 9, pp. 2018–2023, 1981.
217. Egger, K. W., Jola, M. Die NO-katalysierte Isomerisierung von 3-methylen-1,5,5-trimethyl-cyclohexen zu 1,3,5,5-tetramethyl-1,3-cyclohexadien in der gas-phase: Thermodynamische Daten as gleichgewische Daten ans Gleichgewichtsmessungen. *Helv. Chim. Acta*, vol. 52, no. 2, pp. 449–453, 1969.
218. Van Binst, B., Merck, Y. Physico-chemical study of indole alkaloids models. 6. Relative thermodynamic stability of endo- and semi-cyclic olefins in the bicyclo [2,2,2]octane system. *Tetrahedron Lett.*, no. 40, pp. 3897–3899, 1967.
219. Augustine, R. L., Caputo, J. A. The thermal equilibrium of *cis*- and *trans*- decanols. *J. Am. Chem. Soc.*, vol. 86, no. 13, pp. 2751–2752, 1964.
220. Mann, G. Das thermische Gleichgewicht zwischen *cis*- and *trans*-β-dekalon. *Z. Chem.*, vol. 8, no. 8, pp. 301–302, 1968.
221. Allinger, N. L., Siefert, J. H. Conformational analysis. 36. The equilibrium between the *cis*- and *trans*-isomers of 2-decolone. *J. Am. Chem. Soc.*, vol. 94, no. 23, pp. 8082–8084, 1972.
222. Kozina, M. P., et al. Teploty sgoraniya nekotorykh bitscyklanov (The heats of combustion of selected bicyclanes). *Zh. Fiz. Khim.*, vol. 35, no. 10, pp. 8082–8084, 1961.
223. Stolow, R. D., Groom, T. Equilibrium of *cis*- and *trans*-2-*t*-butyl-4-hydroxycyclohexanone. *Tetrahedron Lett.*, no. 38, pp. 4069–4072, 1968.
224. Pasto, D. J., Rao, D. R. Thermodynamic, conformational and chemical reactivity studies of the 2,5-*t*-butylcyclohexyl system: The reversal of the thermodynamic stability with chemical reactivity trends of cyclohexyl derivatives. *J. Am. Chem. Soc.*, vol. 92, no. 17, pp. 5151–5160, 1970.
225. Yoshida, T., Komatsu, A., Indo, M. Isomerization and racemisation of menthols. 2. Isomerization and recemisation with hydrogenation catalysts and sodium mentholates. *Agr. Biol. Chem.*, vol. 29, no. 9, pp. 824–831, 1965.
226. Speros, D. M., Rossini, F. D. Heats of combustion and formation of naphthalene, the two methylnaphthalenes, *cis*- and *trans*-decahydronaphthalene and related compounds. *J. Phys. Chem.*, vol. 64, no. 11, pp. 1723–1727, 1960.
227. Zheglova, D. Kh., Ershov, B. A., Kol'tsov, A. I. Issledovaniye tautomerii β-dikarbonil'nykh soedineniy metodom spektroskopii PMR. 4. Termodinamicheskie kharakteristiki keto-enol'novo ravnovesiya etilovykh efirov zameshchennykh benzoyluksusnykh kislot (PMR spectroscopic investigation of tautomerism of β-dicarbonyl compounds. 4. Thermodynamic functions of the keto-enol equilibrium in the ethyl esters of substituted benzoylacetic acids). *Zh. Org. Khim.*, vol. 10, no. 1, pp. 18–21, 1974.
228. Good, W. D. The enthalpies of combustion and formation of indan and seven alkylindans. *J. Chem. Thermodyn.*, vol. 3, no. 5, pp. 711–717, 1971.
229. Eliel, E. L., Reese, M. C. Conformational analysis. 15. The conformational enthalpy, entropy and free energy of the carboxyl, carboxylate, carbomethyoxy, carbonyl chloride and methyl ketone groups. *J. Am. Chem. Soc.*, vol. 90, no. 6, pp. 1560–1566, 1968.
230. Allinger, N. L., Blatter, H. M. Conformational analysis. 16. The energy of the boat form of

the cyclohexanone. Direct measurement of 2-alkyl ketone effects. *J. Am. Chem. Soc.*, vol. 83, no. 4, pp. 994–995, 1961.
231. Taskinen, E. Thermodynamics of vinyl ethers. 26. Relative stabilities of the geometrical isomers of 8- to 15-membered-1-methoxycycloalkenes. *Acta. Chem. Scand.*, vol. B34, no. 9, pp. 643–646, 1980.
232. Good, W. D. The enthalpies of combustion and formation of 1,8-dimethylnaphthalene, 2,2-dimethylnaphthalene, 2,6-dimethylnaphthalene and 2,7-dimethylnaphthalene. *J. Chem. Thermodyn.*, vol. 5, no. 5, pp. 715–720, 1973.
233. Popov, V. E., et al. Ravnovesiye disproportsionirovaniya izopropilbenzola (Disproportionation equilibrium of isopropylbenzene). *Neftekhimiya*, vol. 14, no. 3, pp. 364–367, 1974.
234. Kosykhin, A. S. Konstanty ravnovesiya i teplovoy effekt reaktsii izomerizatsii 2,2,4,6,6-pentametil-3-geptena v 1,1-dineopentyletilen, rasschitannye po intensivnostyam kharakteristicheskikh liniy spektrov KR sveta (The equilibrium constants and heat effect of 2,2,4,6,6-pentamethyl-3-heptene isomerization into 1,1-dineopentylethylene calculated from the intensities of characteristic lines in RS spectre). *Zh. Prikl. Spectrosc.*, vol. 24, no. 1, pp. 100–106, 1976.
235. Melaugh, R. A., Mansson, M., Rossini, F. D. The energy of isomerization of *n*-dodecane into 2,2,4,6,6-pentamethylheptane. *J. Chem. Thermodyn.*, vol. 8, no. 7, pp. 623–626, 1976.
236. Allinger, N. L., Freiberg, L. A. Conformational analysis. 46. The conformational energies of the simple alkyl groups. *J. Org. Chem.*, vol. 31, no. 3, pp. 894–897, 1966.
237. Ouelette, R. J., Booth, C. E. Conformational analysis. 7. The conformational preference of the nitro and carboethoxyl groups. *J. Org. Chem.*, vol. 31, no. 2, pp. 587–588, 1966.
238. Fischer, G., Muszkati, K. A., Fischer, E. The thermodynamic equilibrium between *cis*- and *trans*-isomers of stilbene and some derivatives. *J. Chem. Soc.*, no. 10, pp. 1156–1158, 1968.
239. Coleman, D. J., Pilcher, G. Heats of combustion of biphenyl, bibenzyl, naphthalene, anthracene and phenanthrene. *Trans. Faraday Soc.*, vol. 62, no. 4, pp. 821–827, 1966.
240. Baba, H., Taskinen, E. Effects of basic substances on the rate and equilibrium of the tautomeric reaction between anthrone and anthronol. *Bull. Chem. Soc. Japan*, vol. 37, no. 8, pp. 1241–1242, 1964.
241. Johnstone, D. E., McKervey, M. A., Rooney, J. J. Equilibration of diamantan-1-ol and diamantan-4-ol: Conformational enthalpy of the hydroxy group, and an unusual example of how entropy and symmetry factors can influence relative thermodynamic stabilities. *J. Chem. Soc. Chem. Commun.*, no. 1, pp. 29–30, 1972.
242. Nesterova, T. N., et al. Issledovaniye ravnovesiya vzaimnykh prevrashcheniy tret. butilbenzolov (An equilibration study of interconversions of *t*-butylbenzenes). *Zh. Fiz. Khim.*, vol. 58, no. 2, pp. 491–493.
243. Allinger, N. L., Karkowski, F. M. The energy of the boat form of a simple cyclohexanone. *Tetrahedron Lett.*, no. 26, pp. 2171–2174, 1965.
244. Walter, W., Meyer, H.-W. Lösungsmittel- und Temperaturabhangigkeit der Thiomidsäureester/keten-S,Nacetal-tautomerie. *Lieb. Ann. Chem.*, no. 1, pp. 36–40, 1975.
245. Volkov, R. N., Yanchuk, V. A. Issledovaniye termodinamicheskoy ustoychivosti prostransvennykh izomerov indanovykh uglevodorodov (A study of thermodynamic stability in the stereoisomers of indane hydrocarbons). *Neftekhimiya*, vol. 11, no. 5, pp. 635–638, 1971.
246. Garbisch, E. W., Patterson, D. B. Conformations. 4. Conformational preference of the phenyl group in cyclohexane. *J. Am. Chem. Soc.*, vol. 85, no. 20, pp. 3228–3231, 1963.
247. Lyttle, M. H., Streitwieser, A., Kluttz, R. Q. Unusual equilibrium between 1,4- and 1,6-di-*tert*-butylcyclooctatetraenes. *J. Am. Chem. Soc.*, vol. 103, no. 11, pp. 3232–3233, 1981.
248. Dulec, E. P., Dulec, G. O. Spectroscopic studies of ketoenol equilibria. 9. N^{15}-substituted anilides. *J. Am. Chem. Soc.*, vol. 88, no. 11, pp. 2407–2412, 1966.
249. Dabbagh, A.-M., et al. Kinetics and thermodynamics of the thermal gas-phase interconversion of hexakis-(penta-fluoroethyl)-benzene and its para-bonded (Dewar) isomer. *J. Chem. Soc. Chem. Commun.*, no. 9, pp. 323–324, 1975.

250. Craig, N. C., Piper, L. G., Wheeler, V. L. Thermodynamics of *cis*-, *trans*- isomerization. 2. The 1-chloro-2-fluoro-ethylenes. 1,2-difluorocyclopropanes and related molecules. *J. Phys. Chem.*, vol. 75, no. 10, pp. 1453–1460, 1971.
251. Sharonov, K. G., Rozhnov, A. M. Ravnovesiye izomerizatsii 1,2-dibrompropana v 1,3-dibrompropan (Equilibrium of 1,2-dibromopropane isomerization into 1,3-dibromopropane). *Zh. Org. Khim.*, vol. 7, no. 10, pp. 2022–2024, 1971.
252. Draeger, J. A. Chemical-thermodynamic properties of molecules that undergo inversion. 3. The aminopyridines. *J. Chem. Thermodyn.*, vol. 16, no. 11, pp. 1075–1079, 1984.
253. Platonov, V. A., Simulin, Yu. N. Eksperimental'noye opredeleniye standartnykh ental'piy obrazovaniya polykhlorbenzolov. 3. Standartnye ental'pii obrazovaniya mono-, 1,2,4-, 1,3,5-tri, 1,2,3,4- i 1,2,3,5-tetrakhlorbenzolov (Experimental determination of the standard enthalpies of formation of polychlorobenzenes. 3. The standard enthalpies of formation of mono-, 1,2,4-, 1,2,5-tri, 1,2,3,4- and 1,2,3,5-tetrachlorobenzenes). *Zh. Fiz. Khim.*, vol. 59, no. 2, pp. 300–304, 1985.
254. Platonov, V. A., Simulin, Yu. N. Eksperimental'noye opredeleniye standartnykh ental'piy obrazovaniya polykhlorbenzolov. 2. Standartnye ental'pii obrazovaniya dikhlorbenzolov (Experimental determination of the enthalpies of formation of polychlorobenzenes. 2. The standard enthalpies of formation of dichlorobenzenes). *Zh. Fiz. Khim.*, vol. 58, no. 11, pp. 2682–2686, 1984.
255. Petrova-Kuminskaya, S. V., Roganov, G. N., Kabo, G. Ya. Ravnovevesiye izomerizatsii 2-metilpentadienov (Isomerization equilibrium of 2-methylpentadienes). *Neftekhimiya*, vol. 24, no. 4, pp. 485–490, 1984.
256. Da Silva, M. D. M. C., da Silva, M. V. A. R., Pilcher, G. Enthalpies of combustion of 1,2-dihydrobenzene and of six alkylsubstituted 1,2-dihydroxybenzenes. *J. Chem. Thermodyn.*, vol. 16, no. 12, pp. 1149–1155, 1984.
257. Draeger, J. A. Chemical-thermodynamic properties of molecules that undergo inversion. 2. The methylanilines. *J. Chem. Thermodyn.*, vol. 16, no. 11, pp. 1067–1073, 1984.
258. Draeger, J. A. The methylbenzenes. 2. Fundamental vibrational shifts, statistical thermodynamic functions and properties of formation. *J. Chem. Thermodyn.*, vol. 17, no. 3, pp. 263–275, 1985.
259. Colomina, M., et al. Thermochemical properties of benzoic acid derivatives. 11. Vapour pressure and enthalpies of sublimation and formation of the six dimethylbenzoic acids. *J. Chem. Thermodyn.*, vol. 16, no. 12, pp. 1121–1127, 1984.
260. Colomina, M., et al. Thermochemical properties of benzoic acid derivatives. 12. Vapour pressures and enthalpies of sublimation and formation of the five dimenthoxybenzoic acids. *J. Chem. Thermodyn.*, vol. 17, no. 11, pp. 1091–1096, 1985.
261. Nesterova, T. N., et al. Molar enthalpies of formation of isopropylchlorobenzenes derived from equilibrium measurements. *J. Chem. Thermodyn.*, vol. 17, no. 7, pp. 649–656, 1985.
262. Nesterova, T. N., et al. Issledovaniye ravnovesiya tretalkilfenolov: Termodinamocheskiy analiz alkilirovaniya fenolov razvetvlennymi olefinami (An equilibration study of *t*-alkylphenols: Thermodynamic analysis of phenol alkylation with branched-chain olefins). *Zh. Prikl. Khim.*, vol. 58, no. 4, pp. 827–883, 1985.
263. Pimerzin, A. A., Nesterova, T. N., Rozhnov, A. M. Equilibria of isomeric transformations and relations between thermodynamic properties of secondary alkylbenzenes. *J. Chem. Thermodyn.*, vol. 17, no. 7, pp. 641–648, 1985.

Errata for *Thermodynamic Properties of Isomerization Reactions.*
Index page numbers have been updated from the original printing.

FORMULA INDEX

Formula	Page	Formula	Page	Formula	Page
F_2N_2	16	C_4H_7Br	23	C_5H_9NS	40
H_8B_5Cl	16	C_4H_7BrO	23	C_5H_{10}	40
CH_3NO_2	16	C_4H_7Cl	23	$C_5H_{10}Br_2$	41
C_2HClF_2	16	C_4H_7ClO	24	$C_5H_{10}Cl_2$	42
$C_2H_2Br_2$	17	C_4H_8	25	$C_5H_{10}N_2$	43
C_2H_2ClF	17	$C_4H_8Br_2$	27	$C_5H_{10}O$	43
$C_2H_2Cl_2$	17	$C_4H_8Cl_2$	28	$C_5H_{10}OS$	44
$C_2H_2F_2$	17	C_4H_8O	28	$C_5H_{10}O_2$	44
$C_2H_2J_2$	17	C_4H_8OS	28	$C_5H_{10}O_2S$	45
$C_2H_3Cl_3$	18	$C_4H_8O_2$	29	$C_5H_{10}S$	46
C_2H_3N	18	$C_4H_8O_2S$	30	$C_5H_{11}Br$	47
$C_2H_4Br_2$	18	C_4H_8S	30	$C_5H_{11}Cl$	47
$C_2H_4Cl_2$	18	C_4H_9Br	31	C_5H_{12}	47
$C_3H_3Cl_3$	18	C_4H_9NO	31	$C_5H_{12}GeO$	48
$C_3H_4F_2$	18	C_4H_9NSi	31	$C_5H_{12}S$	48
C_3H_5Br	18	C_5H_4BrNO	31	C_6F_{10}	48
C_3H_5Cl	19	C_5H_4ClNO	31	$C_6H_2Cl_4$	48
C_3H_5F	19	C_5H_4ClNS	32	$C_6H_2F_4$	49
C_3H_5J	20	C_5H_5NO	33	$C_6H_3Cl_3$	50
C_3H_5N	20	C_5H_5NS	33	$C_6H_3F_3$	50
$C_3H_6Br_2$	20	$C_5H_5N_3O_2$	33	$C_6H_4Br_2$	51
$C_3H_6Cl_2$	20	$C_5H_5N_5O$	33	$C_6H_4Cl_2$	51
C_3H_6O	21	$C_5H_6N_2$	34	$C_6H_4F_2$	52
C_3H_7Br	21	$C_5H_6N_2O$	35	$C_6H_4N_2$	53
C_3H_7Cl	21	C_5H_8	35	$C_6H_6O_2$	53
C_3H_7J	21	C_5H_8O	35	$C_6H_7F_3O_3$	53
$C_3H_7N_5$	21	$C_5H_8O_2$	36	C_6H_7N	54
C_4F_6	21	$C_5H_8O_3$	37	C_6H_7NO	54
C_4F_8	21	C_5H_9BrO	37	C_6H_7NS	55
$C_4H_4F_2$	22	$C_5H_9BrO_2$	37	$C_6H_8N_2O_2S$	55
$C_4H_5Cl_2NO$	22	C_5H_9ClO	38	$C_6H_9ClO_2$	55
C_4H_6	22	C_5H_9FO	39	$C_6H_9ClO_3$	55
$C_4H_6O_2$	23	C_5H_9NO	39	C_6H_{10}	56

227

$C_6H_{10}O$	60	$C_7H_{13}ClO_2$	93	$C_8H_{17}Cl$	131
$C_6H_{10}O_2$	62	$C_7H_{13}FO_2$	93	C_8H_{18}	131
$C_6H_{10}O_3$	64	$C_7H_{13}NO_2$	94	$C_9H_7ClO_3$	132
$C_6H_{11}N$	64	$C_7H_{13}NO_4$	94	$C_9H_7NO_3$	132
$C_6H_{11}NO$	65	C_7H_{14}	94	$C_9H_8O_3$	133
$C_6H_{11}NO_2$	65	$C_7H_{14}O$	97	$C_9H_{10}O$	133
$C_6H_{11}NS$	65	$C_7H_{14}OS$	99	$C_9H_{10}O_2$	133
C_6H_{12}	65	$C_7H_{14}O_2$	100	$C_9H_{10}O_4$	134
$C_6H_{12}O$	71	$C_7H_{14}O_2S$	102	$C_9H_{11}Br$	135
$C_6H_{12}OS$	72	$C_7H_{14}O_3$	103	$C_9H_{11}F$	136
$C_6H_{12}O_2$	73	$C_7H_{14}S$	103	$C_9H_{11}NO$	136
$C_6H_{12}O_2S$	74	$C_7H_{14}S_2$	104	$C_9H_{11}NS$	137
$C_6H_{12}O_3$	75	$C_7H_{15}Br$	104	C_9H_{12}	137
$C_6H_{12}S$	75	$C_7H_{15}Cl$	105	C_9H_{14}	137
$C_6H_{12}S_2$	76	C_7H_{16}	105	$C_9H_{14}BrNO_4S$	138
$C_6H_{13}Cl$	76	$C_7H_{16}N_2S$	106	$C_9H_{14}O_2$	139
C_6H_{14}	78	$C_8H_6F_6O_4$	106	$C_9H_{14}O_3$	139
$C_6H_{14}N_2S$	78	C_8H_9Br	107	C_9H_{16}	139
$C_6H_{14}O$	78	$C_8H_9NO_2$	107	$C_9H_{16}O$	141
$C_6H_{14}S$	79	C_8H_{10}	108	$C_9H_{16}O_2$	142
$C_7H_4N_4S$	79	$C_8H_{10}OS$	108	$C_9H_{16}O_3$	143
$C_7H_4N_4S$	79	$C_8H_{10}O_2$	108	$C_9H_{16}O_4$	145
$C_7H_5ClN_2O_4$	79	$C_8H_{11}N$	109	$C_9H_{17}ClO$	146
$C_7H_5Cl_3N_2O$	79	$C_8H_{11}N_5O$	109	C_9H_{18}	146
$C_7H_6ClNO_2$	79	C_8H_{12}	109	$C_9H_{18}O$	147
$C_7H_6Cl_2N_2O$	80	$C_8H_{12}O_2$	110	$C_9H_{18}OS_2$	149
C_7H_8	80	$C_8H_{12}O_3$	111	$C_9H_{18}O_2$	149
$C_7H_8O_2$	80	$C_8H_{13}NO_2$	111	$C_9H_{18}O_3$	151
$C_7H_8O_2S$	81	C_8H_{14}	111	$C_9H_{18}S$	153
$C_7H_8O_3$	81	$C_8H_{14}O$	114	$C_9H_{18}S_2$	153
C_7H_9N	81	$C_8H_{14}O_2$	117	$C_9H_{18}S_3$	154
$C_7H_9N_5O$	82	$C_8H_{15}ClO$	119	$C_9H_{19}Cl$	154
C_7H_{10}	82	$C_8H_{15}NO_4$	120	C_9H_{20}	154
$C_7H_{10}O_2$	83	C_8H_{16}	120	$C_{10}H_{10}O$	154
$C_7H_{11}ClO$	83	$C_8H_{16}O$	123	$C_{10}H_{10}O_2$	155
C_7H_{12}	83	$C_8H_{16}O_2$	125	$C_{10}H_{10}O_3$	155
$C_7H_{12}O$	86	$C_8H_{16}O_2S$	127	$C_{10}H_{10}O_4$	155
$C_7H_{12}O_2$	88	$C_8H_{16}O_3$	128	$C_{10}H_{11}ClO$	155
$C_7H_{12}O_3$	91	$C_8H_{16}O_3S$	130	$C_{10}H_{11}FO$	156
$C_7H_{12}O_4$	91	$C_8H_{16}O_4S$	130	$C_{10}H_{12}O$	156
$C_7H_{13}BrO_2$	92	$C_8H_{16}S_2$	131	$C_{10}H_{13}Cl$	157

FORMULA INDEX **229**

$C_{10}H_{13}N$	157	$C_{11}H_{20}O$	180	$C_{14}H_{14}N_2O$	195
$C_{10}H_{14}$	157	$C_{11}H_{20}O_2$	181	$C_{14}H_{16}O_2$	195
$C_{10}H_{14}O$	162	$C_{11}H_{22}O$	182	$C_{14}H_{20}O$	196
$C_{10}H_{14}O_2$	163	$C_{11}H_{22}O_2$	182	$C_{14}H_{20}O_2$	196
$C_{10}H_{16}$	163	$C_{11}H_{22}S_2$	183	$C_{14}H_{20}S_2$	196
$C_{10}H_{16}O$	164	$C_{12}H_{12}$	184	$C_{14}H_{22}$	197
$C_{10}H_{18}$	164	$C_{12}H_{12}O_2$	184	$C_{14}H_{22}O$	197
$C_{10}H_{18}O$	165	$C_{12}H_{14}O_3$	185	$C_{14}H_{26}O$	197
$C_{10}H_{18}O_2$	165	$C_{12}H_{14}O_4$	185	$C_{14}H_{26}O_2$	198
$C_{10}H_{18}O_4$	166	$C_{12}H_{15}BrO_2$	186	$C_{14}H_{28}O$	198
$C_{10}H_{20}$	166	$C_{12}H_{15}FO_2$	186	$C_{14}H_{28}O_2$	199
$C_{10}H_{20}O$	167	$C_{12}H_{16}O_2$	187	$C_{15}H_{14}ClN$	199
$C_{10}H_{20}O_2$	169	$C_{12}H_{18}$	187	$C_{15}H_{14}FN$	200
$C_{10}H_{20}O_3$	169	$C_{12}H_{18}N_2O$	187	$C_{15}H_{15}N$	200
$C_{10}H_{20}S_2$	170	$C_{12}H_{22}O$	188	$C_{15}H_{18}BrNO_4S$	200
$C_{10}H_{21}Cl$	170	$C_{12}H_{22}O_2$	188	$C_{15}H_{22}O_2$	201
$C_{11}H_{10}$	171	$C_{12}H_{24}$	189	$C_{15}H_{30}O_2$	201
$C_{11}H_{10}N_2O_2S$	171	$C_{12}H_{24}O_2$	189	$C_{16}H_{14}O_2$	201
$C_{11}H_{10}O_2$	171	$C_{12}H_{24}S_2$	190	$C_{16}H_{16}$	201
$C_{11}H_{11}BrO_3$	172	$C_{12}H_{26}$	190	$C_{16}H_{16}O_2$	202
$C_{11}H_{11}ClO_3$	172	$C_{13}H_{15}F_3O_2$	190	$C_{16}H_{17}N$	202
$C_{11}H_{11}NO_5$	173	$C_{13}H_{18}$	191	$C_{16}H_{17}NO$	203
$C_{11}H_{12}O_2$	174	$C_{13}H_{19}Cl$	191	$C_{16}H_{24}$	203
$C_{11}H_{12}O_3$	174	$C_{13}H_{20}$	191	$C_{16}H_{26}$	203
$C_{11}H_{14}$	175	$C_{13}H_{24}O$	192	$C_{16}H_{30}O$	206
$C_{11}H_{14}O$	175	$C_{13}H_{24}O_2$	192	$C_{17}H_{13}NO$	206
$C_{11}H_{14}O_2$	176	$C_{13}H_{26}O_2$	193	$C_{17}H_{18}$	207
$C_{11}H_{16}$	176	$C_{14}H_8F_4$	193	$C_{17}H_{20}N_2$	207
$C_{11}H_{16}O$	176	$C_{14}H_{10}$	193	$C_{18}F_{30}$	208
$C_{11}H_{16}O_2$	177	$C_{14}H_{10}Br_2$	194	$C_{18}H_{15}NO$	208
$C_{11}H_{16}O_3$	178	$C_{14}H_{10}N_2O_4$	194	$C_{18}H_{20}$	209
$C_{11}H_{19}ClO$	180	$C_{14}H_{10}O$	195	$C_{20}H_{24}$	209
$C_{11}H_{20}$	180	$C_{14}H_{12}$	195		

FORMULA INDEX

F_2N_2	20	C_4H_7Br	27	C_5H_9NS	44
H_8B_5Cl	20	C_4H_7BrO	27	C_5H_{10}	44
CH_3NO_2	20	C_4H_7Cl	27	$C_5H_{10}Br_2$	45
C_2HClF_2	20	C_4H_7ClO	28	$C_5H_{10}Cl_2$	46
$C_2H_2Br_2$	21	C_4H_8	28	$C_5H_{10}N_2$	47
C_2H_2ClF	21	$C_4H_8Br_2$	31	$C_5H_{10}O$	47
$C_2H_2Cl_2$	21	$C_4H_8Cl_2$	32	$C_5H_{10}OS$	48
$C_2H_2F_2$	21	C_4H_8O	32	$C_5H_{10}O_2$	48
$C_2H_2J_2$	21	C_4H_8OS	32	$C_5H_{10}O_2S$	49
$C_2H_3Cl_3$	22	$C_4H_8O_2$	33	$C_5H_{10}S$	50
C_2H_3N	22	$C_4H_8O_2S$	34	$C_5H_{11}Br$	51
$C_2H_4Br_2$	22	C_4H_8S	34	$C_5H_{11}Cl$	51
$C_2H_4Cl_2$	22	C_4H_9Br	35	C_5H_{12}	51
$C_3H_3Cl_3$	22	C_4H_9NO	35	$C_5H_{12}GeO$	52
$C_3H_4F_2$	22	C_4H_9NSi	35	$C_5H_{12}S$	52
C_3H_5Br	22	C_5H_4BrNO	35	C_6F_{10}	52
C_3H_5Cl	23	C_5H_4ClNO	35	$C_6H_2Cl_4$	52
C_3H_5F	23	C_5H_4ClNS	36	$C_6H_2F_4$	53
C_3H_5J	24	C_5H_5NO	37	$C_6H_3Cl_3$	54
C_3H_5N	24	C_5H_5NS	37	$C_6H_3F_3$	54
$C_3H_6Br_2$	24	$C_5H_5N_3O_2$	37	$C_6H_4Br_2$	55
$C_3H_6Cl_2$	24	$C_5H_5N_5O$	37	$C_6H_4Cl_2$	55
C_3H_6O	25	$C_5H_6N_2$	38	$C_6H_4F_2$	56
C_3H_7Br	25	$C_5H_6N_2O$	39	$C_6H_4N_2$	57
C_3H_7Cl	25	C_5H_8	39	$C_6H_6O_2$	57
C_3H_7J	25	C_5H_8O	39	$C_6H_7F_3O_3$	57
$C_3H_7N_5$	25	$C_5H_8O_2$	40	C_6H_7N	58
C_4F_6	25	$C_5H_8O_3$	41	C_6H_7NO	58
C_4F_8	25	C_5H_9BrO	41	C_6H_7NS	59
$C_4H_4F_2$	26	$C_5H_9BrO_2$	41	$C_6H_8N_2O_2S$	59
$C_4H_5Cl_2NO$	26	C_5H_9ClO	42	$C_6H_9ClO_2$	59
C_4H_6	26	C_5H_9FO	43	$C_6H_9ClO_3$	59
$C_4H_6O_2$	27	C_5H_9NO	43	C_6H_{10}	60

$C_6H_{10}O$	64	$C_7H_{13}ClO_2$	97	$C_8H_{17}Cl$	135
$C_6H_{10}O_2$	66	$C_7H_{13}FO_2$	97	C_8H_{18}	135
$C_6H_{10}O_3$	68	$C_7H_{13}NO_2$	98	$C_9H_7ClO_3$	136
$C_6H_{11}N$	68	$C_7H_{13}NO_4$	98	$C_9H_7NO_3$	136
$C_6H_{11}NO$	69	C_7H_{14}	98	$C_9H_8O_3$	137
$C_6H_{11}NO_2$	69	$C_7H_{14}O$	101	$C_9H_{10}O$	137
$C_6H_{11}NS$	69	$C_7H_{14}OS$	103	$C_9H_{10}O_2$	137
C_6H_{12}	69	$C_7H_{14}O_2$	104	$C_9H_{10}O_4$	138
$C_6H_{12}O$	75	$C_7H_{14}O_2S$	106	$C_9H_{11}Br$	139
$C_6H_{12}OS$	76	$C_7H_{14}O_3$	107	$C_9H_{11}F$	140
$C_6H_{12}O_2$	77	$C_7H_{14}S$	107	$C_9H_{11}NO$	140
$C_6H_{12}O_2S$	78	$C_7H_{14}S_2$	108	$C_9H_{11}NS$	141
$C_6H_{12}O_3$	79	$C_7H_{15}Br$	108	C_9H_{12}	141
$C_6H_{12}S$	79	$C_7H_{15}Cl$	109	C_9H_{14}	141
$C_6H_{12}S_2$	80	C_7H_{16}	109	$C_9H_{14}BrNO_4S$	142
$C_6H_{13}Cl$	80	$C_7H_{16}N_2S$	110	$C_9H_{14}O_2$	143
C_6H_{14}	82	$C_8H_6F_6O_4$	110	$C_9H_{14}O_3$	143
$C_6H_{14}N_2S$	82	C_8H_9Br	111	C_9H_{16}	143
$C_6H_{14}O$	82	$C_8H_9NO_2$	111	$C_9H_{16}O$	145
$C_6H_{14}S$	83	C_8H_{10}	112	$C_9H_{16}O_2$	146
$C_7H_4N_4S$	83	$C_8H_{10}OS$	112	$C_9H_{16}O_3$	147
$C_7H_4N_4S$	83	$C_8H_{10}O_2$	112	$C_9H_{16}O_4$	149
$C_7H_5ClN_2O_4$	83	$C_8H_{11}N$	113	$C_9H_{17}ClO$	150
$C_7H_5Cl_3N_2O$	83	$C_8H_{11}N_5O$	113	C_9H_{18}	150
$C_7H_6ClNO_2$	83	C_8H_{12}	113	$C_9H_{18}O$	151
$C_7H_6Cl_2N_2O$	84	$C_8H_{12}O_2$	114	$C_9H_{18}OS_2$	153
C_7H_8	84	$C_8H_{12}O_3$	115	$C_9H_{18}O_2$	153
$C_7H_8O_2$	84	$C_8H_{13}NO_2$	115	$C_9H_{18}O_3$	155
$C_7H_8O_2S$	85	C_8H_{14}	115	$C_9H_{18}S$	157
$C_7H_8O_3$	85	$C_8H_{14}O$	118	$C_9H_{18}S_2$	157
C_7H_9N	85	$C_8H_{14}O_2$	121	$C_9H_{18}S_3$	158
$C_7H_9N_5O$	86	$C_8H_{15}ClO$	123	$C_9H_{19}Cl$	158
C_7H_{10}	86	$C_8H_{15}NO_4$	124	C_9H_{20}	158
$C_7H_{10}O_2$	87	C_8H_{16}	124	$C_{10}H_{10}O$	158
$C_7H_{11}ClO$	87	$C_8H_{16}O$	127	$C_{10}H_{10}O_2$	159
C_7H_{12}	87	$C_8H_{16}O_2$	129	$C_{10}H_{10}O_3$	159
$C_7H_{12}O$	90	$C_8H_{16}O_2S$	131	$C_{10}H_{10}O_4$	159
$C_7H_{12}O_2$	92	$C_8H_{16}O_3$	132	$C_{10}H_{11}ClO$	159
$C_7H_{12}O_3$	95	$C_8H_{16}O_3S$	134	$C_{10}H_{11}FO$	160
$C_7H_{12}O_4$	95	$C_8H_{16}O_4S$	134	$C_{10}H_{12}O$	160
$C_7H_{13}BrO_2$	96	$C_8H_{16}S_2$	135	$C_{10}H_{13}Cl$	161

FORMULA INDEX

$C_{10}H_{13}N$	161	$C_{11}H_{20}O$	184	$C_{14}H_{14}N_2O$	199
$C_{10}H_{14}$	161	$C_{11}H_{20}O_2$	185	$C_{14}H_{16}O_2$	199
$C_{10}H_{14}O$	166	$C_{11}H_{22}O$	186	$C_{14}H_{20}O$	200
$C_{10}H_{14}O_2$	167	$C_{11}H_{22}O_2$	186	$C_{14}H_{20}O_2$	200
$C_{10}H_{16}$	167	$C_{11}H_{22}S_2$	187	$C_{14}H_{20}S_2$	200
$C_{10}H_{16}O$	168	$C_{12}H_{12}$	188	$C_{14}H_{22}$	201
$C_{10}H_{18}$	168	$C_{12}H_{12}O_2$	188	$C_{14}H_{22}O$	201
$C_{10}H_{18}O$	169	$C_{12}H_{14}O_3$	189	$C_{14}H_{26}O$	201
$C_{10}H_{18}O_2$	169	$C_{12}H_{14}O_4$	189	$C_{14}H_{26}O_2$	202
$C_{10}H_{18}O_4$	170	$C_{12}H_{15}BrO_2$	190	$C_{14}H_{28}O$	202
$C_{10}H_{20}$	170	$C_{12}H_{15}FO_2$	190	$C_{14}H_{28}O_2$	203
$C_{10}H_{20}O$	171	$C_{12}H_{16}O_2$	191	$C_{15}H_{14}ClN$	203
$C_{10}H_{20}O_2$	173	$C_{12}H_{18}$	191	$C_{15}H_{14}FN$	204
$C_{10}H_{20}O_3$	173	$C_{12}H_{18}N_2O$	191	$C_{15}H_{15}N$	204
$C_{10}H_{20}S_2$	174	$C_{12}H_{22}O$	192	$C_{15}H_{18}BrNO_4S$	204
$C_{10}H_{21}Cl$	174	$C_{12}H_{22}O_2$	192	$C_{15}H_{22}O_2$	205
$C_{11}H_{10}$	175	$C_{12}H_{24}$	193	$C_{15}H_{30}O_2$	205
$C_{11}H_{10}N_2O_2S$	175	$C_{12}H_{24}O_2$	193	$C_{16}H_{14}O_2$	205
$C_{11}H_{10}O_2$	175	$C_{12}H_{24}S_2$	194	$C_{16}H_{16}$	205
$C_{11}H_{11}BrO_3$	176	$C_{12}H_{26}$	194	$C_{16}H_{16}O_2$	206
$C_{11}H_{11}ClO_3$	176	$C_{13}H_{15}F_3O_2$	194	$C_{16}H_{17}N$	206
$C_{11}H_{11}NO_5$	177	$C_{13}H_{18}$	195	$C_{16}H_{17}NO$	207
$C_{11}H_{12}O_2$	178	$C_{13}H_{19}Cl$	195	$C_{16}H_{24}$	207
$C_{11}H_{12}O_3$	178	$C_{13}H_{20}$	195	$C_{16}H_{26}$	207
$C_{11}H_{14}$	179	$C_{13}H_{24}O$	196	$C_{16}H_{30}O$	210
$C_{11}H_{14}O$	179	$C_{13}H_{24}O_2$	196	$C_{17}H_{13}NO$	210
$C_{11}H_{14}O_2$	180	$C_{13}H_{26}O_2$	197	$C_{17}H_{18}$	211
$C_{11}H_{16}$	180	$C_{14}H_8F_4$	197	$C_{17}H_{20}N_2$	211
$C_{11}H_{16}O$	180	$C_{14}H_{10}$	197	$C_{18}F_{30}$	212
$C_{11}H_{16}O_2$	181	$C_{14}H_{10}Br_2$	198	$C_{18}H_{15}NO$	212
$C_{11}H_{16}O_3$	182	$C_{14}H_{10}N_2O_4$	198	$C_{18}H_{20}$	213
$C_{11}H_{19}ClO$	184	$C_{14}H_{10}O$	199	$C_{20}H_{24}$	213
$C_{11}H_{20}$	184	$C_{14}H_{12}$	199